AF353010

Metal Micro-Forming

Metal Micro-Forming

Editor

Ken-ichi Manabe

MDPI • Basel • Beijing • Wuhan • Barcelona • Belgrade • Manchester • Tokyo • Cluj • Tianjin

Editor
Ken-ichi Manabe
Tokyo Metropolitan University
Japan

Editorial Office
MDPI
St. Alban-Anlage 66
4052 Basel, Switzerland

This is a reprint of articles from the Special Issue published online in the open access journal *Metals* (ISSN 2075-4701) (available at: https://www.mdpi.com/journal/metals/special_issues/Metal_Micro-forming).

For citation purposes, cite each article independently as indicated on the article page online and as indicated below:

LastName, A.A.; LastName, B.B.; LastName, C.C. Article Title. *Journal Name* **Year**, *Article Number*, Page Range.

ISBN 978-3-03943-158-8 (Hbk)
ISBN 978-3-03943-159-5 (PDF)

Contents

About the Editor

Ken-ichi Manabe is currently an Emeritus Professor since 2017, he was awarded his Ph.D. in mechanical engineering in April 1985 from the Tokyo Metropolitan University (TMU), Japan. He has been a Professor at TMU since 2002. Prof. Manabe has undertaken extensive research on the theory and modelling of tube/sheet metal forming processes and intellectualization of their forming processes for over 40 years. Recently his research interest extends toward microforming technology and deformation mechanics in micro/meso scale. He has gained academic recognition in the Japan Society for Technology of Plasticity (JSTP), the Japan Society of Mechanical Engineers (JSME) and so on. He contributes to the dissemination of tube forming technology and its theory and modeling as a chair (2004–2012) of the Tube Forming Research Committee of the JSTP. He was a Conference chair (1993, 2011), Co-chair (1995, 1997, 1999, 2017), and Honorary chair (2019) of the International Conference on Tube Hydroforming (TUBEHYDRO). He received the Best Paper Award in 1989, 2009, and 2012 from the JSTP. He received the JSTP Medal in 2010 and attained the grade of Fellow in 2009 from the JSTP. He was the Vice President of the JSTP in 2011–2012, and the President of the JSTP in 2015–2016. He has published 19 book chapters, over 230 refereed journal papers, and over 230 refereed conference papers. He has accumulated broad knowledge and extensive interdisciplinary experience through his work in Japan, Russia, Australia, and China.

metals

Editorial

Metal Micro-Forming

Ken-ichi Manabe

Department of Mechanical Systems Engineering, Tokyo Metropolitan University, 1-1 Minamiosawa, Hachioji-shi 192-0397, Japan; manabe@tmu.ac.jp

Received: 8 June 2020; Accepted: 16 June 2020; Published: 18 June 2020

1. Introduction

Metal micro-forming is the technological field of micro-manufacturing. It is a microfabrication technology for micro parts that takes advantage of the excellent mechanical and functional characteristics peculiar to metal, and is a vital forming technology that has been scaled down to the microscale range suitable for mass production. Yole Développement [1] predicts that the compound annual growth rate (CAGR) of the micro electromechanical systems (MEMS) market in micro-manufacturing from 2018 to 2024 is 43.6% for telecom applications, 7.5% for medical applications and 11.0% for industry applications. These are mainly related to the semiconductor market. Similarly, the market prediction for metal micro-components with the excellent characteristics of metals is also expected to grow.

Investigations into metal micro-forming technology have been developing over the last 20 years. These studies include research and development activity for micro-bulk forming, micro-sheet metal forming, micro-hydroforming, laser-assisted micro-forming, surface engineering for die and tooling, embossing and micro-sintering. Among them, the elucidation of the size effect associated with scale-down in each processing method is one of the central issues, and its fundamental research and new special processing technology has been developed.

The purpose of this Special Issue is to collect expert views and contributions on the present achievements on metal micro-forming, and through this issue, to find present challenges and to offer ideas for possible solutions.

2. Contributions

In this Special Issue, there are nine papers related to conventional metal forming processes, and two basic testing and evaluation methods (surface roughening and evaluation of micro-tribological characteristics with lubricants); a total of 11 papers. Their processing is categorized into nine methods: tube forming, wire drawing, wire rolling, orbital forging, ring-shape forging, channel forming and embossing. As a basic common characterization, it contains two themes: the micro-tribological characteristics of surface texture on a micro die and the surface roughening evaluation test of raw material. Some techniques are assisted by temperature (high-frequency induction heating and resistance heating) and others by ultrasonic vibration, hydroforming and shock wave liquid medium high-energy-rate processing. Examples of copper, titanium, aluminum, stainless steel and the metallic powders of stainless steel and aluminum are used for medical and communication fields.

So far, relatively few review articles on the forming technology of micro-tubes have been published. It is welcome, therefore, that Hartl [2] reviews and classifies the recent progress in research and development on the fabrication and secondary metalworking methods of micro-tubes. The various micro tube manufacturing methods include various heat assisted dieless drawing methods, micro tube manufacturing by conventional technology, micro tube fabrication for special tubes and the latest manufacturing technology for micro hydroforming technology for T-shapes and cross shapes, laser heating assisted bending processing, micro tube processing technology using severe plastic deformation and micro tube characteristic evaluation methods. A total of 117 papers are surveyed and summarized. Yasui et al. [3] focus on micro T-shape hydroforming of 0.5 mm diameter micro-tubes and examine the effects of the friction and the length of micro-tube on their hydroformability and material flow, using both experiments and finite element analysis. They confirm that, for a short micro-tube, the friction effect is suppressed and high hydroformability is obtained, but by increasing the length of the tube and friction, the hydroformability decreases.

For microwires, Hwang and Liu [4] examine the effects of processing temperature and the oxide layer on the mechanical properties of the drawn wire in microwire dieless drawing by high-frequency induction heating using fine, 1 mm diameter stainless steel wire. Additionally, the influence of the drawing speed and the processing temperature on the drawing limit is examined by the finite element dieless drawing analysis. Xie et al. [5] examine the size effect of microstructure on the deformation characteristics of microwires with a flat rolling outer diameter of 1.5 to 0.8 mm in flat rolling. The effect of the surface layer effect on rolling resistance and the effect of wire diameter on the development of surface roughness.

For micro forging, Presz [6] investigates the material flow related to ultrasonic orbital micro forging, the material flow in different die shapes was clarified in detail by the application of 20 kHz vibration with a 16 μm longitudinal amplitude at surface of punch nose to an aluminum cylinder, with a diameter of 1 mm. Yang and Shimizu [7] develop an innovative resistance heating method using a surface-modified punch and evaluate the heating characteristics. The coating effect of the die in the ring-shape forging is tested by applying it to Ti, Cu, SUS304 materials using a die coated with 0.5 and 1 μm of AlCrSiN film, to improve heating characteristics.

As an application of high-energy rate micro-forming technology, Liu et al. [8] develop a laser impact liquid flexible micro-forming process, which was taken up and applied to the forming of a microchannel of copper foil material. The fabricated shape was compared with finite element analysis results.

In terms of the micro metal compaction for micro sintering, Emadinia et al. [9] perform micro hot embossing using aluminum and stainless steel metal powders with carbon nanotube (CNTs), and examine the optimum conditions of the powder component blending ratio to make the green compact uniform, and to improve the surface properties of the compact.

On the other hand, Shimuzu et al. [10] carry out research on basic micro-tribological characteristics related to metal micro-forming technology, investigating the lubrication effect of dimple texture with a diameter of 10, 50, 100 μm provided on the die surface, which is important for the tribological characteristics in sheet forming, the bending and ironing experiment of 0.1 mm thick SUS304 foil material. In addition to the finite element forming analysis and computational fluid dynamics analysis (CFD) on die contact lubrication condition, an in-situ observation of lubricant flow is performed to confirm the effect of a dimple array on the die. Furushima et al. [11] propose a new compression test method for thin sheet metals, to characterize the surface roughening evolution under compressive strain, and confirm the surface roughening behavior during the straining of thin sheet.

For micro-die manufacturing, Shiratori et al. [12] develop a manufacturing method of micro-meshing punch array for micro-embossing on copper thin sheet by a low-temperature plasma printing process. A 180 μm × 180 μm square prism array of a thin copper plate is formed by micro-embossing.

3. Conclusions

This Special Issue provides a total of 11 articles from seven countries, including one review paper. From these collected papers, it may be seen that micro-forming research in the world is steadily progressing and developing, and it is possible to get a glimpse of the present situation and the academic and technological issues. I hope that advanced metal micro-forming technology will further progress through the elucidation of material deformation behavior in microscale, and that this Special Issue will contribute to prosperous future research.

I would like to express sincere thanks to all the contributors to this Special Issue. Additionally, to the reviewers for making useful comments, and the Metal Editorial staff for prompt publishing.

Conflicts of Interest: The author declares no conflict of interest.

References

1. Status of the MEMS Industry 2019. Available online: https://yole-i-micronews-com.osu.eu-west-2.outscale.com/uploads/2019/06/YD19031_Status_of_the_MEMS_industry_2019_Sample_Yole_Developpement.pdf (accessed on 3 June 2020).
2. Hartl, C. Review on advances in metal micro-tube forming. *Metals* **2019**, *9*, 542. [CrossRef]
3. Yasui, H.; Yoshihara, S.; Mori, S.; Tada, K.; Manabe, K. Material deformation behavior in T-shape hydroforming of metal microtubes. *Metals* **2020**, *10*, 199. [CrossRef]
4. Hwang, Y.-M.; Liu, H.-H. Formability analysis and oxidation layer effects in dieless drawing of stainless steel wires. *Metals* **2019**, *9*, 828. [CrossRef]
5. Xie, H.; Manabe, K.; Jiang, Z. Study of wire deformation characterization and size effects during the micro-flat-rolling process. *Metals* **2020**, *10*, 405. [CrossRef]
6. Presz, W. Material flow in ultrasonic orbital microforming. *Metals* **2019**, *9*, 475. [CrossRef]
7. Yang, M.; Shimizu, T. Development of a novel resistance heating system for microforming using surface-modified dies and evalm, uation of its heating property. *Metals* **2019**, *9*, 440. [CrossRef]
8. Liu, F.; Liu, H.; Jiang, C.; Ma, Y.; Wang, X. Experimental and numerical investigations of a novel laser impact liquid flexible microforming process. *Metals* **2018**, *8*, 599. [CrossRef]
9. Emadinia, O.; Vieira, M.T.; Vieira, M.F. Feedstocks of aluminum and 316L stainless steel powders for micro hot embossing. *Metals* **2018**, *8*, 999. [CrossRef]
10. Shimizu, T.; Kobayashi, H.; Vorholt, J.; Yang, M. Lubrication analysis of micro-dimple textured die surface by direct observation of contact interface in sheet metal forming. *Metals* **2019**, *9*, 917. [CrossRef]
11. Furushima, T.; Aoto, K.; Alexandrov, S. A new compression test for determining free surface roughness evolution in thin sheet metals. *Metals* **2019**, *9*, 451. [CrossRef]
12. Shiratori, T.; Aizawa, T.; Saito, Y.; Wasa, K. Plasma printing of an AISI316 micro-meshing punch array for micro-embossing onto copper plates. *Metals* **2019**, *9*, 396. [CrossRef]

Article

Study of Wire Deformation Characterization and Size Effects during the Micro-Flat-Rolling Process

Haibo Xie [1],*, Ken-ichi Manabe [2] and Zhengyi Jiang [1]

[1] School of Mechanical, Materials, Mechatronic and Biomedical Engineering, University of Wollongong, Wollongong 2522, Australia; jiang@uow.edu.au
[2] Department of Mechanical Engineering, Tokyo Metropolitan University, Hachioji, Tokyo 192-0397, Japan; manabe@tmu.ac.jp
* Correspondence: xie@uow.edu.au; Tel.: +61-2-42215497

Received: 29 February 2020; Accepted: 20 March 2020; Published: 22 March 2020

Abstract: A comprehensive research on the flat rolling deformation characterization of microwire has been conducted systematically through finite element simulation and testified by the results from the experimental analysis. The obtained results are compared in terms of lateral spread, geometrical characteristic, contact area width and surface roughness considering the effects of pass reduction and initial wire diameter. The size effect has been identified and surface layer modeling has been set up based on surface grain share and grain size distribution. The numerical method combined with varied flow stress has been verified by experimental value with a maximum difference of 3.7% for the 1.5 mm wire. With the increase of the height reduction, the curvature radius is decreased while the lateral spread and contact area width are increased. Surface roughness evolution in the range of 0.52–0.85 µm for the rolled wire has also been investigated.

Keywords: size effects; deformation characterization; micro-rolling; wire

1. Introduction

During cold rolling, microwire with a circular cross-section is flat-rolled at room temperature through two rolls to gain the required ratio of thickness-to-width under one or several passes with a specific cross-section due to the barreling effect. The flattened wire products are widely used in medical catheters, electric parts, springs, guide rails and piston rings [1,2] where this side-barrel type cross-section cannot be produced by normal strips or plates through flat rolling. More complex profiles can be manufactured by adding subsequent profile rolling steps. The edge rounding and lateral spreading are two key factors that are related to the rolled wire quality.

Inhomogeneous deformation occurs during the wire flattening rolling deformation process under different height reduction, which influences the contact stress distribution and the deformation force. The methods and techniques applied in the macro-domain cannot be used directly in the micro-domain [3], the size effect due to downscaling should be considered in the micro-forming process [4]. The shear band idea has been used to calculate the contact area width, and the lateral spread and the rolling load can be predicated by slab analysis [5], where the investigation on the effects of height reduction on the effective strain field during flat wire rolling had been conducted and the evolution of the shear bands was predicted as well. The different contact pressure patterns were investigated and confirmed by 3D numerical analysis [6]. The relationship between contact stress distributions and the deformation homogeneity has been identified in flat wire rolling [7–9]. The authors of [10] reported an experimental investigation on the deformation behavior where the influences of the rolling parameter were discussed, and the quantitative relationship of contact area width with rolling reduction was established [11]. The influencing parameters on the spread, residual stresses, loads, and pressure distribution have been reported in references [12–15]. The mechanical anisotropy has played an important role in the geometry evolution of steel wire rolling [13]. Grain size has long been known in influencing the material properties during the forming process, such as surface quality, material flow and yield stress [16–21]. The size effects in the simulation and experiment have been reported in references [22,23]. Although the research mentioned above, the real physical mechanism of size effect is still unclear in the micro-flat-rolling of wires.

In this study, the effects of wire height reduction on the contact area width between the rolls and wire will be investigated and their influence on rolled product characteristics and quality will also be identified. The study will conduct experimentally and numerically on the impact of height reduction on the lateral spread variations during the rolling process. In order to investigate the size effect, flat wire experimental rolling with different wire diameters will be carried out and compared with the finite element (FE) simulation results and this comparison will evaluate their simulation accuracy. The research findings obtained discover the effects of rolling process parameters and further analysis of deformation mechanism during flat microwire rolling will, therefore, provide a parameter optimization method for micro-forming process design.

2. Experimental and Simulation

2.1. Micro-Rolling Mill and Materials Used

The micro-flat-rolling mill system and two-high rolling mill are shown in Figure 1. The rolling mill system includes a roll set, power transmission, a measurement system and a control panel. The two-high roll set consists of two work rolls with 27 mm in diameter and 29.5 mm in barrel length which are made by two different materials: an ultra-strength alloy of cemented carbide (Young's Modulus: 580 GPa) and tool steel SKD 11 (Young's Modulus: 210 GPa), respectively. This kind of work roll with cemented carbide significantly enhances the rolling mill's rigidity and decreases the deflection of the rolls. Turning the screw through the knob to adjust the roll gap and then wire products with the desired thickness can be obtained through the rolling process. By adjusting the electric voltage frequency, various rolling speeds could be applied to investigate the effect of rolling speeds on the micro-rolling process. The rolling capacity of this designed rolling mill is featured with a rolling load of 40 kN. The pure copper wire is used as initially rolled materials, and their chemical composition is shown in Table 1. Two different work rolls are applied and copper wire has three different diameters of 0.8 mm, 1.0 mm and 1.5 mm, respectively.

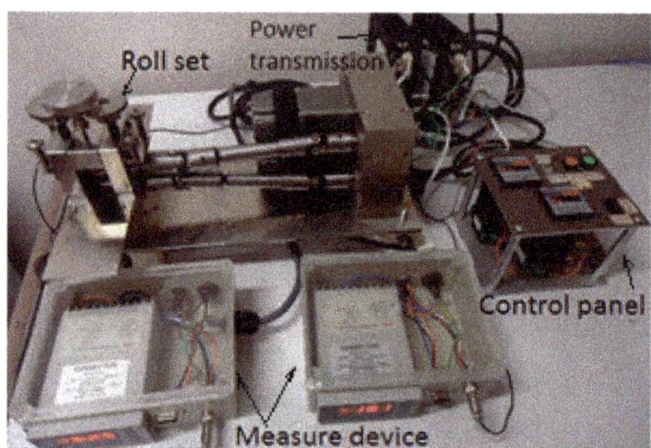

Figure 1. Micro-flat-rolling mill system (two-high).

Figure 2 illustrates the sketch drawing of the deformed wire. It can be seen that the spread is a significant feature in micro-flat-rolling. The main parameters of contact area and spread are included.

Table 1. Chemical composition of copper wire (not more than, wt %) [24].

Bi	Te	As	Fe	Ni
0.001	0.002	0.002	0.005	0.002
Sn	**S**	**Zn**	**O**	**Cu**
0.002	0.002	0.005	0.06	balance

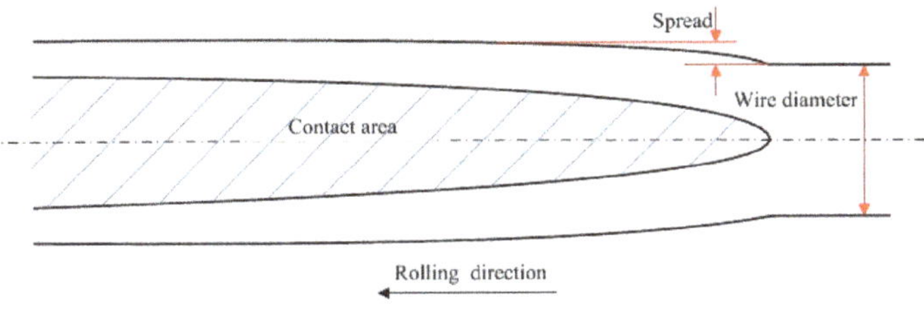

Figure 2. Schematic of the deformed wire (from above).

2.2. Numerical Simulation

In this study, the explicit dynamic finite element software, LS-DYNA R8 (LSTC), is applied to analyze the deformation behavior of the flat microwire rolling process, where the element type solid164 with a three-dimensional 8 node solid element has been used for the rolling simulation. The simulation parameters of the roll and rolled material used in the finite element analysis are the same as the values in the real rolling experiment.

During the simulation of the rolling process, a fixed angular speed is applied on the roll for roll rotation. We can assume the copper wire moves along the rolling direction at a certain speed. When the copper wire moves forward into the bite zone, the friction force between rolls and rolled wire will take effect. The rolling process will go ahead. A coefficient of friction (COF) of 0.18 is applied in rolling simulation after experimental tests. Considering the effect of elastic deformation of work roll, the work roll is considered a linear-elastic material. The rolled copper wire is assumed to be isotropic elastoplastic. Figure 3 shows a simulation model of wire flat rolling on a two-high rolling mill. Figure 3a is the roll meshing in the flat micro-rolling of wire. Figure 3b,c are the cross-section before and after rolling, respectively. In the simulation, the forming resistance, the rolling spread, contact width and rolled wire surface quality are evaluated.

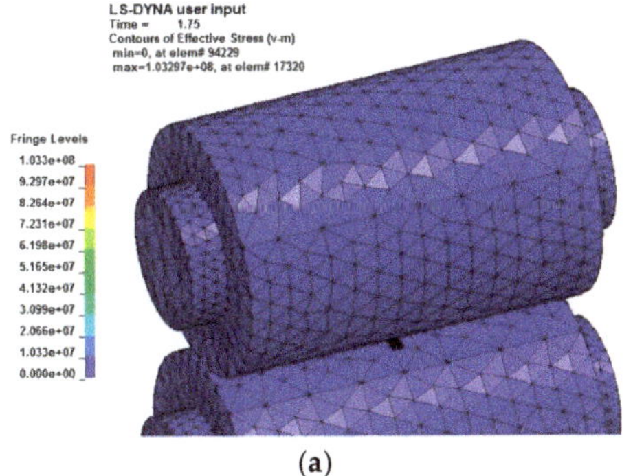

(a)

Figure 3. *Cont.*

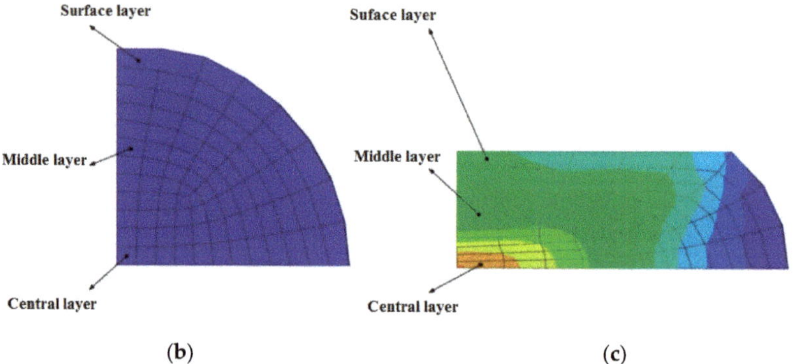

(b)   **(c)**

Figure 3. Simulation of wire flat rolling process. (**a**) Roll meshing in micro-flat-rolling of wire; (**b**) cross-section before rolling; (**c**) cross-section after rolling.

2.3. Micro-Compression Test

In the micro-compression test, the experimental samples are pure copper with three different geometrical dimensions of Φ 0.8 mm × 1.2 mm, Φ 1.0 mm × 1.5 mm, and Φ 1.5 mm × 1.8 mm, respectively. The wire cutting process is applied to prepare for the cylindrical samples as shown in Figure 4, which indicates a 1.5 ratio of height to diameter.

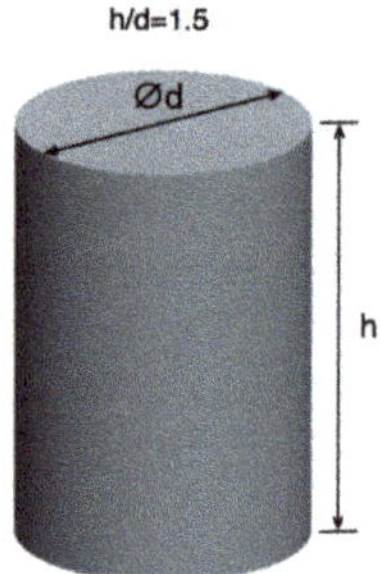

Figure 4. Schematic illustration of cylindrical specimens.

A digital spiral micrometer is used to measure accurately the testing sample geometries. In order to minimize the influence of friction at the interface of tools and samples, the sample cross-sections are rubbed by fine sandpaper [17]. The side surface of these samples was polished using chamois before experimental testing in order to remove any macro-surface defects. Stripping off the disturbance of the initial surface asperity of the specimen is necessary when investigating the variation of specimen side surface during compression deformation. The specimen is annealed under different holding times to obtain various grain sizes. The annealing process of copper wire has been listed in Table 2. The strain–stress curves for various grain sized copper wire are shown in Figure 5 during the micro-compression test. It can be seen that the stress increases with decreased grain size under different strains.

Table 2. Annealing process of copper wire for different holding time under 700 °C.

Annealing	1	2	3	4	5
Holding time, min	10	20	30	90	120
Grain size, μm	32	42	82	160	240

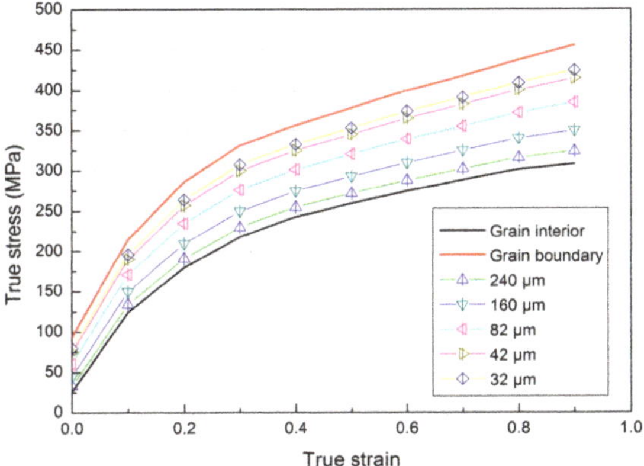

Figure 5. Strain–stress curves.

The microstructure for 32 µm and 240 µm after annealing process are shown in Figure 6.

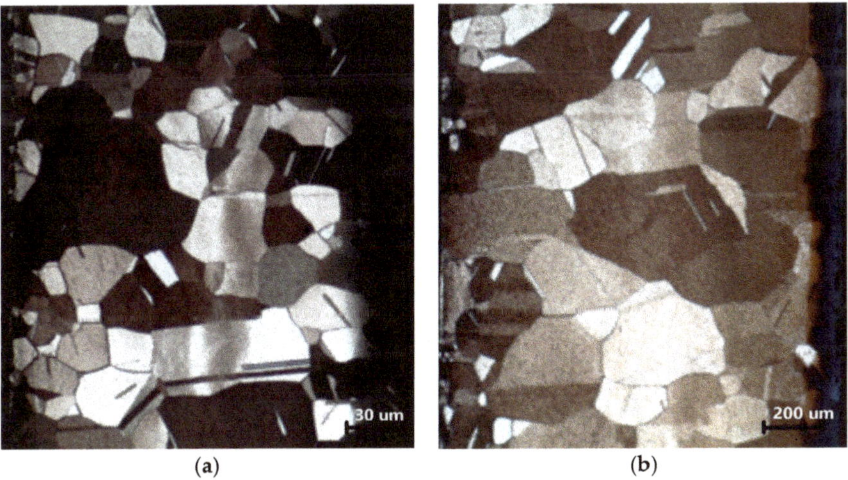

Figure 6. Microstructure distribution after annealing process (**a**) 32 µm and (**b**) 240 µm.

Assuming that there is an elastoplastic behavior involving isotropic deformation hardening, the deformed material was modeled, using the J2 plasticity model which is independent of the deformation rate. A Young modulus $E = 120$ GPa and a Poisson ratio of 0.355 were used to model the elastic deformation zone. The linear hardening law is applied for the plastic deformation zone during simulation,

$$\bar{\sigma} = 183.4 + 1.605\bar{\varepsilon}^{n} \tag{1}$$

where $\bar{\sigma}$ and $\bar{\varepsilon}$ are the equivalent plastic stress and strain, respectively; n is the strain hardening exponent. The values of 183.4 and 1.605 are initial equivalent stress and strength index, respectively, which can be obtained by experiments.

3. Results and Discussion

3.1. Theoretical Analysis

In the flat rolling process of wire, the existence of friction between the wire and rolls will cause the formation of two dead metal zones, resulting in the macroscopic shear bands, which are observed in metallographic cross-section of wire. Their morphology after the pass reduction in height of the wire as shown in Figure 7a–c (initial wire, the first pass and the second pass, respectively). It can be seen that the shear bands are in the form of X as shown in these figures, which may be considered as the slip line in flat rolling of wire.

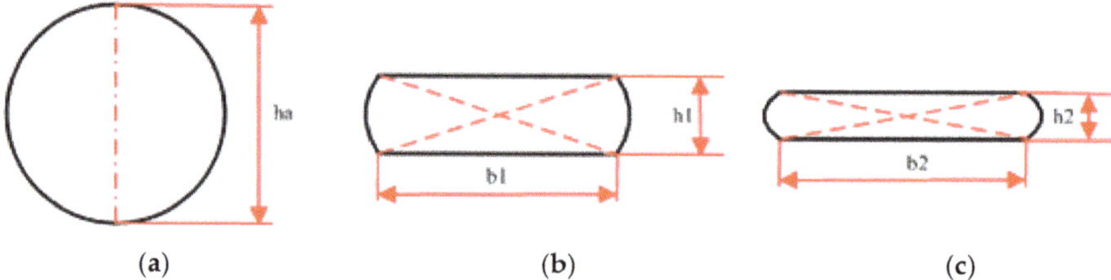

Figure 7. Schematic illustration of the effect of wire height reduction on shear bands. (**a**) Initial wire (**b**) rolled wire after Pass 1 (**c**) rolled wire after Pass 2.

The contact area width between the wire and rolls can be formed below:

$$b = \sqrt{\Delta h(2h_a - \Delta h)} \tag{2}$$

$$\Delta h = h_a - h_1 \tag{3}$$

where Δh is the thickness reduction, h_a and h_1 are the initial diameter and the thickness after flat rolling, respectively. From Equation (2), the wire contact area width is in the proportional relationship with the square root of height reduction.

The wire lateral spread is another important parameter in the deformation process, which can be obtained by the following equation [13,14]:

$$\frac{b_1}{h_0} = ma_1a_2a_3a_4\left(\frac{h_a}{h_1}\right)^p \tag{4}$$

where p is a function of the roll diameter and ratio of the initial width to the height of wire. m, a_1, a_2, a_3 and a_4 are correction factors, which are related to the grain size, material composition, roll material, rolling speed and rolling temperature, respectively, and the values of these parameters can be obtained by experiments.

3.2. Microstructure Evolution and Surface Layer Modeling

Figures 8 and 9 illustrate the microstructure evolution before and after micro-flat-wire rolling. The initial diameter of the rolled wire is 1 mm. From these figures, it is noted that there is a remarkable interface between the microstructures of the inner layer and the outer layer. The ratio of the surface layer in the initial wire is 25%. The ratio of the surface layer is 33.33% and 19.23%, respectively, in the normal direction and transverse direction after a reduction of 45%, while the outer layer thickness of the transverse direction is higher than that of the normal direction. This indicates that the outer layer of the free side surface of rolled samples remains the same before and after rolling.

Figure 8. Microstructure of copper wire before micro-rolling (initial diameter: 1 mm). (**a**) Whole initial wire; (**b**) inner layer.

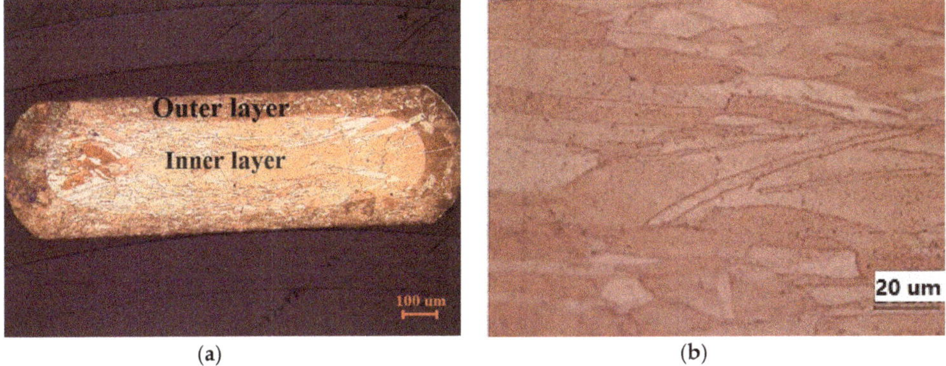

Figure 9. Microstructure of rolled wire after micro-rolling. (**a**) Whole rolled wire; (**b**) inner layer.

As the shape of the component changes, the ratio of surface area to volume changes. The grains on the surface are loosely held compared to the grains in the interior, because of a lesser number of grains surrounding the surface grains. Hence, as the ratio of surface area to volume increases, the flow strength of the material decreases. As the ratio of surface to volume increases, the adhesion of a small-sized component to a gripper is more than a normal-sized component. The size-dependent flow stress can be used to establish a better correlation between the experimental results and numerical ones.

According to surface layer theory [17], the flow stress of polycrystalline aggregate can be obtained through the composite model:

$$\sigma_s = (1 - f)\sigma_i + f\sigma_b \tag{5}$$

where σ_s is the polycrystalline aggregate flow stress. f is the grain boundary volume fraction. σ_i and σ_b are the grain interior flow stress and boundary flow stress, respectively.

3.3. Rolling Parameter Analysis

3.3.1. Forming Resistance

In order to investigate if there exists a size effect during the micro-rolling process of the microstrip from the microwire, a series of rolling experiments have been conducted for different diameter wires in which the rolling process is on the micro-scale. The term of the forming resistance has been introduced in this research, which is the ratio of the rolling force to the contact area between the roll and the rolled wire. The forming resistance during the deformation process is compared between numerical simulation and experiments.

In this study, in order to investigate the size effect of micro-flat-wire rolling, the experimental results have been compared with the numerical simulation ones during the rolling process. Soft annealed pure copper rods of 15 mm length cut out from wire with diameters of 0.8, 1.0 and 1.5 mm are rolled with a height reduction of 10%, 30% and 45%, respectively. The forming resistance of pure copper under various reductions for different wire diameters is shown in Figure 10. It can be seen that for the 1.5 mm wire diameter, the experimental results are in very close agreement with the numerical simulation results with a maximum difference of 3.7% for the 1.5 mm wire, and these results are in the macroscopic domain. The difference between the experiment and numerical simulation results is increasing from 7.5% to 12.8% with a decrease of wire diameter from 1.0 mm to 0.8 mm. Within a reduction range of 10% and 30%, the experimental results are of a different trend from the simulation results. For a 45% reduction, the trend between experimental and simulation results is the same, but the absolute values of the forming resistance are different with a difference of 11.6% for the 0.8 mm wire. The differences between experimental and simulation results indicate that size effects influence the experimental results.

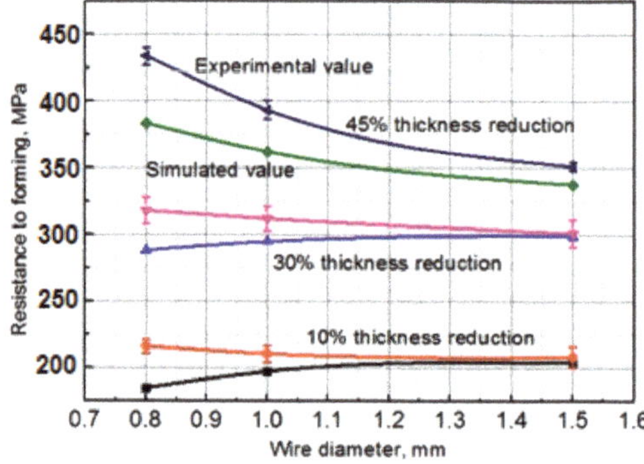

Figure 10. Forming resistance of pure copper wires under various reductions (10%, 30% and 45%) and wire diameters (0.8 mm, 1.0 mm and 1.5 mm).

The different trends between the experimental and simulation results could be illustrated in the following different size effects. The first one is the sample size effect, where the deformation resistance increases with a decrease in sample size. This kind of size effect will be investigated based on surface layer grain share and grain size distribution in the following sections through micro-compression test and microstructure observation. The second one is the COF size effect, where the COF decreases with an increase of sample dimensions [18]. This behavior can be illustrated by the lubricant pockets model. As we know, the wire surface is not smooth and contains many pockets. The closed pockets that are not extended to rolled sample edges can hold some lubricant during deformation. The open pockets that are extended to edges cannot hold some lubricant when the rolling process is conducted. During rolling, with the increase of rolling pressure, the surface peak asperity will be crushed and get flattened, which will increase the friction force. In a lower height reduction within 10–30%, the first size effect can be applied to explain that the forming resistance increases with an increase of wire diameters. When the height reduction increases up to 45%, frictional size effect can be the main factor and will result in forming a resistance to decrease with an increase of wire diameters.

3.3.2. Lateral Spread and Barreling Radius

In wire rolling, there exists plastic deformation along the length direction and side direction in the first several passes. After several passes, the deformation on the flattened sample is similar to the strip rolling; the lateral spread will reduce and can be ignored.

The effects of wire diameter and thickness reduction on the lateral spread are illustrated in Figure 11. The lateral spread increases with thickness reductions while decreases with wire diameters.

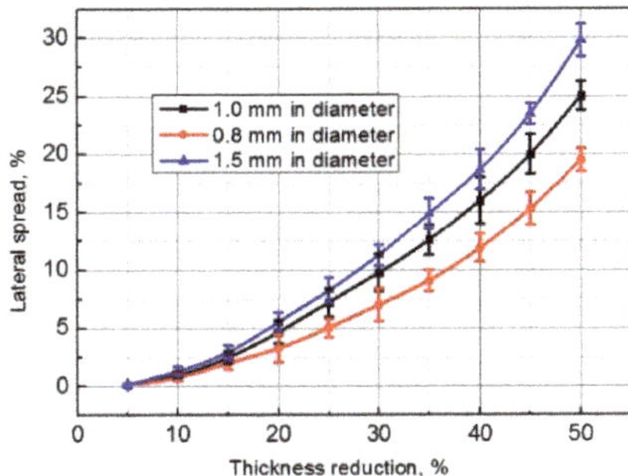

Figure 11. Lateral spread with wire diameter under various rolling thickness reductions.

Figure 12 illustrates the barreling radius versus the thickness reduction in the experiment and FEM simulation. As seen from this figure, the simulation values are in close agreement with the experimental values. In a lower thickness reduction (below 15%), the radius in the FEM simulation is higher than that in the experimental one, while in a higher thickness reduction case, the experimental value is higher than that in the FEM simulation. This is because of the size effect under higher thickness reduction. In addition, this result validates the proposed model with a maximum difference of 7.5% with the experimental one and it can be used to simulate the micro-rolling process.

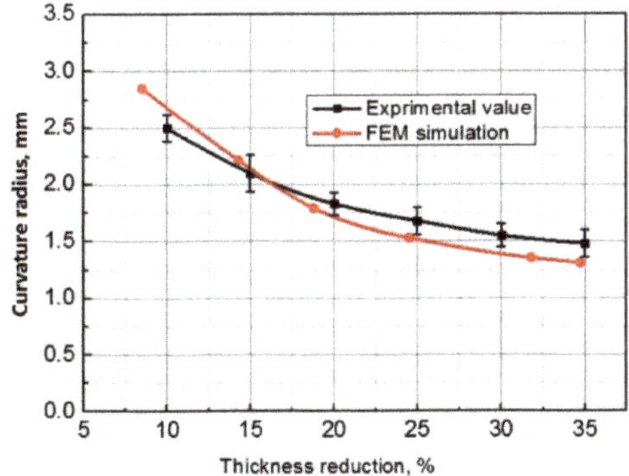

Figure 12. Barreling radius as obtained in the simulations and experiment (1 mm in diameter).

3.3.3. Contact Width

The contact width versus the reduction in thickness under different rolling speeds is shown in Figure 13. It can be seen that the contact width increases with thickness reduction, and the effect of rolling speed is not significant on the contact width. The increased rolling speed will increase the redundant work on the close sample surface, thus increasing the rolling force. Additionally, by increasing the rolling force, the coefficient of friction between the wire and rolls decreases, causing a reduction of friction work, therefore, the rolling force decreases. In addition, it can be seen that the rolling speed in flat wire rolling has less influence on rolling force, thus its effect on contact width is insignificant.

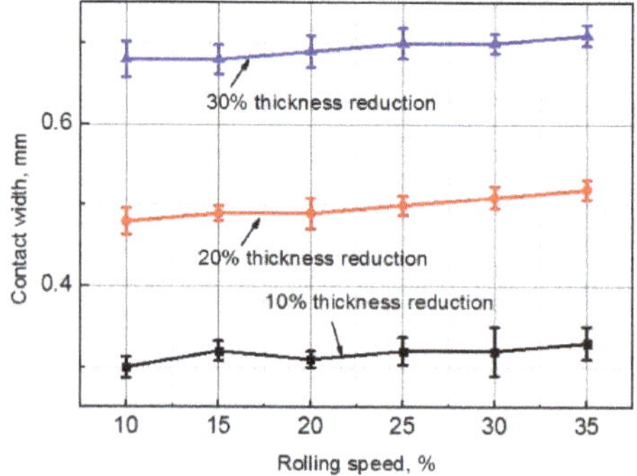

Figure 13. Contact width with rolling reduction and rolling speed (1 mm in diameter).

3.3.4. Surface Roughness Evolution

Figure 14 shows the surface roughness of the initial wire surface and rolled wire surface. It can be seen that the rolling trace is remarkable on the product surface. Normally, the surface roughness of the rolled wire is reduced via compression by the roll because of the smoother surface of the roll. Due to the boundary constraint by the rolls, the surface grains of the rolled sample in the deformation zone cannot deform freely. Therefore, the surface roughness of the contact surface of the rolled wire is reprinted by the roll. Figure 15 illustrates the surface roughness change with wire diameter under a reduction of 45% in the first pass. The initial surface roughness for the three wire diameters is the same, around 1 µm. The roll surface roughness is 0.25 µm. After rolling, the surface roughness is 0.85, 0.73 and 0.52 µm for 0.8, 1.0 and 1.5 mm wire diameters, respectively. The reprinting effect between the roll and wire is less significant for smaller diameter wire, and this is because there are more open pockets on the surface of small diameter wire. These open pockets play an important role in the friction behavior between the roll and wire, and thus influence the deformation of the wire.

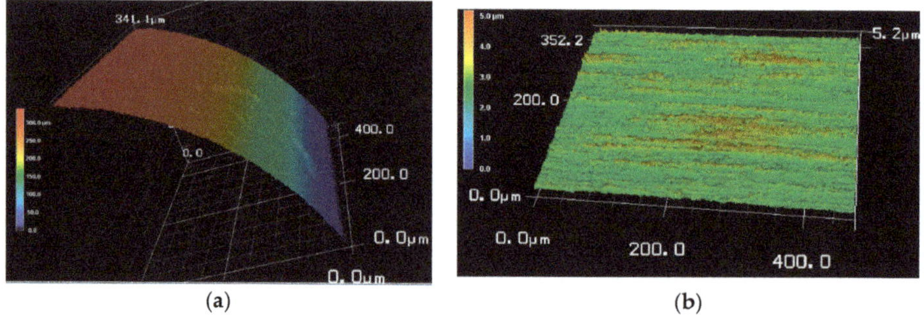

Figure 14. Surface roughness: initial wire surface (**a**), and rolled wire surface (**b**).

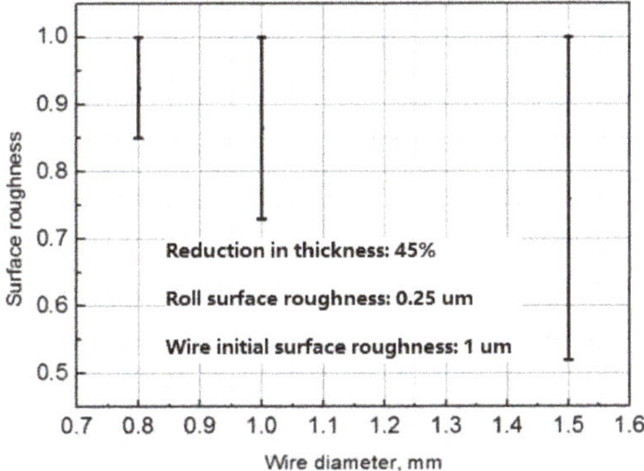

Figure 15. The effect of wire diameter on surface roughness.

4. Conclusions

In this study, the deformation characterization of microwire rolling has been investigated by numerical simulation and rolling experiments via the developed micro-rolling mill. The proposed numerical model has been validated by experimental values in terms of forming resistance and curvature radius with a maximum difference of 3.7% and 7.5%, respectively. The size effect in forming resistance for different wire diameters has been investigated by the comprising rolling experimental results with numerical results. The strain–stress curves have been identified for various grain sizes. The experiment shows the microstructure difference between the outer and inner layers which provide a basis for surface layer modeling with consideration of size effect. Under a higher reduction of small diameter wire, this effect has a significant influence on the deformation behavior of wire flat rolling. The surface roughness evolution from 1 µm to 0.52–0.85 µm before and after rolling for 1.5 mm and 0.8 mm wire, respectively, has been obtained. The findings will provide a theatrical and mechanical basis for the microwire rolling process.

Author Contributions: Conceptualization, H.X., K.-i.M and Z.J.; methodology, H.X. and Z.J.; software, H.X.; validation, H.X., K.-i.M. and Z.J.; formal analysis, H.X. and K.-i.M.; investigation, H.X., K.-i.M. and Z.J.; resources, H.X.; data curation, H.X. and Z.J.; writing—original draft preparation, H.X.; writing—review and editing, H.X. and Z.J.; visualization, H.X.; supervision, Z.J.; project administration, Z.J.; funding acquisition, Z.J. All authors have read and agreed to the published version of the manuscript.

Funding: This research was funded by Australian Research Council (ARC) Discovery Projects, grant number DP190100408.

Conflicts of Interest: The authors declare no conflicts of interest.

References

1. Kazeminezhad, M.; Taheri, A.K. A theoretical and experimental investigation on wire flat rolling process using deformation pattern. *Mater. Design.* **2005**, *26*, 99–103. [CrossRef]
2. Utsunomiya, H.; Hartley, P. Pilingen. *J. Manuf. Sci. Eng.* **2001**, *123*, 397–404. [CrossRef]
3. Vollertsen, F.; Biermann, D.; Hansen, H.N.; Jawahir, I.S.; Kuzman, K. Size effects in manufacturing of metallic components. *CIRP Ann. Manuf. Technol.* **2009**, *58*, 566–587. [CrossRef]
4. Xie, H.B.; Manabe, K.; Furushima, T.; Tada, K.; Jiang, Z.Y. Deformation characterization of micro rolling for stainless steel foil. *Procedia Eng.* **2014**, *81*, 179–184. [CrossRef]
5. Kazeminezhad, M.; Taheri, A.K. Calculation of the rolling pressure distribution and force in wire flat rollingprocess. *J. Mater. Process. Technol.* **2006**, *171*, 253–258. [CrossRef]
6. Vallellano, C.; Cabanillas, P.A.; Garcia-Lomas, F.J. Analysis of deformations and stresses in flat rolling of wire. *J. Mater. Process. Technol.* **2008**, *195*, 63–71.14. [CrossRef]

7. Kazeminezhad, M.; Taheri, A.K. An experimental investigation on the deformation behavior during wire flat rolling process. *J. Mater. Process. Technol.* **2005**, *160*, 313–320. [CrossRef]

8. Kim, N.; Lee, S.M.; Shin, W.; Shivpuri, R. Simulation of square to oval single pass rolling using a computationally effective finite and slab element method. *J. Eng. Ind.* **1992**, *114*, 329–335. [CrossRef]

9. Kobayashi, M. Influence of rolling conditions on spreadingin flat rolling of round wire. *J. Jpn. Soc. Technol. Plast.* **1978**, *19*, 630–637. (In Japanese)

10. Francesco, L.; Antoniomaria, D.I. A parametric study on residual stresses and loads in drawing process with idle rolls. *Mater. Des.* **2011**, *32*, 4832–4838.

11. Bjorn, C. The contact pressure distribution in flat rolling of wire. *J. Mater. Process. Technol.* **1998**, *73*, 1–6.

12. Massé, T.; Chastel, Y.; Montmitonnet, P.; Bobadilla, C.; Persem, N.; Foissey, S. Impact of mechanical anisotropy on the geometry of flat-rolled fully pearlitic steel wires. *J. Mater. Process. Technol.* **2011**, *211*, 103–112. [CrossRef]

13. Helmi, A.; Alexander, J.M. Geometric factors affecting spread in hot flat rolling of steel. *J. Iron Steel Inst.* **1968**, *206*, 1110–1117.

14. Saito, Y.; Takahashi, Y.; Kato, K. Calculation of spread, elongation, effective roll radius, roll force and torque when rolling in the square-diamond, square-oval, and round-oval passes. *J. Iron Steel Inst. Jpn.* **1983**, *2*, 1070–1077.

15. Chitkara, N.R.; Johnson, W. Some experimental results concerning spread in the rolling of lead. *ASME J. Basic Eng.* **1966**, 489–499. [CrossRef]

16. van Putten, K.; Hirt, G. Size effects on miniaturised rolling processes. *Ironmak. Steelmak.* **2010**, *37*, 283–289. [CrossRef]

17. Lu, H.N.; Wei, D.B.; Jiang, Z.Y.; Liu, X.H.; Manabe, K. Modelling of size effects in microforming process with consideration of grained heterogeneity. *Comput. Mater. Sci.* **2013**, *77*, 44–52. [CrossRef]

18. Xie, H.B.; Manabe, K.; Furushima, T.; Tada, K.; Jiang, Z.Y. An experimental and numerical investigation on micro rolling for ultra-thin strip. *Int. J. Mater. Form.* **2016**, *9*, 405–412. [CrossRef]

19. Hum, K.M.; Zhang, W.M.; Zhong, Z.Y.; Peng, Z.K.; Meng, G. Effect of surface layer thickness on buckling and vibration of nonlocal nanowires. *Phys. Lett. A* **2014**, *378*, 650–654.

20. Zhu, T.T.; Bushby, A.J.; Dunstan, D.J. Materials Mechanical size effects: A review. *Mater. Technol.* **2008**, *23*, 196–209. [CrossRef]

21. Liu, J.G.; Fu, M.W.; Chan, W.L. A constitutive model for modeling of the deformation behaviour in microforming with a consideration of grain boundary strengthening. *Comput. Mater. Sci.* **2012**, *55*, 85–94. [CrossRef]

22. Corrado, M.; Paggi, M.; Carpinteri, A. A multi-scale numerical method for the study of size-scale effects in ductile fracture. *Metals* **2014**, *4*, 428–444. [CrossRef]

23. Wang, C.J.; Wang, S.; Wang, S.T.; Chen, G.; Zhang, P. Investigation on springback behavior of Cu/Ni clad foils during flexible die micro V-bending process. *Metals* **2019**, *9*, 892. [CrossRef]

24. Lu, H. A Study on the Micro Cross Wedge Rolling of Metals. Ph.D. Thesis, School of Mechanical, Materials and Mechatronic Engineering, University of Wollongong, Wollongong, Australia, 2013.

Article

Material Deformation Behavior in T-Shape Hydroforming of Metal Microtubes

Hajime Yasui [1], Shoichiro Yoshihara [2,*], Shigeki Mori [3], Kazuo Tada [4] and Ken-ichi Manabe [4]

[1]   Faculty of Engineering, Integrated Graduate School of Medical, Engineering, and Agricultural Sciences, University of Yamanashi, 4-3-11 Takeda Kofu-shi, Yamanashi 400-8511, Japan; g18tma01@yamanashi.ac.jp

[2]   Department of Engineering and Design, Shibaura Institute of Technology, 3-9-14 Minato-ku, Tokyo 108-8548, Japan

[3]   Metal forming unit, Polytechnic University, 2-32-1 Ogawanishi-Machi, Kodaira-shi, Tokyo 187-0035, Japan; morisgk3@gmail.com

[4]   Department of Mechanical System Engineering, Tokyo Metropolitan University, 1-1 Minami-Osawa, Hachioji-shi, Tokyo 192-0397, Japan; tada-kazuo@jmj.tmu.ac.jp (K.T.); manabe@tmu.ac.jp (K.-i.M.)

*   Correspondence: yoshi@shibaura-it.ac.jp; Tel.: +81-3-6722-2741

Received: 21 December 2019; Accepted: 26 January 2020; Published: 30 January 2020

Abstract: In this study, the material behavior in the T-shape microtube hydroforming (MTHF) of pure copper and stainless-steel SUS304 microtubes with an outer diameter of 500 μm and wall thickness of 100 μm was examined experimentally and numerically. This paper elucidates the basic deformation characteristics, the forming defects, and the forming limit as well as the effects of lubrication/friction and tube length. The hydroformability (bulge height) of the SUS304 microtube was shown to be higher than that of the copper microtube because of the high buckling resistance of SUS304. Good lubrication experimentally led to the high hydroformability of T-shape forming. The length of the microtube significantly affects its hydroformability. Friction resistance increases with increasing tube length and restricts the flow of the microtube material into the die cavity. By comparing the T-shape and cross-shape MHTF characteristics, we verified the hydroformability of the T-shape microtube to be superior to that of the cross-shape microtube theoretically and experimentally. In addition, the process window for T-shape MTHF had a narrower "success" area and wider buckling and folding regions than that for cross-shape MTHF. Furthermore, conventional finite element (FE) modeling without consideration of the grains was valid for MTHF processes owing to the many grains in the thickness direction.

Keywords: microtube; hydroforming; T-shape bulging; tube materials; friction; tube length; micro hydroformability; process window; FE analysis; microstructure

---

1. Introduction

In recent years, the demand for extremely small tubular parts has increased with the miniaturization of equipment and devices in medical, communication, and electronic fields [1]. To fabricate these microscale metallic tubular parts, tube hydroforming (THF) technology is expected to be applied [2]. However, the hydroforming technology for microtube components requires high precision and fine tooling, a sealing system that can withstand ultrahigh pressure, and a forming machine suitable for microscale processing [3]. In addition, the THF of microtubes might be impossible to simply scale down because of elastic deformation of the tool compared with the normal scale [4], and the size effect [5], friction [6], and other effects will become prominent. In addition, the wall thickness of the microtube increases with downscaling, and handling of the microtube becomes difficult [7]. Therefore, to achieve microtube hydroforming (MTHF), it is necessary to solve these problems. So far, the following investigations have been carried out on MTHF.

Hartl et al. [4,5] successfully deformed an SUS304 stainless-steel tube with an outer diameter of 0.8 mm and a wall thickness of 0.04 mm, where the tubes were expanded to an outer diameter of 1.04 mm. In this process, an internal pressure of 120 MPa, clamping force of 16 kN, and axial compressive force of 800 N were loaded on the SUS304 tube. At the same time, the dimensional error, which was affected by the elastic deformation of the tool, and the cracks caused by the increase in crystal grain size relative to the dimensions of the tube blank were examined. Ngaile et al. [6] developed a simple sealing system with a floating die that could load a maximum internal pressure of 140 MPa on the tube and a maximum die clamping force of 1500 kN. MTHF was carried out using SUS304 tubes with an outer diameter of 1 mm and wall thickness of 0.1 mm, and an outer diameter of 2 mm and wall thickness of 0.2 mm. These tubes were deformed experimentally into T and Y-shapes. In addition, the effects of friction and tube dimensions on the axial compressive force were evaluated by finite element method (FEM) analysis. Cross-shape forming has been investigated by Shirayori et al. [8,9]. Aluminum alloy tubes and copper tubes with outer diameters of 8 mm and thicknesses of 0.5 and 0.8 mm were used in these studies. An aluminum alloy tube loaded with an internal pressure of 25 MPa was deformed to a bulge height of 18.5 mm. Similarly, a copper tube loaded with an internal pressure of 62 MPa was deformed to a bulge height of 20.75 mm. Furthermore, for a fine small magnesium alloy tube with 2.0 mm outer diameter, a bulging test was carried out by Yoshihara et al. [10]. They conducted the bulging test numerically to evaluate the bulging deformation of a fine small tube, which is employed as a medical stent. However, few studies on microtubes with outer diameters of less than 0.8 mm have been reported.

Mori and Manabe, among the present authors, developed a new ultrahigh-pressure MTHF system [11] and applied it to cross-shape MTHF of a phosphorus-deoxidized copper tube with an outer diameter of 0.5 mm. The following results were obtained. (1) To operate the MTHF equipment reliably at the required precision of the microscale under even ultrahigh pressures of over 200 MPa, a new microdie structure was developed with a simplified structure of left and right split dies for easy handling. (2) To reduce the friction resistance between the tube and the die, which would be increased by the size effect, a special axial compressive punch was developed, which has eight grooves cut on the punch surface and end. At the same time, the supplying and sealing of the pressure medium under ultrahigh pressure for the microtube became possible because of the grooves on the punch. (3) A simple ultrahigh-pressure generator and an MTHF device were developed and used in an experiment on cross-shape MTHF for a microtube of 0.5 mm outer diameter. The process windows, forming limit, and forming defects were clarified and classified. (4) In the case of cross-shape MTHF, an excellent bulge height (h_{ave} = 798 μm) of 1.6 times the initial tube outer diameter was successfully attained. In addition, it was revealed that a higher hydroformability of approximately 80% of the theoretical bulge height was obtained.

However, the possibility of an MTHF process for T-shape and other tube materials for a 0.5-mm outer-diameter microtube was not studied; thus, the material deformation behavior in the T-shape MTHF process has not yet been evaluated. Furthermore, the effects of the difference in the deformation characteristics between T- and cross-shape forming processes and the influential factors such as the tube material and the length of the tube have not been clarified.

For the conventional THF process, it is known that the length of tubes and lubrication condition significantly affect the hydroformability. In the earlier studies on the hydroformability (bulge height) using different tube lengths, it was reported that a longer tube has a lower bulging height, indicating lower hydroformability [12]. However, even for conventional THF, few fundamental investigations on the THF of a long tube have been reported. On the other hand, for medical applications, there has been increasing demand for long integrated complex tubular microcomponents without welded or joined portions in recent years.

In this study, a micro-T-shape forming experiment is performed using the MTHF system developed previously [11]. The objective of the current study is to confirm the deformation behavior and evaluate the hydroformability of the microtube in T-shape forming experimentally and numerically. In addition, the effects of the tube material, tube length, and lubrication condition on the hydroformability of microtubes are clarified. Furthermore, by comparison with the previous cross-shape MTHF results, the featured forming characteristics for T-shape MTHF are clarified in detail.

2. Materials and Methods

2.1. Overview of MTHF System

Figure 1a shows the MTHF system used in the experiment. This system was successfully for cross-shape MTHF in a previous study [11]. It is constructed from an MTHF device comprising mounted dies for T-shape forming and an ultrahigh-pressure generator. The MTHF system is compact, as shown in Figure 1b. The axial compressive punch in the figure has a drive system with manually rotated screws, and the screws on each end are synchronized by a gear mechanism, as shown in Figure 1c. The axial compressive displacements at each end were measured by dial gauges connected to the screws. The ultrahigh-pressure generator has length L_u = 120 mm, width W_u = 120 mm, and height H_u = 150 mm. In addition, the MTHF device has length L_d = 200 mm, width W_d = 87 mm, and height H_d = 114 mm. Figure 2 shows the structure and dimensions of the die and the axial compressive punch for MTHF. In consideration of the ease of removing micro-T-shaped samples, the divided die, which has left and right parts as shown in Figure 2a, was employed. Both the microdie and the axial compressive punch were made of cemented carbide. The die cavity has a diameter of $0.51^{+0.010}_{0}$ mm and die shoulder radius of r_1 = 0.3 mm, and the outer diameter of the axial compressive punch is $0.51^{-0.001}_{-0.005}$ mm. The axial compressive punch has grooves (R = 0.025 mm) machined on the outer periphery and end. A hydraulic pressure medium flowing on the grooves reduces the friction, which increases owing to the size effect between the tube and the die. At the same time, the axial compressive punch enables the supply and sealing of the hydraulic pressure medium under ultrahigh pressure (Figure 2b). Figures 3 and 4 respectively show the appearance of one of the divided dies and the profiles of the die shoulder for the designed die and measured machined die. The shoulder profile in the machined die was measured using a confocal laser microscope VK-100 (KEYENCE Corp., Osaka, Japan). It can be seen that the die shoulder was cut precisely.

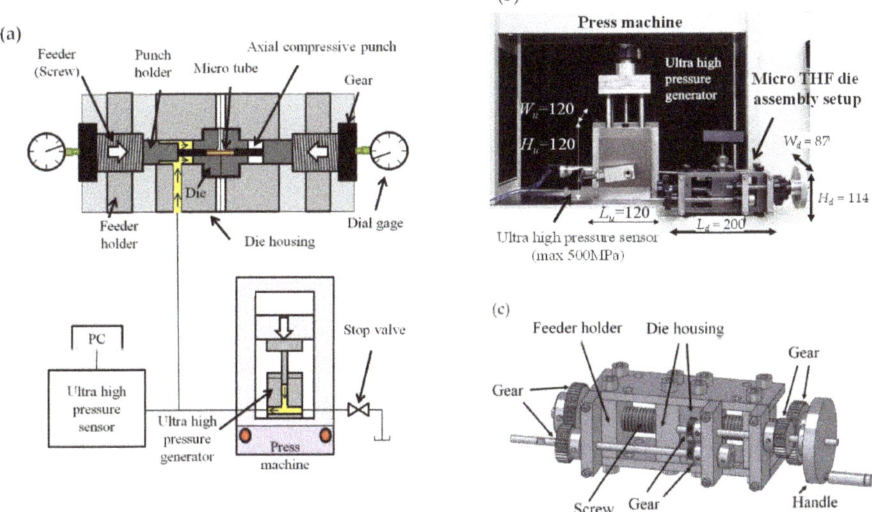

Figure 1. Microtube hydroforming (MTHF) system used for T-shape forming: (**a**) schematic of MTFH system; (**b**) appearance of MTHF system; (**c**) axial feeding mechanism in MTHF device.

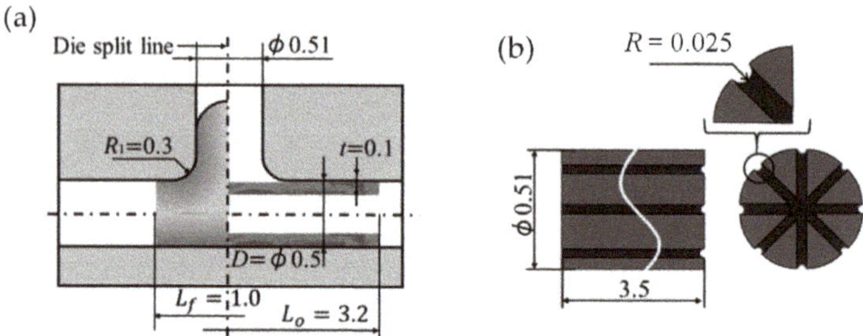

Figure 2. Geometry and dimensions of die and axial punch used for a microtube hydroforming (THF) system (unit, mm): (**a**) die; (**b**) axial punch with groove.

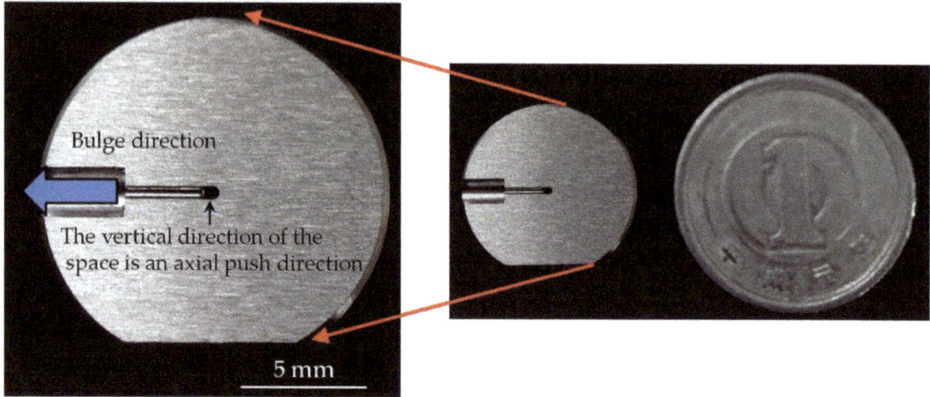

Figure 3. Split die for T-shape microcomponent (comparison of one-yen coin and microdie).

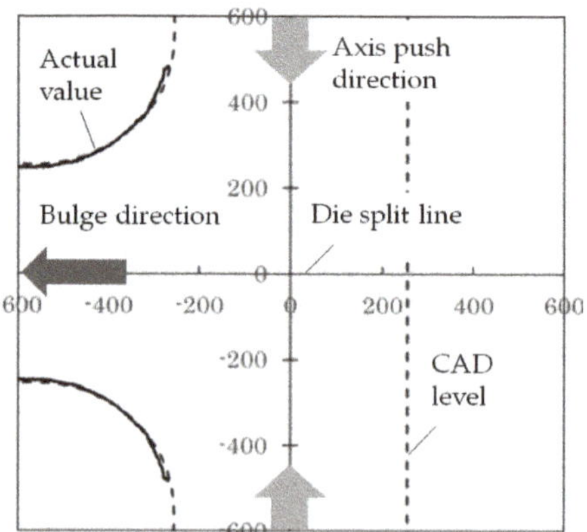

Figure 4. Design profile (dotted line) and measured profile (solid line) nearby die shoulder part for T-shape MTHF (unit, μm).

The ultrahigh-pressure generator (maximum hydraulic pressure 400 MPa) generates ultrahigh pressure by compressing the hydraulic pressure medium, which was filled into the cylinder with a piston attached to a precision universal testing machine. Loading an ultrahigh pressure of 250 MPa is possible owing to this mechanism. The hydraulic pressure was measured with a pressure sensor having a rated capacity of 500 MPa.

2.2. Materials Used and Experimental Methods

The materials used are phosphorus-deoxidized copper tubes (C1220-H) and SUS304 tubes with an outer diameter D = 0.5 mm and wall thickness t = 0.1 mm. A test specimen was cut to an initial length L_0 of 3.2 mm. The end surface of the specimen for experiments was polished with #4000 abrasive paper using a jig to increase the uniformity of the loaded axial compressive force and ensure the sealability in the pressurization stage. The perpendicularity on the tube end was ensured to be 90 ± 2°. The inner and outer diameters of the tubes were measured at cross-sections of the tubes using a digital microscope VHX-900 (KEYENCE Corp., Osaka, Japan).

For the tensile test, a precision universal testing machine AG-50kN (SHIMADZU Corp., Kyoto, Japan) was used. The tubular test specimen was chucked between the upper and lower chucks with a distance between each chuck of approximately 14 mm. Elongation of the test specimen was measured by a video-type noncontact extensometer DVE-101 (SHIMADZU Corp., Kyoto, Japan) on the basis of the gage length (approximately 10 mm). The speed of the applied tensile load was 0.5 mm/min. Figure 5 shows the nominal stress–nominal strain diagram of specimens obtained from the tensile test. It was confirmed that the test results of several specimens had material properties with little variation. A C1220 tube has an almost constant stress after yielding, and the breaking strain is 0.056, whereas the yield point of an SUS304 tube is about 2.25 times larger than that of a C1220 tube, and the subsequent strain-hardening characteristic is outstanding. Its tensile strength exceeds 1200 MPa, which is more than three times that of copper. No detailed data were obtained on the elongation from the tensile test of SUS304 in this study. In addition, by comparing with the tensile test results obtained with a different loading speed of 0.2 mm/min, no marked strain rate dependence was observed.

Figure 6a shows the basic loading path of the internal pressure p_h and axial feeding ΔL adopted in the micro-T-shape forming experiment. In the loading path, Stage (1) is the sealing of the hydraulic pressure applied, whereby the tube is compressed by the punch with distance ΔL_1 from the two ends. ΔL_1 is the distance required to seal the pressure medium flowing from the axial compressive punch to the microtube and is determined below the limit at which the microtube buckles due to axial compression. In this experiment, ΔL_1 = 150 μm for the copper tube (C1220) and ΔL_1 = 450 μm for the SUS304 tube were set on the basis of the results of previous studies [11,13]. Next, in Stage (2), the hydraulic pressure is loaded to the predetermined internal pressure p_h. This pressure is maintained after the pressure medium (oil) is filled and the air in the tube is vented. p_h is set in the range of 0 to 280 MPa. After that, in Stage (3), axial feeding to deform the tube is carried out to the maximum axial feeding length ΔL_2 of 1000 μm from each end while maintaining the internal pressure p_h.

The loading path used to confirm the process window is shown in Figure 6b. In Stage (1), the path is selected to reveal the buckling limit of the tube ΔL_b in the sealing stage. In Stages (2) and (4), the paths are for determining the bursting limit of the tube p_f. In Stage (3), the path is a combination of those in Stages (2) and (4). The path in Stage (4) traces a complicated route, and it is difficult to conduct experiments with the present system. Therefore, p_f was determined by finite element method (FEM) analysis. In this MTHF machine, machine oil was used as the hydraulic pressure medium. Fluorocarbon, which was applied by dry spraying, was employed for lubrication between the tube and the die.

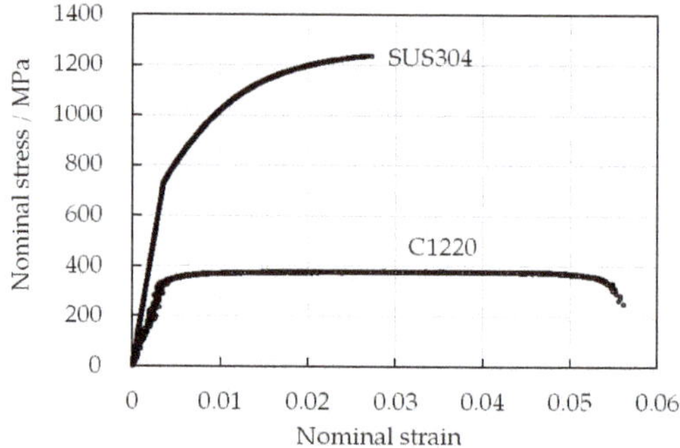

Figure 5. Stress–strain curves obtained from tensile test results of microtubes used.

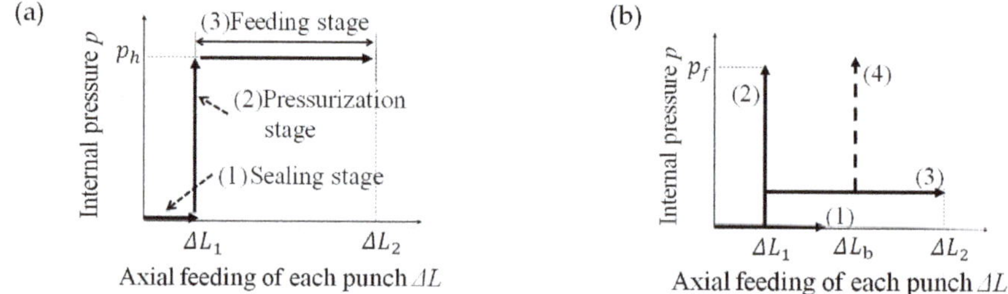

Figure 6. Loading path used in T-shape MTHF: (**a**) basic loading path; and (**b**) loading path for determining the process window.

2.3. FEM Model

The MTHF process was analyzed by FEM to evaluate the deformation behavior and characteristics of the tube. The FEM code used was the dynamic explicit LS-DYNA 3D (ver. 19. 2) (ANSYS Inc. Canonsburg, PA, USA). The FE model was constructed for one-quarter of the entire part considering the geometrical symmetry of the T-shape forming model, as shown in Figure 7. The model was based on the dimensions of the microtube and the die (Figure 2a). Solid elements were used for the tube, the die, and the axial compressive punch. A rigid body was employed for the die and the punch model. The tube model was meshed with eight-node elements and an isotropic elastoplastic material was assumed. The minimum element size of the tube was 20 µm × 20 µm × 15.7 µm. The mechanical properties of the C1220 copper microtube used in the simulations are shown in Table 1. The material characteristics of the tube were determined using the true stress–true strain diagram in Figure 5 in the uniform deformation region. The stress–strain data were input as the values approximated by a multiline approximation approach. The data in the large-strain region were extrapolated and input. In the simulation, the contact state was assumed using the normal Lagrange method. The Coulomb friction model was used in the simulation, and the friction coefficients μ_s (static) and μ_d (kinetic) between the tube and the die used in the simulation were 0.01 and 0.005, 0.025 and 0.015, 0.1 and 0.05, and 0.2 and 0.15, respectively. In the simulation, the internal pressure and the axial feeding were linearly loaded along the steplike loading path in Figure 6a.

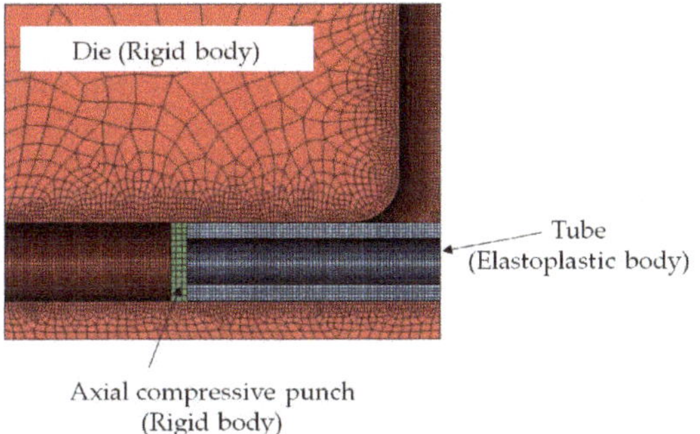

Figure 7. Schematic of finite element (FE) model for T-shape MTHF.

Table 1. Mechanical properties of C1220 copper microtube used in the finite element method (FEM).

Material	C1220
Material model	Elastoplastic body
Mass density ρ/g cm^{-3}	8.89
Elastic modulus E/GPa	118
Yield stress σ_y/MPa	324.47
Poisson's ratio v	0.3

3. Results and Discussion

3.1. Deformation Behavior in T-Shape MTHF Process

Figure 8 shows the appearance of a successful micro-T-shape sample of a C1220 tube on a finger and the magnified microsample with a rice grain [14]. Figure 9 shows the deformation behavior of the microtube during MTHF and a comparison of deformation profiles obtained by experiment and FE analysis. The input conditions of MTHF in FE analysis are p_h = 50 MPa, μ_s = 0.01, and μ_d = 0.005. The bulge height h grows with increased axial feeding ΔL_2 in both the experiment and analysis. From the comparison, it is seen that the two sets of results are in good agreement qualitatively. Using the data on bulge height shown in Figure 9, we examined the relationship between bulge height and axial feeding. Figure 10a shows the bulge height behavior in T-shape MTHF for the copper tube. The bulge height grows linearly with the increased axial feeding amount ΔL_2 in the early stage. However, in the late stage, the experimental height gradually becomes larger than the simulated bulge height. In particular, when ΔL_2 is 744 µm, a significant difference arises in the bulge height between the experimental and FEM results. This variation in the bulge height with the increasing axial feeding amount is shown in Figure 10a. From this figure, this behavior of bulge deformation growth can be well understood. In the early deformation stage, the simulated result agrees very well with the experimental result up to ΔL_2 of about 430 µm. However, after that, the experimental result increases gradually and then rapidly, but the simulated result continues to increase steadily. Figure 10b shows the shape of samples obtained in the experiment and analysis at ΔL_2 = 431 µm. Not only the bulge height but also the shape of the samples was almost in agreement. However, the shape of the samples differed between the experiment and analysis at ΔL_2 = 744 µm, as shown in Figure 10c; the branch was higher in the experiment.

Therefore, to investigate the cause of the rapid increase in the bulge height seen in Figure 10, we investigated what happens inside the tube when ΔL_2 is about 430 µm or more. Figure 11 shows the deformation state inside the tube observed from the axial cross-sections of T-shape samples with different loading path conditions. Figure 11 shows the axial cross-section of the microtube with the loading conditions of (a) p_h = 50 MPa and ΔL = 450 µm and (b) p_h = 120 MPa and ΔL = 700 µm. Under conditions (a), wrinkling/buckling occurs at the bottom side opposite the branch side. On the other hand, under conditions (b), despite the higher internal pressure applied, the buckling at the bottom of the tube is not suppressed. Moreover, since axial feeding is advanced further, the buckling deformation occurs and progresses to folding deformation. These results suggest that the difference in bulge height behavior arising in the later stage of the process in Figure 10a is due to the occurrence of buckling at the bottom of the tube, localized variation in the wall thickness distribution, and the change in the lubrication conditions during the process.

Figure 12 shows the overall appearance and axial cross-section of the T-shaped SUS304 tube. Even though the bulge height is 500 µm or more, which is greater than the bulge height in Figure 11, there is no buckling at the bottom of the tube, and the entire bottom thickens more uniformly. Thus, the thickening and buckling behaviors on the bottom of the SUS304 tube is different from those of the copper tube. The cause is considered to be the difference in the buckling resistance of the tube materials used.

The empirical buckling stress σ_K of a cylindrical tube under axial compression was formulated by Miyagawa et al. [15] on the basis of Geckeler's plastic buckling theory [16]. When the ratio of the diameter to the wall thickness of the tube is expressed as d_0/t_0, the plastic buckling stress of the tube is expressed by

$$\frac{\sigma_k}{\sigma_B} = \left(1 - \frac{0.026\left(\frac{d_0}{t_0}\right)}{0.5\left(\frac{d_0}{t_0}\right) - 1}\right) \tag{1}$$

where σ_B is the tensile strength. From Equation (1), the plastic buckling stress of the C1220 and SUS304 microtubes with a diameter of 0.5 mm and a wall thickness of 0.1 mm can be calculated by using the tensile strengths of the copper tube, σ_B = 375 MPa, and the stainless-steel tube, σ_B = 1235 MPa, which was obtained from Figure 5. The calculated plastic buckling stresses of the copper microtube and stainless-steel microtube were 342.5 MPa and 1127.97 MPa, respectively. Thus, the nominal strain at the onset of plastic buckling can be obtained from the stress–strain diagram in Figure 5 using the calculated plastic buckling stresses. The nominal strains of the copper and stainless-steel tubes are determined to be approximately 0.00413 and 0.0144, respectively. Thus, it is seen that the strain at the onset of plastic buckling strongly depends on the material properties; the plastic buckling onset strain of SUS304 is much larger than that of C1220, and the plastic buckling resistance of SUS304 is high. Since SUS304 has a higher plastic buckling resistance than C1220, it is concluded that for SUS304, the occurrence of plastic buckling at the bottom side opposite the branch is suppressed and delayed. Consequently, it can be concluded that a microtube material with high buckling resistance has high micro-hydroformability in T-shape MTHF.

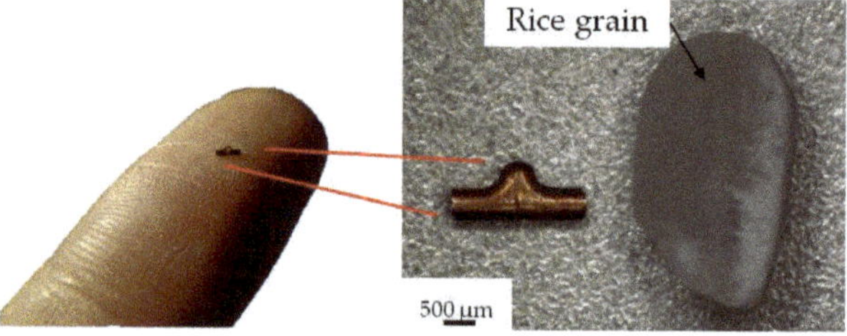

Figure 8. Micro-T-shape sample on the finger and magnified sample compared with a rice grain (C1220).

ΔL_2 / μm	269.5	431	580	744
Experiment				
h / μm	201	365	588	878
FEM μ_s, μ_d = 0.01, 0.005				
h / μm	229.187	361.901	466.42	606.32

Figure 9. Deformation behavior of microtube in T-shape MTHF (p_h = 50 MPa, μ_s = 0.01, μ_d = 0.005) (C1220).

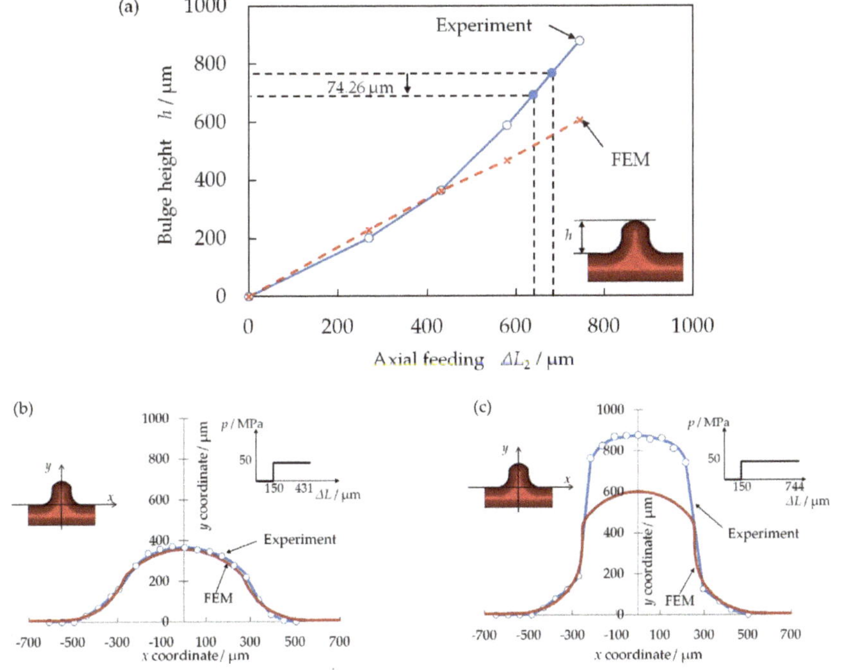

Figure 10. Bulging behavior and bulge shape in experiment and simulation (μ_s = 0.01 and μ_d = 0.005) (C1220, fluorocarbon): (**a**) bulging behavior; (**b**) bulge shape at ΔL_2 = 431 μm; (**c**) bulge shape at ΔL_2 = 744 μm.

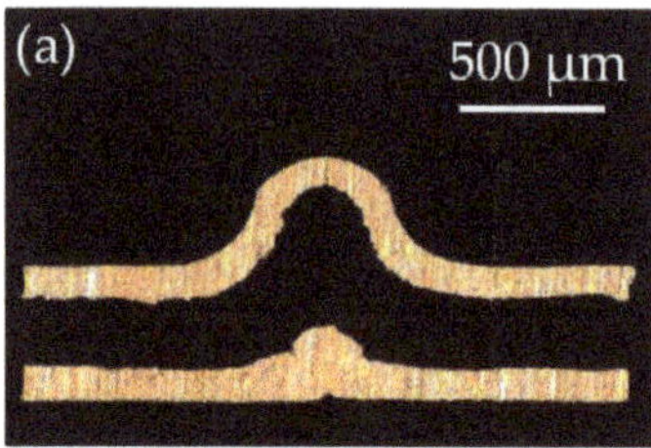 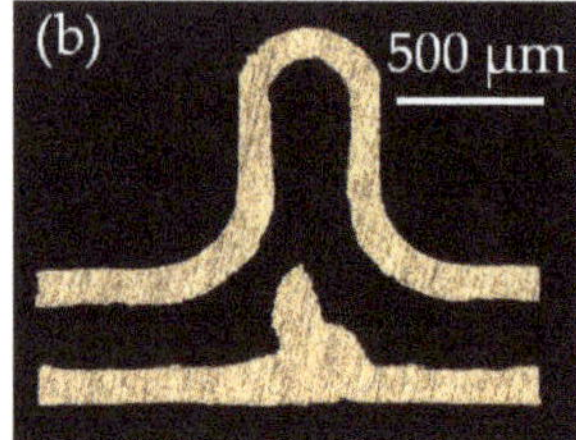

Figure 11. Axial cross-section of T-shape samples (C1220) under different MTHF conditions: (**a**) sample hydroformed at p_h = 50 MPa, ΔL = 450 μm; (**b**) sample hydroformed at p_h = 120 MPa, ΔL = 700 μm.

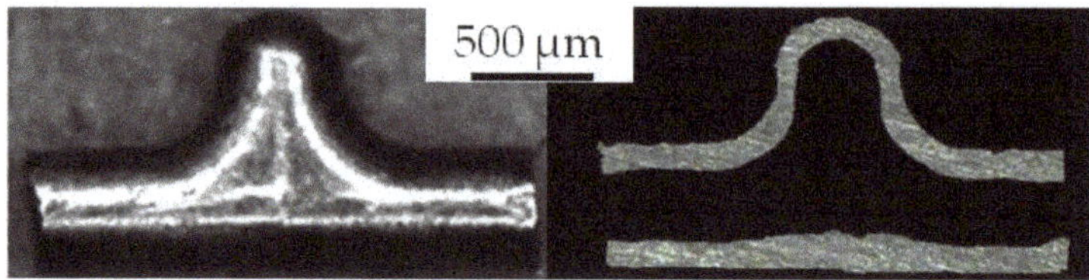

Figure 12. Appearance and axial cross-section of T-shape sample of SUS304 (p_h = 70 MPa, ΔL = 750 μm).

3.2. Effect of Lubrication/Friction on Bulge Height and Material Behavior in Micro-T-Shape Forming

The size effects of lubrication/friction are that the tribological characteristics increase with the miniaturization of process dimensions [17], and with the microscaling of dimensions, the lubrication effect decreases [18]. In the deformation of microscale parts, friction affects the formability of parts because the ratio of the volume to surface area of parts increases [6,18]. Therefore, it is necessary to evaluate the effect of lubrication/friction on hydroformability in MTHF. To determine confirm such effects, the bulge height h after the T-shape MTHF of a C1220 tube was evaluated under different lubrication conditions: no lubrication (degreasing treatment) and fluorocarbon lubrication. Figure 13 shows the appearance of the initial test specimens and the T-shape samples under the different lubrication conditions with the internal pressure p_h and the axial feeding amount ΔL_2 during loading unchanged. The bulge height h obtained with fluorocarbon lubricant is considerably larger than that with no lubrication. However, under the fluorocarbon condition (Figure 13a), the amount of axial feeding is slightly larger than that in the case of no lubrication (Figure 13b). As a result, the bulge height h with fluorocarbon lubricant was twice that of the sample without the fluorocarbon coating shown in Figure 13b. If both of these experiments had been conducted with the same amount of axial feeding (ΔL_2 = 639 μm), the bulge height h with fluorocarbon lubrication would be expected to be reduced by 74.26 μm judging from Figure 10a. However, it is possible that the lubrication state on the friction interface is complicated, and the no-lubrication condition greatly changes owing to adherence of leaked pressure medium to microtube because of the sealing mechanism mentioned below. In any case, it can be confirmed that the lubrication condition greatly affects the deformation behavior and the bulge height, or hydroformability, in T-shape MTHF. Furthermore, as can be seen from the microtooling of the MTHF system in Figure 2b, some pressure medium may flow over and adhere to the outer surface of the tube blank owing to the structure of the apparatus. Therefore, it has not been determined whether a no-lubrication state can be realized.

Next, the effects of the friction coefficient between the tube and the die on the deformation and material behavior of the tube in the micro-T-shape forming were evaluated by FEM. The sample used was a 3.2-mm-long C1220 tube. The friction coefficients μ_s and μ_d were 0.01 and 0.005, 0.025 and 0.015, 0.1 and 0.05, and 0.2 and 0.15, respectively. The internal pressure p_h applied to the tube was 50 MPa, and the amount of axial feeding ΔL_2 was 744 μm.

(a) Bulge height: Figure 14 shows the effect of the friction coefficient on the bulge height h of the T-shape samples. When the friction coefficient is low, the bulge height increases, but when the friction coefficient is large, the bulge height decreases. This is because the greater friction prevents the material flow into the die cavity and suppresses the bulge height and hence, the hydroformability. These results confirm the experimental results in Figure 13 described above. To enhance the hydroformability, the improvement of friction and lubrication conditions is essential.

(b) Wall thickness at different portions: Figure 15 shows the effect of the friction coefficient on the wall at the crown of the bulge. The wall at the crown of the T-shape tube becomes thick during MTHF with this loading path. As the friction coefficient increases, the wall thickness at the crown increases and reaches the maximum when μ_s and μ_d are 0.1 and 0.05; then, it decreases considerably when μ_s and μ_d are 0.2 and 0.15, respectively. This behavior can be explained as follows. When the friction coefficient is small, there is a large amount of material flow into the die cavity, and at this time, the overall wall thickness increases uniformly as a result of axial feeding, so that the wall at the crown can thicken gradually with the increasing friction coefficient. Meanwhile, when the friction coefficient increases, the compressive deformation and wall thickening become concentrated near the tube end area so that the wall at the crown can only deform in the early stage and not thicken so much under a high friction coefficient. Furthermore, when the friction coefficient increases, the material flow itself into the die is also suppressed. Then, since the bulge height is reduced, as shown in Figure 14, the thickness at the crown increases only slightly. Therefore, maximum thickness of the wall at the crown is the net effect of the increase in the thickness owing to the axial feeding and the decrease in the thickness owing to the decrease in the bulge height resulting from the increased friction coefficient.

Figure 16 shows the effect of the friction coefficient on the wall thickness at the bottom opposite the T-shape bulged part. The wall thickness at the bottom decreases with the increasing friction coefficient. In the case of a low friction coefficient, most of the axial compressive force due to axial feeding can be transmitted at the center of the tube bottom where the tube material cannot move, and the axial compressive force causes a significant increase in the wall thickness. However, when the friction coefficient increases and the values of μ_s and μ_d are 0.2 and 0.15, respectively, the compression force transmitted to the center of the tube bottom is too small; thus, the amount of tube material supplied to the branch cavity is not enough for a sufficient increase in the wall thickness.

Figure 17 shows the effect of the friction coefficient on the wall thickness of the T-shape samples at the contact surface between the tube and the axial punch. The wall thickness at the end of the tube increases with increasing friction coefficient. This is apparently because the frictional resistance also increases towing to the increase in the friction coefficient and because the end of the tube that receives the greatest frictional resistance is greatly compressed, thereby increasing the wall thickness. On the other hand, when the friction coefficients μ_s and μ_d are 0.2 and 0.15, respectively, the wall thickness is slightly increased, and a peak t_{max} is observed in the inner portion from the contact surface with the axial punch. The high coefficient of friction on the contact surface between the tube and the axial punch prevents the increase in wall thickness at the end of the tube. Thus, the tube material that could not flow toward the die cavity owing to a high coefficient of friction accumulates in the inner portion from the contact surface with the axial punch, and a peak is observed. On the basis of these results obtained from experiments and FE analysis, it was confirmed that the friction/lubrication condition is an extremely important processing factor affecting the flow and behavior of the tube material to fabricate highly accurate and high-quality microcomponents and to enhance the hydroformability.

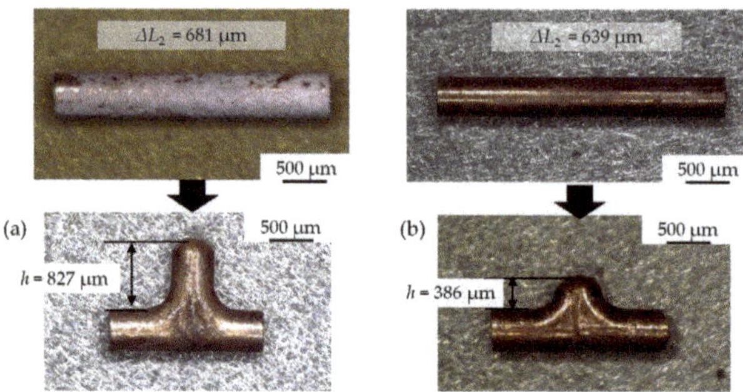

Figure 13. Effect of lubrication condition on the hydroformability (bulge height) of T-shape MTHF: (**a**) fluorocarbon; (**b**) no lubrication (p_h = 120 MPa, ΔL_1 = 150 μm) (C1220).

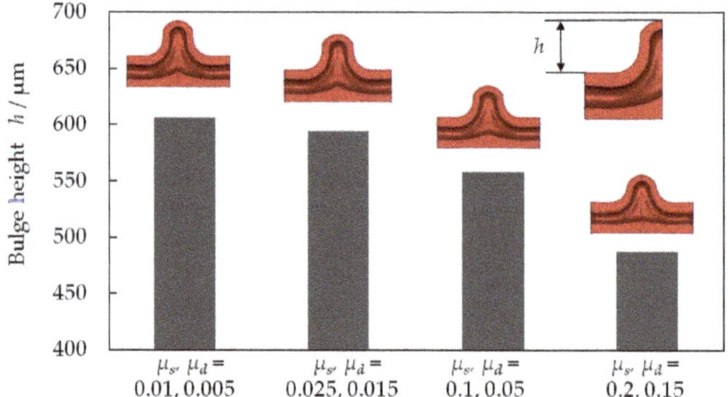

Figure 14. Effect of friction coefficient on bulge height of a micro-T-shape tube (p_h = 50 MPa, ΔL_1 = 150 μm).

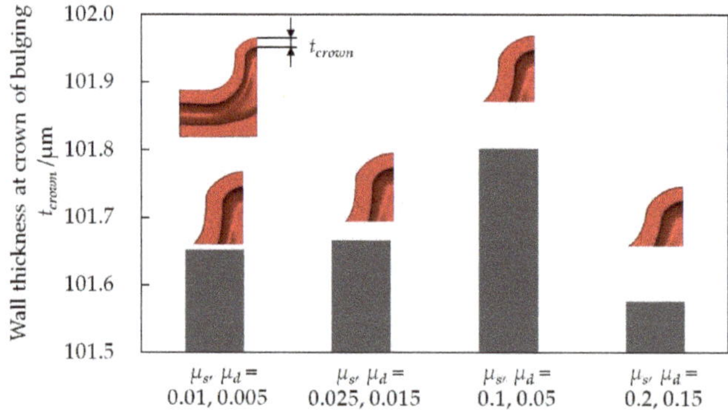

Figure 15. Effect of friction coefficient on wall thickness at bulged crown of a micro-T-shape tube (p_h = 50 MPa, ΔL_1 = 150 μm).

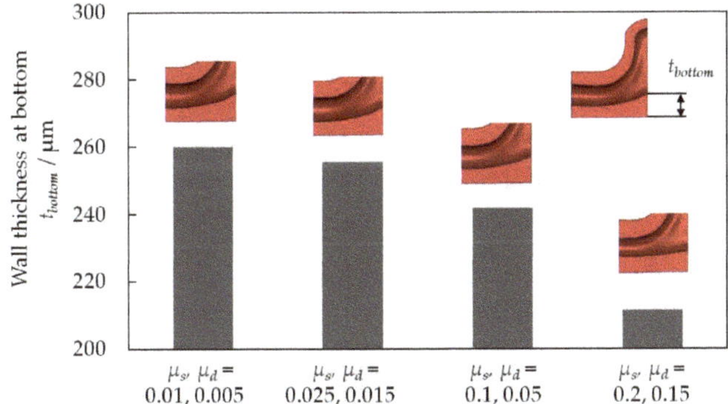

Figure 16. Effect of friction coefficient on wall thickness at bottom opposite branch part of a micro-T-shape tube (p_h = 50 MPa, ΔL_1 = 150 μm).

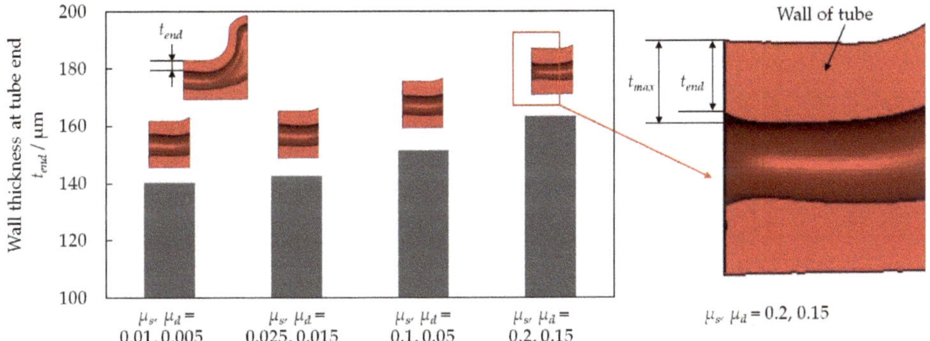

Figure 17. Effect of friction coefficient on wall thickness at end of a micro-T-shape tube (p_h = 50 MPa, ΔL_1 = 150 μm).

3.3. Effect of Tube Length on Hydroformability in T-Shape MTHF

The length of the tube is known to affect hydroformability in connection with friction and lubrication. However, there has been little research on the effects of tube length on hydroformability so far. Here, the effect of the length of the tube on its deformation behavior in the T-shape MTHF process is investigated by FE simulation. The length of the tube ranged from a short tube with L_0 = 3.2 mm (L_0/d_0 = 6.4) to a long tube with L_0 = 20 mm (L_0/d_0 = 40). The friction coefficients μ_s and μ_d are 0.1 and 0.05, respectively. The loading conditions are an internal pressure p_h = 50 MPa and axial feeding amount ΔL_2 = 750 μm.

Figure 18 shows the effect of the tube length on the deformed profile and equivalent plastic strain distribution of the micro-T-shape tube. The bulge height decreases with the increasing initial tube length ($L_0 = 2L_h$). In terms of the equivalent strain distribution of the T-shape sample, when the amount of axial feeding is the same, the longer the tube, the lower the overall average strain, but the strain distribution state differs greatly. In the case of L_0 = 3.2 mm (L_h = 1.6 mm), a sufficient bulge height is obtained, and the entire region is deformed and strained. Regarding the characteristic strain distribution, a particularly large equivalent strain with shear deformation is concentrated along the centerline of the bulge. However, when L_0 = 10 mm (L_h = 5 mm), the height of the bulge formed is insufficient, so that the strain in the bulge region is low, and the area around the tube end is larger. The highest strain occurs at the end of the tube, and it gradually decreases toward the center of the tube. It can be seen that the length of the axial feeding-affected region is about 4.5 mm in the axial direction, but there is less deformation from there toward the bulge die-cavity entrance. In addition, it is suggested that when the bulge height is smaller, the displacement increment vectors at the center of the tube are small, and the vectors hardly rotate toward the bulge direction. This result shows that the amount of material flowing in the bulge region decreases when a long microtube is deformed, and insufficient material flow could lead to a smaller bulge height h.

Moreover, when $L_0 = 20$ mm ($L_h = 10$ mm), the bulge-forming process is incomplete. Hardly any deformation occurs at the center of the bulge, and thereby, only the area near the end of the tube is strained by axial feeding compression. From the figure, the axial-feeding-affected zone (AFAZ) has a length of about 4.5 mm, which is similar to that in the case of $L_0 = 10$ mm. Generally, the factors affecting this AFAZ are considered to be the processing internal pressure, friction coefficient, tube strength, and wall thickness, all of which are related to material flow. However, according to Figure 18, the length of this AFAZ does not vary greatly and is considered to be a measure of the initial length of the tube, which indicates the possibility of bulge deformation with axial feeding. For a tube with a length close to that of this AFAZ, the material flow due to axial feeding and the material flow into the die cavity causing bulge deformation may occur at the same time. It is estimated that the initial length is slightly greater than the length of the AFAZ.

From these results, it is suggested that for longer tubes, increasing the axial feeding does not necessarily increase the bulge height and only compresses a certain area from the end of the tube, resulting in material flow only in that area. In addition, not only the strain distribution but also the displacement increment vector from the initial position at each node of the finite element mesh was observed; thus, the material flow was examined. As a result, it was also confirmed that the material flow state almost corresponds to the equivalent strain distribution in Figure 18.

As Figure 19 shows, it is clear that the bulge height decreases rapidly with the increasing initial length of the tube. Therefore, in the case of different values of the friction coefficient, it is predicted from the results in Figure 14 that the bulge height curve is below the curve in Figure 19 when the friction coefficient is larger and slightly above it when the friction coefficient is smaller. As a result, it is suggested that bulge forming becomes difficult if friction is increased, even with a shorter tube ($L_0 = 10$ mm ($L_h = 5$ mm)).

Here, the theory on the axial compressive force required in T-shape forming is formulated as the following equation referring to Figure 20 [19].

$$F_X = F_P + F_\mu + F_f = \frac{\pi(d_0 - 2t_0)^2}{4}p + \mu\pi d_0 l_\mu p_{die} + \frac{\pi t_0(d_0 - t_0)k_f}{2} \tag{2}$$

Here, F_P is the axial compressive force against internal pressure, F_μ is the axial compressive force against the friction between the tube and the die, F_f is the axial compressive force related to the plastic deformation of the tube, p is the internal pressure for deforming, p_{die} is the contact surface pressure on the die cavity surface, d_0 is the outer diameter of the tube, l_μ is half the contact length between the tube and the die, t_0 is the wall thickness of the tube, k_f is the yield stress of the tube during forming, and μ is the friction coefficient. Then, $l_\mu = l_0 - x$, where l_0 is half the initial tube length and x is the axial displacement of the tube (one side).

From the second term of Equation (2), we see that the axial compressive force required in the hydroforming of the microtube depends on the tube length l_μ. That is, the axial compressive force against the friction resistance F_μ between the tube and the die increases in proportion to the current tube length l_μ. When T-shape forming was carried out with a longer tube, the deformation of the tube occurred only at the end of the tube, as shown in Figures 17 and 18. As mentioned above, the cause of the reduction in the bulge height h is the effect of friction on the material flow into the die cavity. If the tube is longer, the material cannot flow toward the branch because of the increased friction effect. It can be concluded that insufficient material flow leads to lower bulge height h. For the deformation of a long microtube in MTHF, not only reduced friction between the tube and the die but also a higher axial compressive force is required. Additionally, the axial compressive force during the MTHF, which depends on the normal force (pressure p), is a product of p_{die} and the friction coefficient μ between the tube and the die. An excellent friction state might lead to an increase in the size of the region that can propagate the axial compressive force. Therefore, it is necessary to develop a processing technique that suppresses frictional (flow) resistance due to friction and promotes the plastic flow of the material by maintaining satisfactory friction and lubrication states as much as possible.

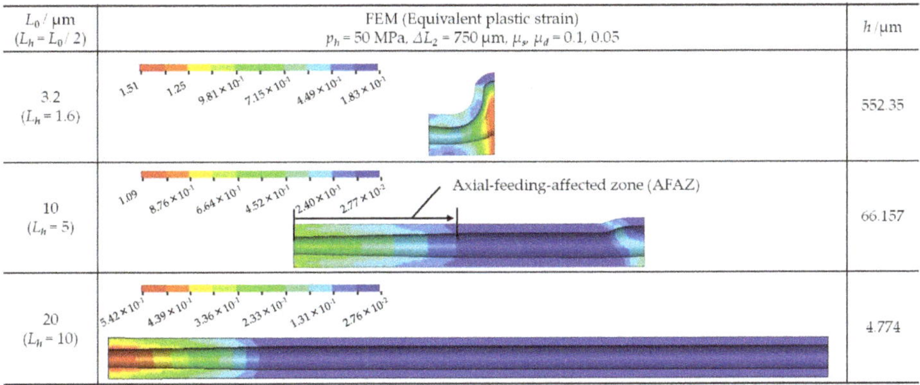

L_0 / μm ($L_h = L_0 / 2$)	FEM (Equivalent plastic strain) $p_h = 50$ MPa, $\Delta L_2 = 750$ μm, μ_s, $\mu_d = 0.1$, 0.05	h / μm
3.2 ($L_h = 1.6$)		552.35
10 ($L_h = 5$)		66.157
20 ($L_h = 10$)		4.774

Figure 18. Effect of the initial tube length on the deformation and equivalent plastic strain of a micro-T-shape tube.

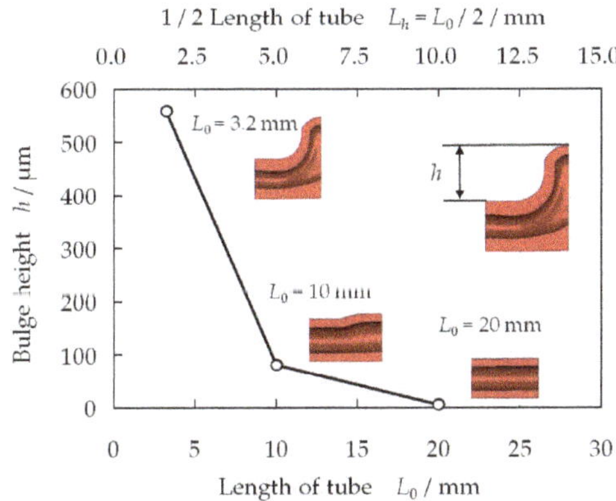

Figure 19. Effect of the initial tube length on the bulge height of a micro-T-shape tube.

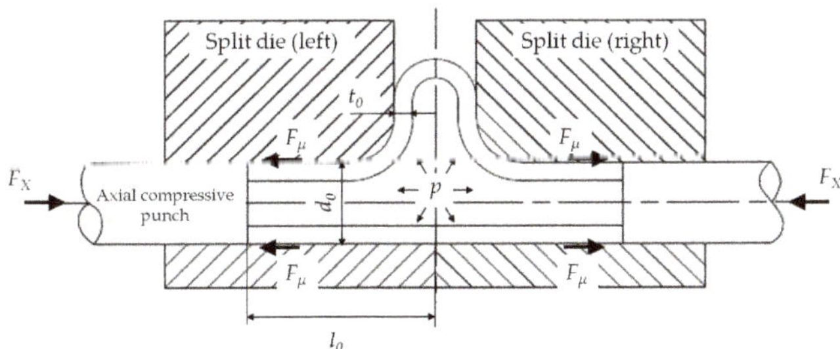

Figure 20. Axial compressive force acting against friction between tube and die F_μ during processing.

3.4. Microforming Defects and Process Windows

Figure 21 shows typical microforming defects of T-shape samples in both the T-shape forming experiment and FE analysis. There are three typical modes of the forming defects: (1) thinning and fracture of the branch wall (Figure 21a), (2) thickening and folding at the bottom side opposite the bulge (Figure 21a,b), and (3) rupture at the crown of the bulging branch (Figure 21c). The type of defect depends on the lubrication condition and loading path.

On the basis of these forming defects, a process window for T-shape MTHF is constructed. Figure 22 shows the process window created using all data obtained from experiments and FE analysis. The bursting pressure of the microtube increased with the amount of axial feeding. Similar results were obtained in previous studies on T-shape forming [20]. It can be understood that the wall thickness of the microtube was increased by the axial-feeding-affected bursting pressure of the microtube. In the low internal pressure region of the window, thickening occurred on the bottom side at the initial stage of T-shape forming. Then, this thickening led to buckling and folding with an increasing amount of axial feeding as shown in Figure 11. On the other hand, the defect of the deformed branch appeared as headband-shaped thinning or rupture in the low internal pressure region where buckling or folding wrinkles occurred. In the T-shape MTHF experiment, it is not easy to constantly maintain a stable lubrication state between the tube and the die. Even with forming at low internal pressure, different defects occur at different locations because of the variation in the lubrication state. This is a phenomenon that was not observed in the cross-shape MTHF experiment described in the previous report [11]. In the forming of a T-shape, which is a geometrically asymmetrical shape, it is difficult to obtain a process window with clear divisions between the defects such as buckling and rupture, unlike a process window for the forming of a cross-shape, which is symmetrical. The material flow in T-shape forming occurs toward the die cavity in only one direction. Therefore, the plastic flow of material in T-shape forming is not smooth because it is more complex than that in cross-shape MTFH, which has two opposing flow directions.

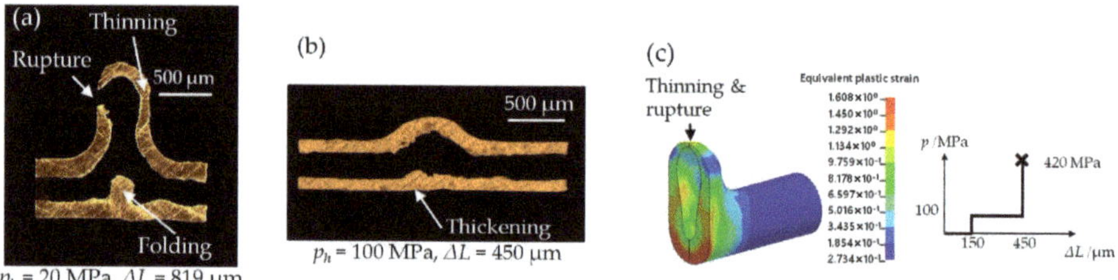

Figure 21. Typical failure modes that occur in the T-shape MTHF process observed in experiments and FE simulation (C1220): (**a**) folding, thinning and rupture; (**b**) thickening; (**c**) rupture at crown.

Figure 23 shows the process window of T-shape forming with the results of cross-shape forming for comparison. There was a "success" region that suggests an optimal process design between the rupture region with high internal pressure and the low-internal-pressure region where complex defects occurred. The "success" area in T-shape forming shifted toward the higher pressure direction compared with that in cross-shape forming. Thinning of the wall and bursting occurred at the crown of the bulge in the high-internal-pressure region in the process window. On the other hand, these defects were different from the headband-shaped thinning and rupture occurring in the low-internal-pressure region. Figure 24 shows the thinning behavior of the T-shape and cross-shape samples loaded only with internal pressure and analyzed by FEM. In cross-shape forming, thinning, which led to cracking of the tube, occurred at the border of the two branch tubes. In T-shape forming, thinning, which led to rupture of the tube, occurred at the crown of the bulge. As shown in Figure 24, the location of cracks depends significantly on the shape of the formed microproduct. The difference in their process windows is influenced by not only the difference in the flow of the material in the die cavity but also the location of the crack.

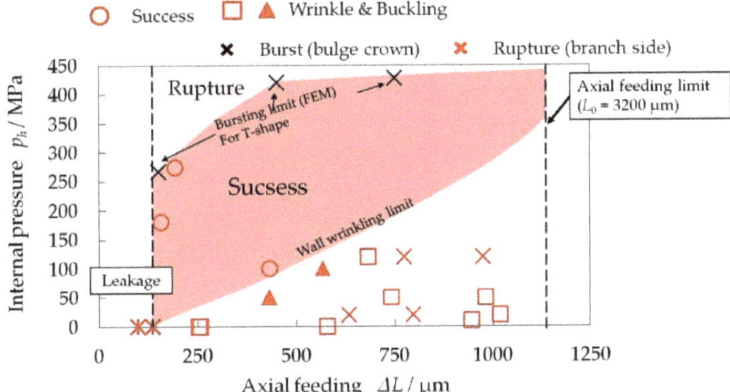

Figure 22. Process window for T-shape MTHF (C1220).

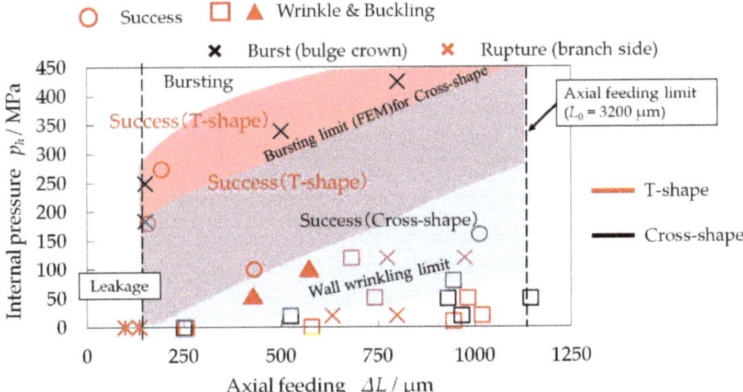

Figure 23. Process windows for T-shape MTHF with results of cross-shape MTHF plotted for comparison (C1220).

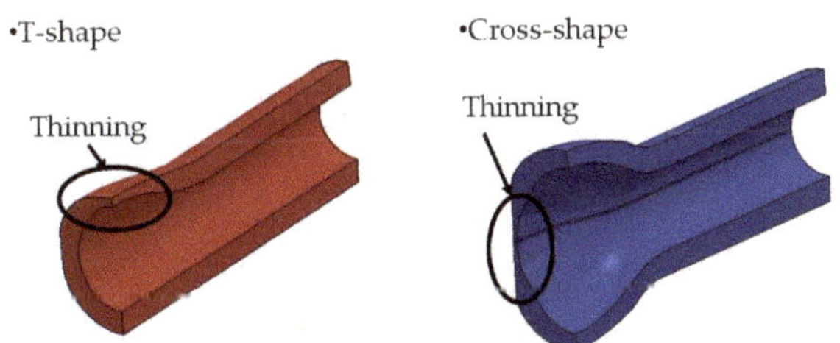

Figure 24. Sections where thinning occurs for T-shape and cross-shape MTHFs.

3.5. Theoretical and Experimental Bulge Height of T-Shape and Cross-Shape MTHF

Figure 25 shows the relationship between the bulge height h and the amount of axial feeding ΔL_2 for a C1220 microtube with fluorocarbon lubrication plotted using data from T-shape MTHF experiments with internal pressures p_h of 20 to 274 MPa. The results for cross-shape forming are also plotted in the figure for comparison. The bulge height h of the cross-shape sample was measured as half of the bulge heights on the two sides ($h = (H_f - 500)/2$), where H_f is the height between two branches, so that it could be directly compared with the bulge height h of the T-shape samples. The regression lines of both experimental results are represented by

$h = 1.3437\Delta L - 186.61$ for T-shape forming, and

$h = 0.852\Delta L - 120.58$ for cross-shape forming.

The bulge height h of the T-shape sample was higher than that of the cross-shape sample. It can be understood that the flow of the material at the bottom was led toward the branch in the sample in T-shape forming. Equations (3) and (4) respectively present the theoretical equations for the bulge height h in T-shape and cross-shape forming. In both formulae, the bulge height is proportional to the axial displacement, and the coefficient is different between them. The volume and wall thickness of the tube are assumed to be unchanged after MTHF forming, on the basis of the relation shown in Figure 26. The applicable range of the axial feeding amount ΔL for the theoretical formula is equal to the range of displacement from the value when the height of the bulge having a hemispherical branch tube head with a radius of $D/2$ is added to the die shoulder height r_1 to the value when the axial punch head reaches the boundary of the die shoulder in the axial direction.

- T-shape forming

$$h = 4\Delta L + A - \frac{4(V_a + V_c + V_e)}{\pi(D^2 - d^2)} \tag{3}$$

- Cross-shape forming

$$h = 2\Delta L + A - \frac{4(V_a + V_c + V_e)}{\pi(D^2 - d^2)} \tag{4}$$

$$A = 4r_1 + 3D \tag{5}$$

$$V_a = \frac{\pi(D^3 - d^3)}{6} \tag{6}$$

$$V_c = \left(1 - \frac{\pi}{4}\right)(2r_1 + D)^2(D - d) \tag{7}$$

$$V_e = \pi^2\left(\frac{D}{8} + \frac{r_1}{4}\right)(D^2 - d^2) - \frac{\pi(D^3 - d^3)}{6} \tag{8}$$

Figure 27 shows the experimental and theoretical results of the bulge height in the process window for T-shape and cross-shape forming. From Equations (3) and (4), it can be seen that the bulge height h of the T-shape sample can theoretically be twice that of the cross-shape sample under the same amount of axial feeding ΔL. However, in the experimental data, the bulge height h obtained with one-way material flow in T-shape forming is approximately 150% that of the cross-shape with thinning of the wall in two-way material flow. In addition, the FE analysis results are also plotted in the figure. From this figure, the results of FE analysis are in good agreement with the experimental data for both T-shape and cross-shape MTHF; together with Figures 9 and 10a; the validity of FE modeling for this MTHF process is consequently verified.

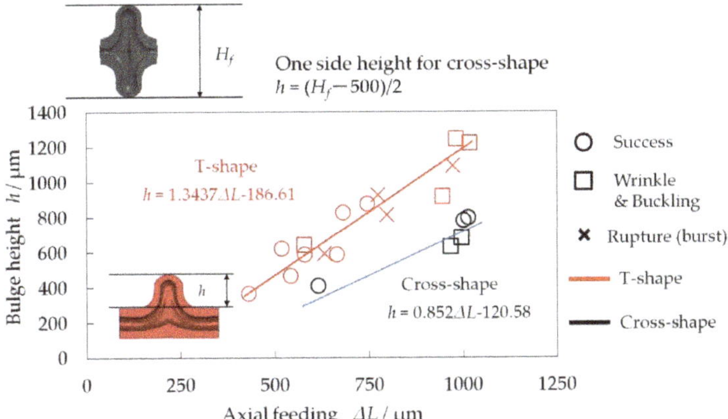

Figure 25. Relationship between bulge height and amount of axial feeding for T-shape and cross-shape hydroforming (C1220, experiment).

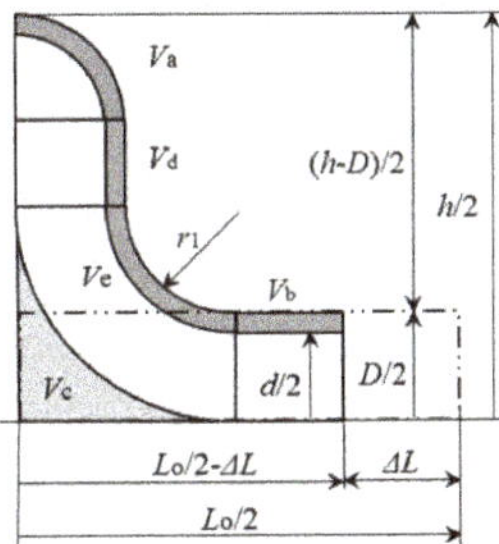

Figure 26. Model of geometric relationship between bulge height h and amount of axial feeding ΔL in MTHF process.

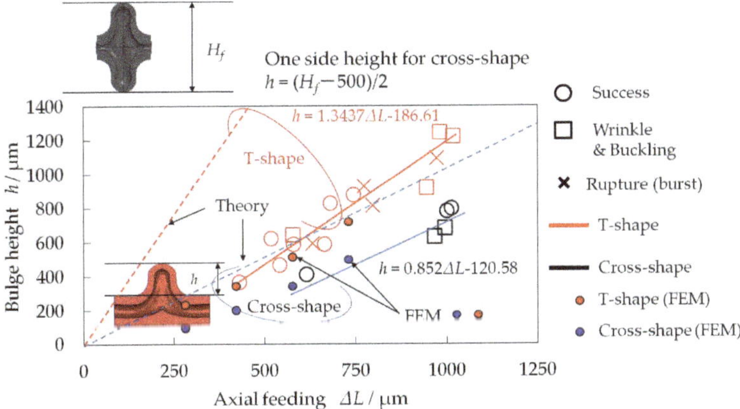

Figure 27. Theoretical and experimental results showing relationship between bulge height and amount of axial feeding for T-shape and cross-shape hydroforming (C1220).

3.6. Microstructure of Microtube in T-shape MTHF

Figure 28 shows the microstructure of the initial microtube of C1220 copper. The original microtube has approximately 20 fine-grained layers in the wall thickness direction with a lamellar texture and average grain size of approximately 4.0 μm. This lamellar texture has a circular shape in the cross section of the hoop. Therefore, it is clarified that the microstructure in the wall thickness of the microtube is stretched in the axial direction by drawing. On the other hand, to observe the microstructure after T-shape MTHF, a T-shape sample was fabricated under the forming conditions of p_h = 50 MPa, ΔL_1 = 150 μm, and ΔL_2 = 580 μm. Figure 29 shows the microstructures at different areas of the micro-T-shape sample [14]. The grain size of the T-shape sample is almost the same as that of the initial microstructure. Therefore, the evaluation of the hydroformability and deformation behavior of the microtube is possible by FE analysis, as shown previously, without considering the grain boundaries. In contrast, the microstructure of the T-shape sample changes from the initial state. As shown in Figure 29, the crystal orientation of the grains at the branch wall rotated. However, there was no rotation of the grains at the crown of the bulge, where the biaxial tensile stress was loaded. In addition, no rotated grains were observed at the area where compressive stress was loaded. In this area, the rotation of the grains was suppressed. The rotation of the grains may have been caused by uniaxial tensile stress at the branch wall.

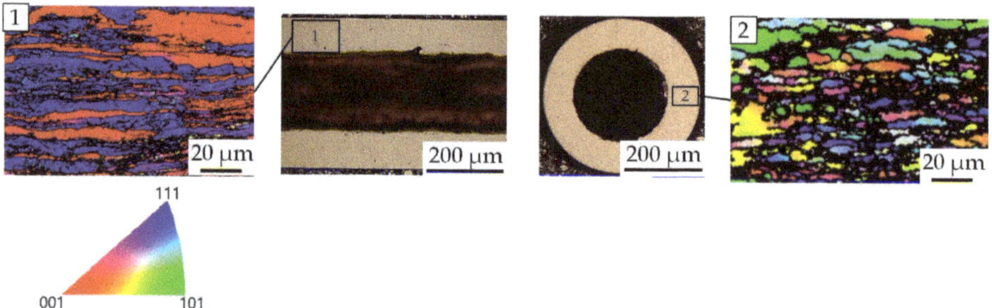

Figure 28. Microstructure in the initial microtube (C1220).

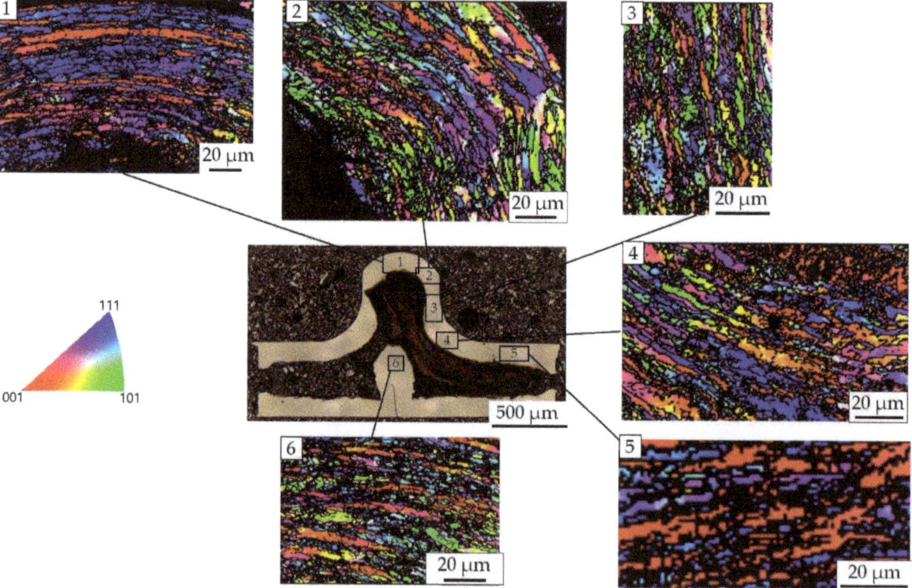

Figure 29. Microstructure at different positions of micro-T-shape sample fabricated by the MTHF process (C1220).

On the other hand, the microstructure in the folding portion at the bottom opposite the bulge had a different texture. There were grains with an average size of 3.4 µm. These grains were approximately 14% smaller than the initial grains. Therefore, it is considered that the material at the bottom opposite the bulge was loaded with higher compression and underwent greater folding deformation than the other areas, presumably as a result of grain refinement. FEM could not simulate the buckling and folding at the bottom. One of the reasons for this disagreement could be that the refinement of grains in the material was not considered in FEM.

4. Conclusions

In this study, micro-T-shape forming for phosphorus–deoxidized copper and SUS304 stainless-steel tubes with an outer diameter of 0.5 mm and a wall thickness of 0.1 mm was examined by ultrahigh-pressure MTHF and FEM analysis. The deformation behavior and forming characteristics, the effects of lubrication/friction, tube material, and length on hydroformability, and the forming defects of the tubes were examined and evaluated. The process window of T-shape MTHF was compared with that of cross-shape MTHF. The results obtained in this study are summarized below.

1. The deformation behavior and forming characteristics of T-shape MTHF were clarified from experiments and FEM. The two sets of results showed relatively good agreement. In a late forming stage, a discrepancy between experimental and simulated bulge height results was seen and is thought to be due to the occurrence of irregular deformation at the bottom opposite the bulge in the experiment, and the experimental bulge height was larger than the FEM result.
2. The hydroformability of C1220 copper microtubes is lower than that of SUS304 because the SUS304 microtubes having high buckling resistance and the occurrence of bottom buckling in the T-shape MTHF process being suppressed.
3. The microtube length, which is related to the friction resistance, greatly affects the hydroformability. Even when T-shape MTHF of a short microtube is successful, for a longer tube, even with an extremely good friction/lubrication state, the flow of material into the die cavity becomes difficult, and the process cannot be successful. Compared with cross-shape MTHF, the hydroformability (bulge height) in T-shape MTHF is superior, and increased bulge height can be obtained. Since the tube material in T-shape MTHF flows asymmetrically in one direction into the die cavity, there is a high possibility of buckling deformation easily occurring at the bottom of the bulge, and the buckling region in the process window becomes wider under low internal pressure, and the success region moves to one of higher internal pressure.
4. Lubrication and friction between the tube and the die greatly affect the hydroformability (bulge height) of the microtube. When the friction resistance increases, the amount of material flowing into the die cavity is restricted and decreases, and then the bulge height decreases. Correspondingly, the microtube greatly thickens in the vicinity of its end.
5. The asymmetric material behavior in T-shape MTHF causes a variety of undesired deformation defects in low internal pressure loading paths. This is because the plastic flow into one side of the bulge die cavity causes irregular complex deformation behavior, whereby a variety of forming defects easily occur in the early stages of T-shape MTHF compared with cross-shape MTHF. A featured typical forming defect for T-shape MTHF is buckling/folding at the bottom of the bulge.
6. The success region in the process window for T-shape MTHF shifts toward higher internal pressure compared with that for cross-shape MTHF. There were many defects in the low internal pressure region, and the area where the microtube could be deformed was also narrowed. Since the thickening of the wall was greater than that in cross-shape forming, the forming efficiency considering the bulge height deteriorated by approximately 25%.
7. From observation of the metal microstructure, 20 or more fine grains arranged in the thickness direction were observed for the C1220 copper microtube. The deformation behavior of a T-shape tube was roughly estimated by FEM analysis that did not consider the structural size effect. The

microstructure in buckling and folding wrinkles that occurred at the bottom opposite the bulge showed a decrease of approximately 14% in average grain size compared with the surrounding grains. This refinement could be caused by the difference in the buckling and folding behavior in FEM described in Equation (1) above.

Author Contributions: Conceptualization, K.-i.M.; Methodology, K.-i.M. and S.M.; experimental work and methodology, S.M.; numerical analysis, H.Y.; writing—original draft preparation, H.Y. and S.M.; final review and editing, K.-i.M.; supervision of numerical analysis, S.Y.; funding acquisition, K.-i.M.; experimental setup, K.T. All authors have read and agreed to the published version of the manuscript.

Funding: This work was supported by JSPS KAKENHI Grant Number JP18K04781 and The Die and Mold Technology Promotion Foundation.

Acknowledgments: The authors would like to express their appreciation for the implementation of this project in cooperation with Marko Vilotic of Novi Sad University and Kenta Itai, Kanta Sasaki and Naoya Kikuchi, who were the students of Tokyo Metropolitan University.

References

1. The Japan Society for Technology of Plasticity. *Tube Hydroforming Keiryoukanotameno-Seikei-Gijutsu*, 1st ed.; Morikita-Shuppan: Tokyo, Japan, 2015; pp. 1–288.
2. Manabe, K.; Fuchizawa, S. Tube Forming Technology: Toward the Highest Card for Weight Savings. *J. Jpn. Soc. Technol. Plast.* **2011**, *52*, 33 41. [CrossRef]
3. Geiger, M.; Kleiner, M.; Eckstein, R.; Tiesler, N.; Engel, U. Microforming. *CIRP Ann.* **2001**, *50*, 445–462. [CrossRef]
4. Hartl, C.; Anyasodor, G. Experimental and numerical investigations into micro-hydroforming processes and machine design. *Steel Res. Int.* **2010**, *81*, 1193–1196.
5. Zhuang, W.; Wang, S.; Lin, J.; Hartl, C. Experimental and numerical investigation of localized thinning in hydroforming of micro-tubes. *Euro. J. Mech. A/Solids* **2011**, *31*, 67–76. [CrossRef]
6. Ngaile, G.; Lowrie, J. New micro tube hydroforming system based on floating die assembly concept. *Trans. ASME J. Micro Nano-Manuf.* **2014**, *2*, 041004. [CrossRef]
7. Manabe, K. Micro forming processing of metal tubes. *B. Jpn. Soc. Technol. Plast.* **2019**, *2*, 8–13.
8. Shirayori, A.; Ando, T.; Narazaki, M.; Usui, M. Effect of Initial thickness on cross-shaped-tube hydroforming of small-diameter tubing. *J. Jpn. Soc. Technol. Plast.* **2013**, *54*, 988–992. [CrossRef]
9. Shirayori, A.; Ando, T.; Usui, M.; Narazaki, M. Effect of Initial Thickness on Cross-Shaped Tube Hydroforming of Small Diameter Tubing. In Proceedings of the 6th International Conference on Tube Hydroforming (TUBEHYDRO 2013), Jeju, Korea, 25–28 August 2013; pp. 64–67.
10. Yasui, H.; Yoshihara, S.; Yamada, R.; Ito, Y. Deformation Behavior on Small-Diameter ZM21 Magnesium Alloy Tube in Warm Tube Hydroforming. *J. Jpn. Soc. Technol. Plast.* **2019**, *60*, 346–351. [CrossRef]
11. Mori, S.; Sato, H.; Itai, K.; Manabe, K. Development of microtube hydroforming system and its application to cross-shape microtube. *J. Jpn. Soc. Technol. Plast.* **2017**, *58*, 72–77. [CrossRef]
12. Jirathearanat, S.; Hartl, C.; Altan, T. Hydroforming of Y-shapes—product and process design using FEA simulation and experiments. *J. Mater. Process. Technol.* **2004**, *146*, 124–129. [CrossRef]
13. Manabe, K.; Sato, H.; Itai, K.; Vilotic, M.; Tada, K. Factors influencing the forming characteristics in micro tube hydroforming by ultra high-forming pressure. *Procedia Eng.* **2017**, *207*, 2334–2339. [CrossRef]
14. Mori, S.; Itai, K.; Sato, H.; Tada, K.; Manabe, K.; Takahashi, S. Micro Tee Hydroforming of Copper Tube. In Proceedings of the 68th Japanese Joint Conference for the Technology of Plasticity, Fukui, Japan, 9–12 November 2017; pp. 409–410.
15. Miyagawa, M.; Back, C. The buckling of the circular cylindrical shells under uniaxial compression and nose gorming. *J. Jpn. Soc. Technol. Plast.* **1962**, *3*, 397–405.
16. Geckeler, J.W. Plastic buckling of the wall of hollow cylinders and some other folding phenomena on trays and sheets. *J. Appl. Math. Mech.* **1928**, *8*, 341–352.
17. Vollertsen, F.; Hu, Z. Tribological size effects in sheet metal forming measured by a strip drawing test. *CIRP Ann.* **2010**, *55*, 1193–1196. [CrossRef]

18. Putten, V.K.; Franzke, M.; Hirt, G. Size effect on friction and yielding in wire flat rolling. In Proceedings of the 2nd International Conference on New Forming Technology, Bremen, Germany, 20–21 September 2007; pp. 583–592.
19. Shirayori, A.; Tamanoi, S.; Narazaki, M. Elementary Analysis Modeling for Branch Forming of Small Diameter Thick-Walled Tube. In Proceedings of the 63rd Japanese Joint Conference for the Technology of Plasticity, Kitakyushu, Japan, 4–6 November 2012; pp. 183–184.
20. Fuchizawa, S.; Kitamura, K.; Narazaki, M.; Sukimoto, M.; Watanabe, I. Deformations of aluminum alloy tubes under forming of T-fitting. *J. Jpn. Soc. Technol. Plast.* **1995**, *30*, 80–86.

Article

Lubrication Analysis of Micro-Dimple Textured Die Surface by Direct Observation of Contact Interface in Sheet Metal Forming

Tetsuhide Shimizu [1,*], Hiroyuki Kobayashi [1], Jochen Vorholt [2] and Ming Yang [1]

[1]   Division of Mechanical Systems Engineering, Graduate School of System Design, Tokyo Metropolitan University, Japan 6-6, Asahigaoka, Hino, Tokyo 191-0065, Japan
[2]   Faculty of Production Engineering-Mechanical Engineering & Process Engineering, University of Bremen, Bibliothekstraße 1, 28359 Bremen, Germany
*   Correspondence: simizu-tetuhide@tmu.ac.jp; Tel.: +81-42-585-8650; Fax: +81-42-585-8440

Received: 16 July 2019; Accepted: 15 August 2019; Published: 22 August 2019

Abstract: To investigate the underlying mechanism of the effects of surface texturing on lubricated sliding friction in the metal forming operation, an in-situ observation system using transparent silica glass dies and a high speed recording camera was newly developed. To correlate the dimensional parameters of micro-dimple textured structures and tribological properties in the metal forming operation, the in-situ observation was performed during bending with the ironing process of the stainless steel sheet with a thickness of 0.1 mm. The lubrication behavior were compared between the different lubricant viscosities and the micro-dimple textures with different diameters of 10 μm, 50 μm, 100 μm fabricated by using femto-/pico-second laser processing. As a result, the textured die with dimple diameters of 10 μm and 50 μm showed the lubricant flow transferred from one to the other dimples owing to the lubricant reservoir effect, while that of 100 μm indicated the less supply of the lubricant. However, the textured die with a dimple diameter of 10 μm demonstrated higher ironing force than that of 50 μm, due to the severe adhesion of work materials inside the dimple structures. Based on these experimental findings, the dimple size dependencies on lubricant reservoirs effects and the generation of the hydrodynamic pressure were discussed by correlating with the in-situ observation results, a fluid-flow analysis and a laminar two-phase flow analysis using the finite element method.

Keywords: surface texturing; sheet metal forming; in-situ observation; micro-dimple; lubricant

---

1. Introduction

Surface texturing has attracted a great attention as a geometrical modification approach to improve tribological performances in the metal forming operation [1]. In particular, the development of laser processing technology has expanded its possibility in applying micro-textured array patterns on complex three-dimensional structures. Although the initial studies were mainly focused on the surface texturing of random dimple patterns on work materials, e.g., by shot-blast texturing [2] or by electrical discharge texturing [3], research activities are gradually shifting either for the application to metal forming tools within a great advancement in efficiency and varieties of laser processing [4].

Geiger et al. applied the micro-texture surface by using an excimer laser radiation onto a TiN coated cold forging tool surface. They have shown the significant improvement of tool life by applying the surface texture [5]. Costa and Hutchings investigated the effect of surface texturing by strip drawing tests under a lubricated sliding contact including plastic deformation of stainless steel work sheets. The impact of groove orientation to the drawing direction was demonstrated indicating the significant decrease of the coefficient of friction when it oriented perpendicularly to the drawing direction [6].

Xu et al. applied the micro-channel patterns by using laser surface texturing, and demonstrated the friction reduction during the single point incremental forming process [7].

General understanding for the functions of surface textures is the creation of the load carrying capacity owing to the hydrodynamic lift effect [8]. In addition, it acts as a lubricant reservoir, which enhances the lubricant entrainment and retention under boundary lubrication [9]. Moreover, it is also known as the structure for the wear debris entrapment to minimize the effect on damaging to the contact pair [10]. However, these effects are strongly dependent on the contact state and operating conditions, so that the optimal geometrical parameters for the design of surface texturing patterns are still under discussion.

In view of the design of surface texturing patterns, one of the most commonly used geometrical patterns are circular dimple structures sized in the range from a few microns to hundreds of microns [11]. In particular, dimple diameters, texture densities, and the relative dimple depth are the important design parameters for surface textures in contact. Most of the studies have been contributed to find optimal values for those parameters to obtain the best tribological performance to achieve the maximum load carrying capacity under hydrodynamic lubrication [12]. In comparison with the conventional studies in the field of tribology for bearing [13] or automotive piston [14], characteristic features of the contact state in metal forming is the contact pair loaded under plastic deformation. Thus, the amount of strain induced in the work material will strongly affect the real area of contact and so as the lubrication between tools and materials [15]. In such a contact state, it is well known that the role of the plasto-hydrodynamic lubrication (PHL) becomes dominant under certain specific contact conditions [16]. As the surface deformation proceeds under the mixed lubrication regime, the lubricant trapped in the valley of the surface asperities act as a hydrostatic medium [17]. In particular, this increase of the effect of PHL by the intentional formation of lubricant pockets in the work material or tool surface is called the microplastohydrodynamic lubrication (MPHDL) [18]. Thus, the surface texturing patterns for the metal forming die surface should be designed based on those specific contact states. However, the basic mechanisms of the contribution of the surface texture and design guideline optimized for metal forming operations are still under discussion, due to its difficulties in observing the lubricant flow behavior between workpieces and tools.

Within these backgrounds, the present study aims to clarify the basic mechanism and relationship between the geometrical parameters of dimple surface textures and lubrication properties during metal forming operations. In sheet metal forming, for example, the surface is loaded under contact pressure from few hundreds of MPa in the deep drawing process to more than 1 GPa such as for the ironing process, with the ratio of surface enlargement less than 2% [19]. As a first approach to investigate under these contact states, the in-situ observation system using transparent silica glass dies and a high speed recording camera was developed to directly observe the dynamic variation of the contact interface during ironing of the sheet material under the mixed lubrication regime. To correlate the dimensional parameters of surface textures and tribological properties, micro-dimple array patterns were fabricated on silica glass dies by using femto-/pico-second laser processing. By using this laser surface textured silica glass die, the in-situ observation was performed during ironing of the stainless steel sheet with a thickness of 0.1 mm. The difference in the lubricant flow behavior and its effect on the tribological properties between the different dimple structures under the different contact state was thoroughly investigated.

2. Experimental

2.1. Development of In-Situ Observation System

A new in-situ observation system was developed with a high-speed recording camera system, VW-9000, equipped with a microscope lens VH-Z75 (Keyence Corp., Osaka, Japan) attached to a sheet metal forming die assembly. A load cell and a high-precision stroke sensor were built into the sheet metal forming die assembly, which enables to determine the forming load and stroke by using a data logger. The magnification of the lens can be adjusted from 75 to 750. The whole setup is installed on a high-precision desktop-size servo screw press machine, type DT-J312, Micro Fabrication Laboratory LLC. The in-situ observation system and its schematic illustration are shown in Figure 1. To observe the contact interface between the textured die and work sheet materials, a transparent silica glass die with an elastic modulus of 72 GPa was used for bending with the ironing process. An initial arithmetic mean surface roughness (Sa) of the silica glass die was 0.3 μm. The in-situ observation was achieved by observing from the back side of this silica glass die as shown in the figure. The area of the observation was decided based on the contact state analysis using the finite element (FE) analysis described in Section 3. Another bending die opposed to this silica glass die and a punch were made of tool steels (JIS: SKD11).

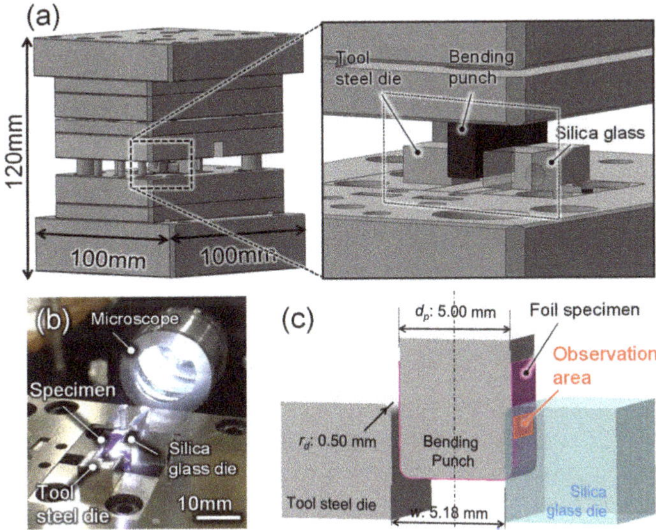

Figure 1. Schematic illustrations of the in-situ observation system, (**a**) bending with ironing die assembly, (**b**) appearance of the system and (**c**) dimensions of bending with ironing dies and punch.

2.2. Fabrication of Surface Texturing

Micro-dimple array patterns were fabricated on a silica glass die surface by using the femto-/pico-second laser in view of the great advantage in its high accuracy and ability to process the μm-scale dimple textures [20]. The geometrical parameters for the micro dimple textures are summarized in Table 1. A texture density is defined as the ratio of the textured to the non-textured area. The dimple aspect ratio is defined as the ratio of the dimple depth to its diameter [11]. The micro dimple diameters were changed from 10 μm, 50 μm to 100 μm, keeping both density and aspect ratio constant. Figure 2 shows surface profiles and cross sectional profiles of dimples with different diameters fabricated by laser processing. The dimples with semi-ellipse shapes and smooth hole edges were successfully manufactured.

Table 1. Geometrical parameters of micro dimple arrays.

Texture Density ρ_d	Aspect Ratio λ	Diameter d [mm]	Depth h_d [mm]	Pitch b [mm]
0.1	0.1	0.1	0.01	0.18
0.1	0.1	0.05	0.005	0.09
0.1	0.1	0.01	0.001	0.02

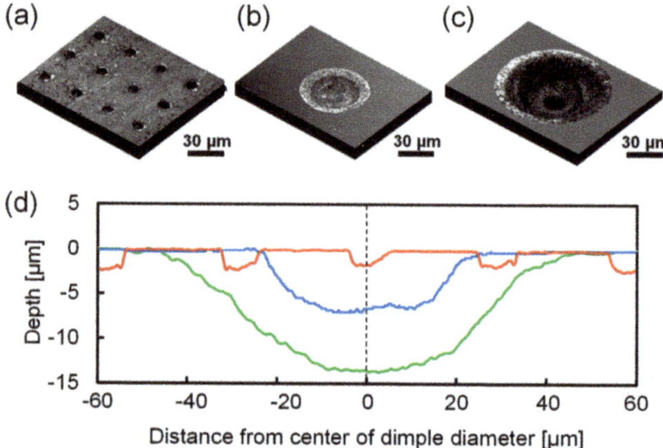

Figure 2. Confocal laser scanning microscope images of the laser textured dimple cell with a diameter of (**a**) 10 μm, (**b**) 50 μm, (**c**) 100 μm and (**d**) its surface line profiles.

2.3. Experimental Conditions

Operation conditions of bending with the ironing process are summarized in Table 2. The material used was the stainless steel JIS: SUS304-H with a thickness of 0.1 mm, a width of 15 mm and a length of 3 mm. The clearance between the punch and die was set to 0.09 mm, so that the ironing was achieved for the sheet thickness of 0.1 mm. The punch velocity was 50 mm/s with a maximum stroke of 4 mm without acceleration during the process. To investigate the effect of the lubricant viscosity on the flow behavior on the textured surface, commercially available lubricants (G-6311, G-6030M, G-3116HS, Nihon Kohsakuyu Co., Ltd., Tokyo, Japan) with three different viscosities of 1.27 mm/s^2, 2.53 mm/s^2 and 4.51 mm/s^2 were used. The in-situ observation during bending with the ironing process was carried out using silica glass dies with different dimple diameters as above. The non-textured silica glass die was also used as a reference.

Table 2. Operation conditions of bending with the ironing process.

Work Material	Stainless Steel	(JIS:SUS304-H)
Sheet thickness t	[mm]	0.1
Bending die radius r_d	[mm]	0.5
Punch corner radius r_p	[mm]	0.5
Clearance c	[mm]	0.09
Maximum stroke h_s	[mm]	4
Punch Velocity v_p	[mm/s]	50
Lubricant viscosity η	[mm^2/s]	1.27, 2.53, 4.51

3. Finite Element (FE) Simulation of Contact State

To decide the in-situ observation area along the sliding surface, the apparent contact length and contact pressure were estimated by using the FE-analysis. The analysis was done with an implicit dynamic finite element model using ABAQUS 6.16. A 2D plane-strain model was used as schematically shown in Figure 3. The mechanical properties of the work material and bending dies are summarized in Table 3 [21,22]. The model of the work sheet material was meshed with linear quadrilateral 4-node plane stress elements and was assumed as an isotropic elasto-plastic body. The material properties were obtained by uniaxial tensile tests of SUS304-H with a thickness of 0.1 mm. A total number of 77,278 elements were meshed in the whole FE model. The silica glass die and tool steel die were modeled as the elastic body and the punch was modelled as the rigid body. In the analysis, the contact state was assumed using the penalty method with a friction coefficient μ of 0.1. The validation of the simulation was confirmed by the comparison with a load-stroke curve of the non-textured silica glass die obtained by the experiment.

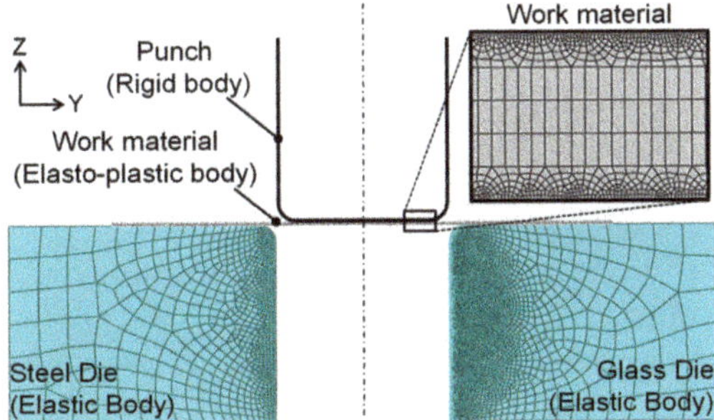

Figure 3. Schematic illustration of the finite element (FE) model for the contact state analysis of bending with the ironing process.

Table 3. Analytical conditions of material models for the contact state analysis.

Material		Stainless Steel	Silica Glass Die	Tool Steel Die	(JIS:SUS304-H)
Material model		Elasto-plastic body	Elastic body	Elastic body	
Mass density ρ	[g/cm^3]	7.93	2.2		7.72
Elastic modulus E	[GPa]	200	72		206
Yield stress σ_y	[MPa]	1170	-		-
Poisson's ratio ν		0.3	0.17		0.3

4. FE Based Computational Fluid Dynamics (CFD) Analysis

To analyze the lubricant flow behavior for different dimple dimensions obtained by the in-situ observation, two different CFD analyses were performed by using the COMSOL Multiphysics 5.2a. One is a laminar two-phase (air and lubricant) analysis to simulate the flow-in behavior of the lubricant to the inside of the dimple structures, and another is a fluid-flow analysis to simply simulate the steady state flow behavior during the sliding surface. Figure 4 shows a schematic illustration of the numerical model. A single unit cell of the texture patterns was chosen for the modelling domain. It consists of the moving wall at the top surface as the sliding work sheet and the lubricant film filled in between the textured tool and the work sheet. As geometrical parameters of the texture patterns, dimple diameter d, pitch b, lubricant film thickness h_l were defined. To reduce the calculation time, a 2D computational domain with the textured die in cross-section was used. Details of each analysis are described in the following sections.

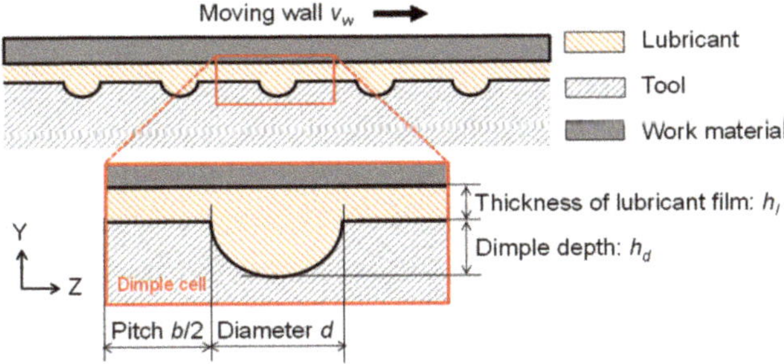

Figure 4. Schematic illustration of the model of micro dimple cells for the CFD (computational fluid dynamics) analysis.

4.1. Laminar Two-Phase Flow Analysis

To analyze the inflow behavior of the lubricant to the inside of the dimple structure with different dimensions, the laminar two-phase flow analysis at the fluid interface between the lubricant and air was carried out. In this simulation, the phase field were time dependent since the position of the interface was dependent of its history. Two-dimensional incompressible Navier-Stokes equations were used. In addition, the Level Set method was used to describe the convection of the fluid interface. The level set function ϕ was defined below;

$$\phi = 0 \text{ (Gas phase)} \quad \phi = 0.5 \text{ (Interface)} \quad \phi = 1 \text{ (Liquid phase)} \tag{1}$$

In a transition layer of liquid close to the interface between liquid and gas, ϕ varied smoothly from zero to one. Details for the level set function and modified Navier-Stokes equations are described elsewhere [23,24]. The sliding velocity of v_w was defined at the moving wall boundary conditions of the top surface, while the bottom wall with the dimple structure are modelled as a stationary wetted wall. In addition, the lubricant inflow from the inlet at a constant flow velocity of v_f was modelled and the outlet of the lubricant flow was defined as the zero gauge pressure. Table 4 summarizes the analytical conditions for the laminar two-phase flow analysis.

Table 4. Analytical conditions for the laminar two-phase flow analysis.

Mass Density of Lubricant ρ_{liquid}	[kg/m^3]	899.8
Lubricant viscosity of η_{lub}	[Pa·s]	0.04
Sliding speed v_w	[mm/s]	50
Flow velocity v_f	[mm/s]	50
Lubricant film thickness h_l	[μm]	2
Density of gas ρ_{gas}	[kg/m^3]	1.2
Viscosity of gas η_{gas}	[Pa·s]	1.8×10^{-5}
Atmosphere temperature T	[K]	293.15
Contact angle θ	[degrees]	20

4.2. Fluid-Flow Analysis

To roughly estimate the generation of the hydrodynamic pressure in the steady state lubricant flow for each texture dimension during the sliding of the work sheet surface, the simple fluid flow analysis was performed. Following two-dimensional incompressible Navier-Stokes equations were used to simulate the steady state lubricant flow in between the work material and tool surface.

$$\rho v \cdot \nabla v = -\nabla p + \eta \nabla^2 v \tag{2}$$

$$\nabla \cdot v = 0 \tag{3}$$

v is the velocity vector, p is the pressure, ρ is the mass density of the fluid and η is the viscosity of the lubricant. As well as the above laminar two-phase flow analysis, the constant velocity of v was defined as the moving wall boundary conditions of the top surface of lubricant films, while the bottom wall with dimple structure are modelled as the stationary body. In addition, the pressure at the inlet and outlet of the lubricant flow, where the end of the modelled domain of the lubricant is observed, was defined as the zero gauge pressure. Analytical conditions are summarized in Table 5.

Table 5. Analytical conditions for the fluid-flow analysis.

Mass Density ρ	[kg/m^3]	899.8
Viscosity η_l	[Pa·s]	0.04
Sliding speed v_w	[mm/s]	50
Lubricant film thickness h_l	[μm]	0.5, 1, 2

5. Results

5.1. Contact State Analysis

Figure 5a shows a calculated result of the variation of the contact pressure and apparent contact length during bending with the ironing process for the in-situ observation. Contour figures as shown in Figure 5b show the distribution of the equivalent stress state at the corresponding punch strokes denoted in Figure 5a. The contact pressure increases with the increasing punch stroke during bending and it reaches the maximum of 1600 MPa at a stroke of 1.2 mm. Afterwards, it decreases until a stroke of 2.5 mm and it becomes almost constant until the process is completed. On the other hand, the contact length linearly increases from 0.1 mm at a stroke of 1.2 mm and it reaches the maximum contact length of 2.8 mm at a stroke of 4.0 mm. To visualize the variation in the contact state and to observe the lubricant flow of the contact interface during ironing processing, the area directly under the die radius was chosen as the in-situ observation area, where the maximum contact pressure was observed.

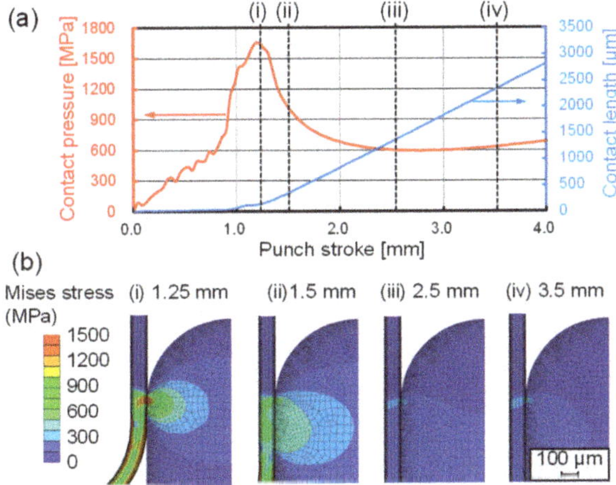

Figure 5. Results of the contact state analysis. (**a**) Variations of the contact pressure and apparent contact length during bending with the ironing process, (**b**) variation of the equivalent stress distribution at each punch stroke denoted in (**a**).

5.2. In-Situ Observation of Lubricant Flow

Figure 6 shows captured images from the recorded in-situ observation movies during the ironing process using a silica glass die with a dimple of 50 μm in the diameter at a stroke of (a) 1.25 mm, (b) 1.5 mm, (c) 2.5 mm and (d) 3.5 mm. The sliding direction of the work sheet is indicated by an arrow directed from top to bottom. The inset figure at each stroke shows magnified views of one dimple cell. A line below the first row of the dimples in Figure 6a indicates the edge of the radius of the work sheet bended by the punch corner radius, showing that it slides over the first row of the dimples. The area around this edge shows a stronger light reflection from the surface of the work sheet and it appears brighter. Considering from the results of the FE-analysis, the brightness of the captured images indicates the high pressure contact leading to a thin lubricant film thickness. While, the small air bubbles are generated inside the dimples directly after the radius edge of the work sheet slides over the dimples. When it slides over the second dimple row, this brighter area shifts downwards as shown in Figure 6b. The diameter of the air bubbles increases with increasing of the punch stroke. Note that brightness at the area below the first row of the dimples in the sliding direction became darker, indicating the lubricant transfer from the first to the second row. At a stroke of 2.5 mm as shown in Figure 6c, the diameter of the air bubble inside the dimples further increases. When the air bubble diameter reaches the diameter of the dimple, an area with the interference color is observed, moving from the first to the second row of the dimples along the sliding direction. Afterwards, the air bubble diameter at the first row dimples decreases as shown in the inset figure of Figure 6d indicating a refill of the lubricant inside the dimple.

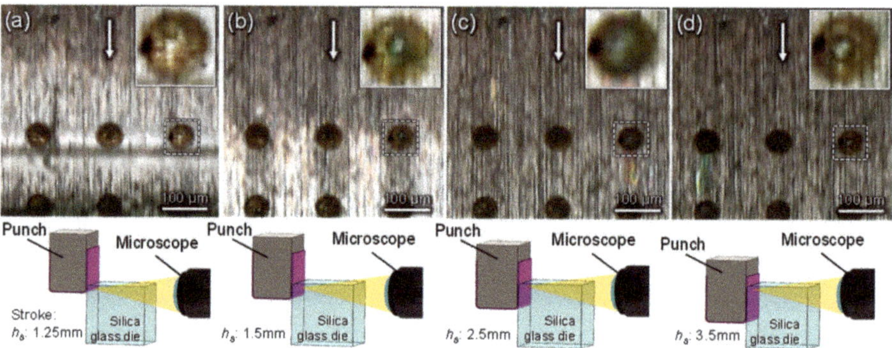

Figure 6. Captured images from in-situ observations at various punch strokes of (**a**) h_s:1.25 mm, (**b**) h_s:1.5 mm, (**c**) h_s:2.5 mm and (**d**) h_s:3.5 mm in bending with the ironing process using the surface textured die with a dimple diameter of 50 µm.

A conceivable explanation for this generation of air bubbles is the air carried into the contact interface or to the local cavitation, which is caused by rapid pressure drops [25]. However, considering from the range of the relative sliding speed and the observation of the remaining air bubbles even without the movement of the punch stroke, that the cavitation effect seems to be less effective [26]. Instead, the air bubbles seem to be a flow from the surface of work material or inside of the initial lubricant. Therefore, the generation of the air bubbles and the variation of its diameter can be identified as an indicator of the lubricant supply from the dimples during the ironing process.

5.3. Effect of Micro-Dimple Diameters

Averaged punch force–stroke curves for the non-textured die and the textured die with different dimple diameters of 10 µm, 50 µm, 100 µm are plotted in Figure 7. The error bars in the figure show the standard deviations for seven processes for each condition. In comparison with the non-textured tool surface, lower punch forces were obtained for a 50 µm and 100 µm dimple diameter, while the die with a dimple diameter of 10 µm shows higher force and it increases over seven processes.

Figure 8 shows the captured images from the in-situ observation movies during the ironing process at a stroke of 1.5 mm with (a) the non-textured die and the dies with the dimple size of (b) 10 µm, (c) 50 µm and (d) 100 µm in diameter. In view of the brightness of the observation images, which shows the less lubricant under higher brightness region, the non-textured silica glass die has the brighter area at the latter half of the image, while the other textured die has the darker region below the dimple arrays. In particular, the textured dies with a dimple size of 10 µm and 50 µm in diameter show clear images of the sufficient supply of the lubricant below the dimple structure, where the lack of the lubricant was observed for the non-textured die at the same stroke of 1.5 mm. As for the textured die of 100 µm in diameter, however, the darker area between the dimples in the sliding direction do not reach the second row of dimple arrays. Additionally, the generation of the air bubbles inside the dimple structures is not clearly observed.

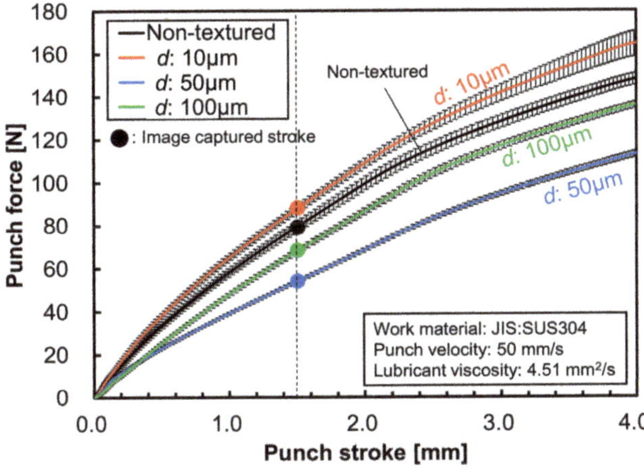

Figure 7. Punch force–stroke curves of bending with the ironing process using the non-textured die and surface textured die with different dimple diameters of 10 μm, 50 μm, and 100 μm.

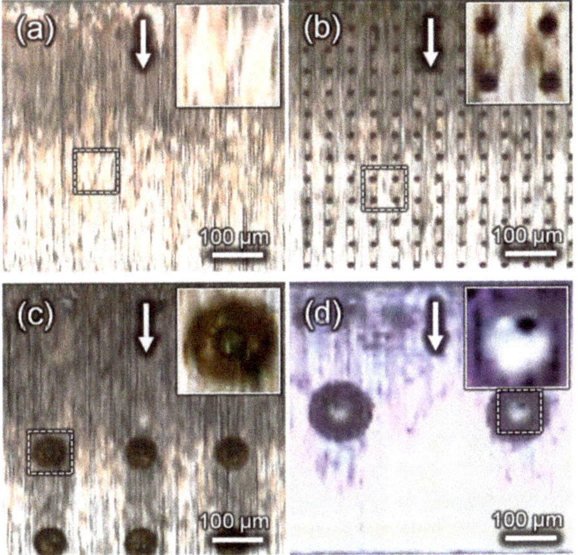

Figure 8. Captured images from in-situ observations at a punch stroke of 1.5 mm in bending with the ironing process using (**a**) the non-textured die and surface textured die with different dimple diameters of (**b**) 10 μm, (**c**) 50 μm and (**d**) 100 μm.

5.4. Effect of Lubricant Viscosity

Figure 9 shows the punch force–stroke curves of bending with the ironing processes using a 50 μm textured silica glass die with different lubricant viscosities of 1.27 mm/s², 2.53 mm/s² and 4.51 mm/s². As mentioned above, error bars in the figure show the standard deviations for seven processes for each viscosity condition. The lubricant with the highest viscosity of 4.51 mm/s² shows the lowest punch force with less deviation, while the lubricant with an intermediate viscosity of 2.53 mm/s² indicates the slightly larger punch force. On the other hand, the lubricant with the lowest viscosity of 1.27 mm/s² results in the highest punch force with a large deviation, which is caused by the significant increase over the series of seven processes.

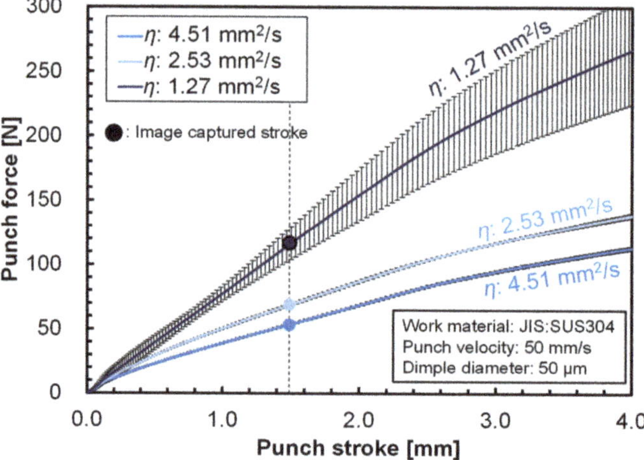

Figure 9. Comparison of punch force–stroke curves between different lubricant viscosities in bending with the ironing process using the surface textured die with dimple diameters of 50 μm.

Figure 10 shows the comparison of the captured images at a 1.5 mm punch stroke with the viscosities of (a) 4.51 mm/s^2, (b) 2.53 mm/s^2, and (c) 1.27 mm^2/s, using the silica glass die with dimple diameters of 50 μm. Since the brightness between the first and second row of the dimples were dark for the viscosity of 4.51 mm/s^2 and 2.53 mm^2/s, the lubricant seems to be well transferred between the dimples. On the other hand, for the lowest viscosity of 1.27 mm/s^2, the brightness at the area between the first and the second row dimple arrays becomes slightly brighter, suggesting a less supply of lubricant between the dimples.

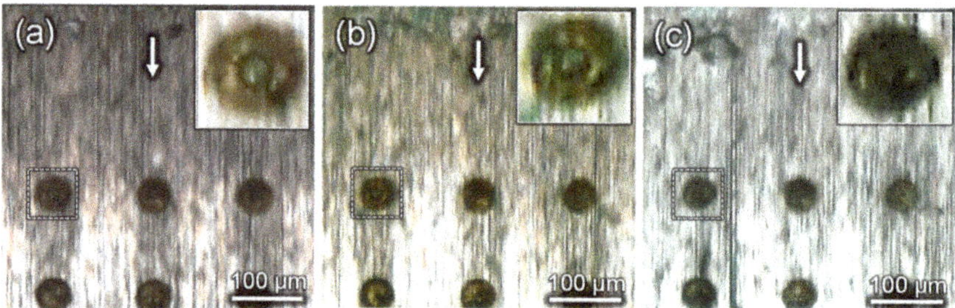

Figure 10. Captured images from in-situ observations at a punch stroke of 1.5 mm in bending with the ironing process with a 50 μm-dimple textured die using lubricants with different viscosities of (**a**) 4.51 mm/s^2, (**b**) 2.53 mm/s^2, and (**c**) 1.27 mm^2/s.

From the above observations, the sufficient lubricant supply and its transfer between the dimple arrays appear to be dominant for the difference in punch forces, as described in Sections 5.3 and 5.4. The cause of this difference in the punch force and related lubricant flow between the different dimple dimensions are further discussed based on the results of the CFD analysis in the following sections.

6. Discussion

From the observation of the lubricant flow behavior shown in Figure 6a–d, dimple array patterns seem to act as lubricant reservoirs transferring the lubricant at the interface, which leads to a fluid film lubrication from boundary lubrication. To analyze the difference in the flow-in behavior between different dimple sizes, as shown in Figure 8, the laminar two-phase flow analysis was performed.

Figure 11 shows the lubricant inflow behavior at each time step of the calculations for different dimple diameters of 10 μm, 50 μm and 100 μm. The colored side bar shows the level set value, which means the transition degree of the two phases of the lubricant and air; one for the lubricant phase and zero means the air phase. As can be seen from the transition at each time step, the time delay flowing into the dimple structure is more significant for the larger dimple size. Although the smallest dimple with 10 μm was filled with the lubricant at a step time of 0.66, the one with 100 μm was not filled with the lubricant even at a step time of 1.0. This is apparently due to the relative volume of the lubricant that flowed into the dimple to the volume of dimple structure. When we assume that the same amount of the lubricant will flow into the interface, the relative volume of the lubricant reserved into the dimple cells and the original volume of the dimple structure with different diameters can be roughly estimated by geometrical calculations.

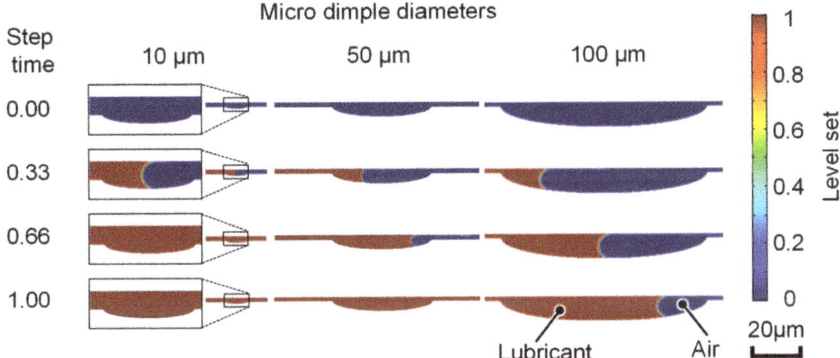

Figure 11. Comparison of the variations of the flow-in behavior of the lubricant at different time steps in the laminar two-phase flow analysis for different dimple diameters of 10 μm, 50 μm and 100 μm.

Figure 12 shows a schematic illustration of the calculation area of the lubricant volume. The geometry of the dimple cell was defined as half of the ellipsoid, and expressed as the following equation,

$$V_d = \frac{1}{6}\pi d^2 h_d + d^2 h_l \tag{4}$$

$$V_l = dbh_l \tag{5}$$

where V_d denotes the volume at the area of dimples and V_l is the volume of the lubricant at the area of the pitch between the dimples flowing into the dimples. According to a report of Azushima et al., the range of the thickness of lubricant films, h_l, was assumed as 0.5 μm to 2.0 μm [27].

Figure 13 shows the variations of the relative volume, V_l/V_d, as a function of the lubricant film thickness. Dotted lines in the figure are quadratic approximation lines for the calculated data plots for each dimple diameter. Since the amount of lubricant flowing into the dimple structure decreases with decreasing of the lubricant film thickness, V_l/V_d indicates the decreasing tendency. In particular, the dimple size with a 100 μm in diameter shows the lowest V_l/V_d under all lubricant film thickness conditions, due to its relative large volume of the dimple cell structure compared to the amount of lubricant flow into the structure. Thus, the lubricant will not be sufficiently reserved during sliding and it will be rather difficult to be supplied to the next row of dimple arrays along the sliding direction. Those tendencies are well corresponding to the in-situ observation results, which showed the less lubricant supply between the dimple array for the 100 μm dimple size as shown in Figure 8d. In view of the inflow and retention behavior of the lubricant into the dimple structure, the relative volume of the inflow lubricant to the dimple structure is of great importance to have a sufficient retention of lubricant and its supply to the next row of the dimple arrays. Therefore, the volume of the dimple structure needs to be designed corresponding to the amount of lubricant transferred by the sliding of work materials, which is related to the lubricant viscosity, contact pressure and sliding velocity at the interface.

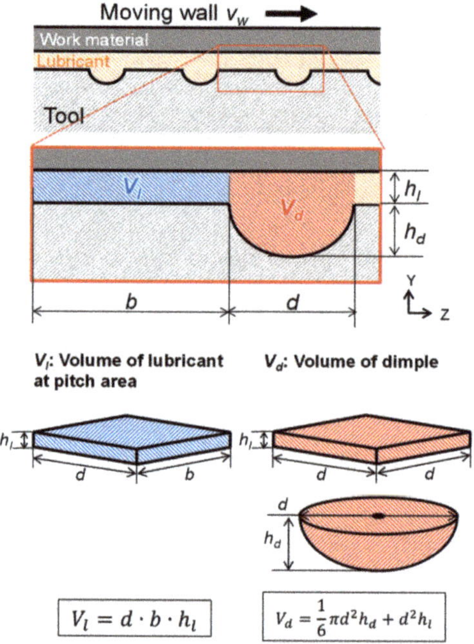

$$V_l = d \cdot b \cdot h_l$$

$$V_d = \frac{1}{6}\pi d^2 h_d + d^2 h_l$$

Figure 12. Schematic illustration for geometrical calculations of the relative volume of lubricant flow into the dimple cell V_l and the volume of dimple cell structure V_d.

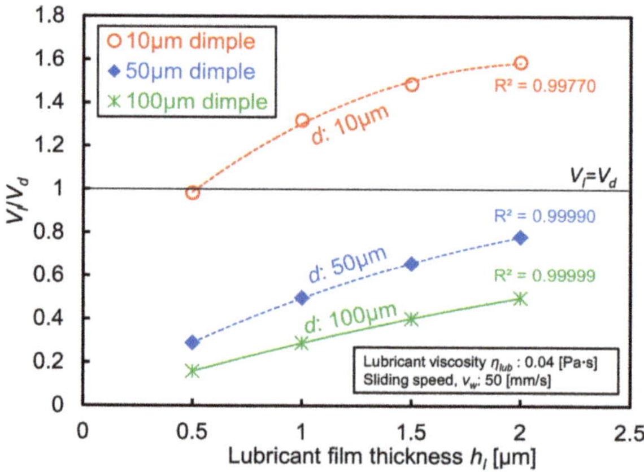

Figure 13. Variations of the relative volume V_l/V_d as a function of the lubricant film thickness h_l for different dimple diameters of 10 μm, 50 μm, and 100 μm. Dotted lines are quadratic approximated lines for the data of each dimple diameter.

Another observation is the results of the dimple with a 10μm in diameter, which shows the highest punch forces even under the sufficient supply of lubricant between the dimple arrays as seen in Figure 8c. To further discuss the cause of this tendency, a steady state fluid flow analysis was carried out focusing on the effect of the hydrodynamic pressure. Figure 14 shows the comparison of the hydrodynamic pressure distribution between different dimple dimensions under different lubricant film thicknesses of (a) 0.5 µm, (b) 1.0 µm and (c) 2.0 µm. As a general tendency, the hydrodynamic pressure significantly decreases at the inlet of the dimple structure, and it increases at the outlet for all dimple diameters. It is well known that those significant decrease and increase of the hydrodynamic pressure at the edge of the dimples contribute to the inlet suction effect to introduce the lubricant into the dimple structure [28]. However, it can be seen that those tendencies of pressure drop and rise at the dimple cell are relatively small for the dimple with 10 µm at a film thickness of 1 µm and 2 µm. This is due to the relative small depth of the dimple cell structure compared to the lubricant film thickness. Figure 15 shows the distribution of the flow velocity and its flow lines for each dimple size of 10 µm, 50 µm and 100 µm under a lubricant film thickness of 2 µm. As can be seen from the direction of the flow lines, the vortex-like flow behavior was observed inside the dimple structure with 50 µm and 100 µm in diameter, while the coquette flow oriented to the sliding direction was obtained for that with 10 µm. The generation of the vortex flow inside the dimple might be attributed to this pressure drops, and it can be also related to the generation of the air bubbles as seen in the in-situ observations [29].

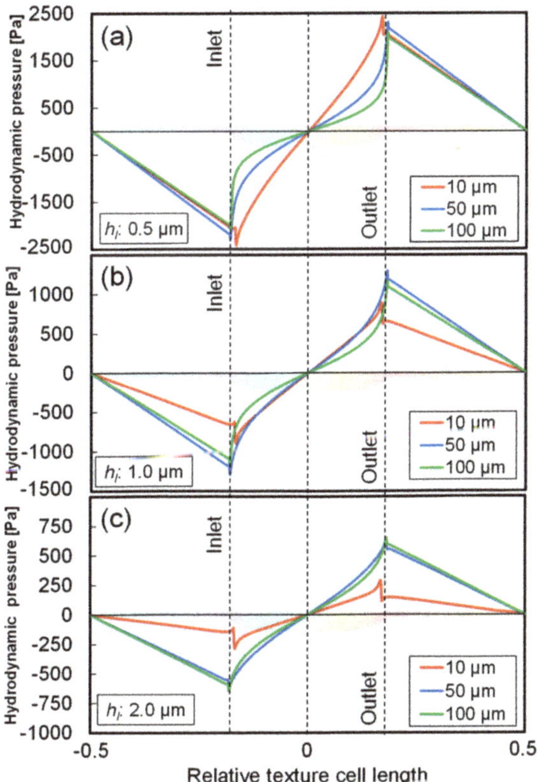

Figure 14. Comparisons of the calculated hydrodynamic pressure distribution between different dimple diameters of 10 µm, 50 µm and 100 µm under different lubricant film thicknesses h_l of (**a**) 0.5 µm, (**b**) 1.0 µm and (**c**) 2.0 µm.

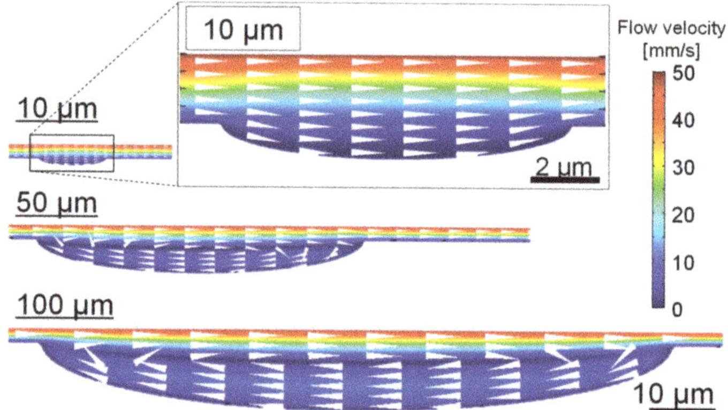

Figure 15. Comparison of the distribution of the lubricant flow velocity in the dimple cells with different diameters of 10 µm, 50 µm, 100 µm under a lubricant film thickness of 2.0 µm.

Due to this lower hydrodynamic pressure for the dimple diameter of 10 µm, the sliding contact pair can be directly contacted with each other. In fact, from the observation by using the optical microscopy as shown in Figure 16, it was found that a metallic material was accumulated inside the dimples, on which the work sheet edges slid over. This accumulation was found to wear the debris from the work material, by detecting the iron (Fe) elements from the energy dispersive X-ray (EDX) analysis as shown in Figure 16. This accumulation of wear debris may cause the instability and higher load during the ironing process.

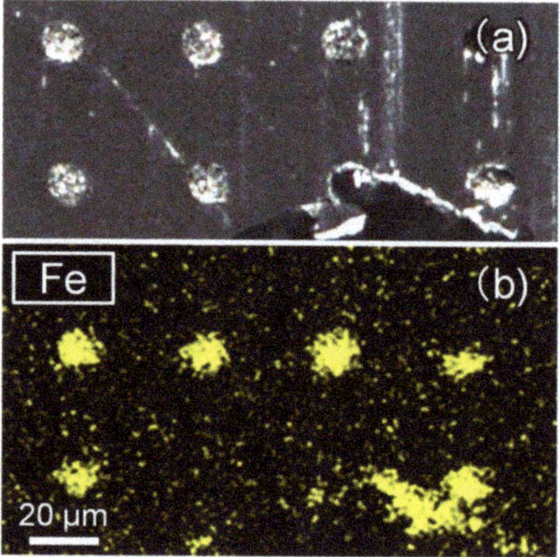

Figure 16. Surface images of silica glass die with textured dimple with 10 µm in diameter after bending with ironing processes, (**a**) optical microscopic image, (**b**) EDX (energy dispersive X-ray) mapping of Fe (Iron).

From the viewpoint of load carrying capacity of inside the dimples, the hydrodynamic pressure drops at the inlet and rapid increase at the outlet should be taken into account, which strongly relates to the relative depth of the dimple structure to the thickness of the lubricant. Even the texture has the lubricant reservoir effect as seen in the dimple with a 10 µm in diameter, it may cause the direct contact with work materials due to the less hydrodynamic pressure, and it promotes the wear debris generation to worsen the friction properties. The similar tendency was also observed in comparison between the different viscosities of the lubricant as shown in Figure 10. The highest punch force with a large deviation at low viscosity lubricant seems to be attributed to the less hydrodynamic pressure, resulting in the flow out of the lubricant from the contact interface. Therefore, the lubricant transfer to the next dimple array patterns became difficult, as leading the direct contact with the work materials.

7. Conclusions

In the present study, to investigate the underlying mechanism of the effect of surface texturing on the lubricated sliding friction in the metal forming operation, an in-situ observation system which enables to observe the contact interface between tools and work materials during the sheet metal forming process was newly developed. The in-situ observation results under the different texture dimensions and the different lubricant viscosities are compared with the load stroke measurements and additional CFD analysis for the single cell of dimple texture. From these experimental and analytical results, the following conclusions can be drawn;

A series of inflow, retaining and supply of the lubricant at the contact interface was successfully observed during bending with the ironing process, by referring the brightness of the captured images and the generation and growth of the air bubbles inside the dimple structure as an indicator.

In comparison between the laser textured dies with different dimple diameters, the lowest punch force was obtained for the dimple diameter of 50 μm, which can be proper dimensions for the sliding contact state with a maximum pressure of 1.6 GPa at a contact length of 0.1 mm.

Although the dimple with a diameter of 100 μm also indicated the lower punch force than the non-textured die, less transfer of lubricant between dimple array patterns due to the large volume of dimple structure resulted in the higher punch force than that with 50 μm.

While for the dimple with a diameter of 10 μm, less hydrodynamic pressure due to the relative shallow depth of the structure caused the direct contact with work materials, and it resulted in the highest punch force with larger deviations.

The tendency between the different lubricant viscosities also supported the above observation that the lowest viscosity lubricant showed less lubricant transfer under low hydrodynamic pressure than the other higher viscosity lubricants.

Consequently, to obtain the sufficient load carrying capacity of the lubricant during the metal forming operation by applying the surface texturing, the contribution by the lubricant reservoir effect will be a main dominant role. To have the sufficient lubricant supply, the surface texture should be designed based on the relative volume of the dimple structure for the lubricant retention, thickness of lubricant films and the amount of lubricant transferred by the work materials during the operation. Further investigation is required to correlate those dimensional parameters of the surface texture and pitch of patterning with the distribution of the contact pressure, thickness of the lubricant film, and resulting friction properties during the forming operation.

Author Contributions: T.S. and M.Y. made the experimental concept and design, J.V. carried out the experiments, H.K. executed the FE simulation.

Funding: This work has been funded by the Tokyo Metropolitan University Co-Tutorial Program between Bremer Institute für Angewandte Strahtechnik (BIAS) and Graduate School of System Design, Tokyo Metropolitan University.

Acknowledgments: The authors gratefully acknowledge Frank Volletsen, Lukas Heinrich and Hendrik Flosky and Hamza Messaoudi at BIAS for the fruitful discussion regarding the lubricant flow behavior on dimple surface texturing. We would also like to thank Tomomi Shiratori and Yasuaki Ohsawa at Komatsu-Seiki Kosakusho. Co., Ltd. and Yoshihiro Sagisaka in Industrial Research Institute of Shizuoka Prefecture for the fabrication of surface dimple textures on silica glass die, and the Nihon-Kousakuyu Company for their support in providing lubricant with different viscosities.

Conflicts of Interest: The authors declare no conflicts of interest.

Nomenclature

List of symbols

b	Pitch	[mm]
c	Clearance between punch and die	[mm]
d	Dimple diameter	[mm]
d_p	Punch with	[mm]
E	Elastic modulus	[GPa]
h_d	Dimple depth	[mm]
h_l	Lubricant film thickness	[μm]
h_s	Maximum stroke	[mm]
l_c	Contact length	[mm]
n_p	Number of forming processes	(dimensionless)
p_c	Contact pressure	[MPa]
r_d	Bending die radius	[mm]
r_p	Punch radius	[mm]
t	Work sheet thickness	[mm]
T	Temperature	[°C]
u	Kinematic viscosity	[mm^2/s]
v_p	Punch velocity	[mm/s]
v_w	Sliding wall velocity	[mm/s]
v_f	Flow velocity	[mm/s]
V_d	Volume of dimple cell	[mm^3]
V_l	Volume of lubricant at pitch area	[mm^3]
w	Width of die gap	[mm]

Greek letters

μ	Friction coefficient	(dimensionless)
η	Kinematic viscosity of lubricant	[mm^2/s]
η_{lub}	Lubricant viscosity	[Pa·s]
η_{gas}	Gas viscosity	[Pa·s]
λ	Dimple aspect ratio	(dimensionless)
ρ	Mass density of fluid	[kg/m^3]
ρ_{gas}	Gas density	[kg/m^3]
ρ_{liquid}	Liquid density	[kg/m^3]
ρ_d	Texture density ratio	(dimensionless)
σ	Surface tension coefficient	[mN/m]
σ_y	Yield stress	[MPa]
ν	Poisson's ratio	(dimensionless)
θ	Contact angle	[degree]
φ	Level set function	(dimensionless)

References

1. Wagner, K.; Volkl, R.; Engel, U. Tool life enhancement in cold forging by locally optimized surfaces. *J. Mater. Process. Technol.* **2008**, *201*, 2–8. [CrossRef]
2. Steinhof, K.; Rasp, W.; Pawelski, O. Development of deterministic-stochastic surface structures to improve the tribological conditions of sheet forming processes. *J. Mater. Process. Technol.* **1996**, *60*, 355–361. [CrossRef]
3. Zhou, R.; Cao, J.; Wang, Q.J.; Meng, F.; Zimowski, K.; Xia, Z.C. Effect of EDT surface texturing on tribological behavior of aluminum sheet. *J. Mater. Process. Technol.* **2011**, *211*, 1643–1649. [CrossRef]
4. Evans, C.J.; Bryan, J.B. "Structured", "Textured", or "Engineered" Surfaces. *CIRP Ann.* **1999**, *48*, 541–556. [CrossRef]
5. Geiger, M.; Popp, U.; Engel, U. Excimer laser micro texturing of cold forging tool surfaces-influence on tool life. *CIRP Ann.* **2002**, *51*, 231–234. [CrossRef]
6. Costa, H.L.; Hutchings, I.M. Effects of die surface patterning on lubrication in strip drawing. *J. Mater. Process. Technol.* **2009**, *209*, 1175–1180. [CrossRef]

7. Xu, D.; Wu, W.; Malhotra, R.; Chen, J.; Lu, B.; Cao, J. Mechanism investigation for the influence of tool rotation and laser surface texturing (LST) on formability in single point incremental forming. *Int. J. Mach. Tools Manuf.* **2013**, *73*, 37–46. [CrossRef]

8. Hamilton, D.B.; Walowit, J.A.; Allen, C.M. A theory of lubrication by micro- irregularities. *J. Fluids Eng.* **1966**, *88*, 177–185.

9. Wakuda, M.; Yamauchi, Y.; Kanzaki, S.; Yasuda, Y. Effect of surface texturing on friction reduction between ceramic and steel materials under lubricated sliding contact. *Wear* **2003**, *254*, 356–363. [CrossRef]

10. Varenberg, M.; Halperin, G.; Etsion, I. Different aspects of the role of wear debris in fretting wear. *Wear* **2002**, *252*, 902–910. [CrossRef]

11. Ibatan, T.; Uddin, M.S.; Chowdhury, M.A.K. Recent development on surface texturing in enhancing tribological performance of bearing sliders. *Surf. Coat. Technol.* **2015**, *272*, 102–120. [CrossRef]

12. Gropper, D.; Wang, L.; Harvey, T.J. Hydrodynamic lubrication of textured surfaces: A review of modeling techniques and key findings. *Tribol. Int.* **2016**, *94*, 509–529. [CrossRef]

13. Ramesh, A.; Akram, W.; Mishra, S.P.; Cannon, A.H.; Polycarpou, A.A.; King, W.P. Friction characteristic of microtextured surface under mixed and hydrodynamic lubrication. *Tribol. Int.* **2013**, *57*, 170–176. [CrossRef]

14. Profito, F.J.; Vlădescu, S.J.; Reddyhoff, T.; Dini, D. Transient experimental and modelling studies of laser-textured micro-grooved surface with a focus on piston-ring cylinder liner contacts. *Tribol. Int.* **2017**, *113*, 125–136. [CrossRef]

15. Ike, H.; Makinouchi, A. Effect of lateral tension and compression on plane strain flattening processes of surface asperities lying over a plastically deformable bulk. *Wear* **1990**, *140*, 7–38. [CrossRef]

16. Schey, J.A. *Tribology in Metalworking. Friction, Lubrication and Wear*; ASM, Metals Park: Novelty, OH, USA, 1983.

17. Azushima, A.; Uda, M.; Kudo, H. An interpretation of the speed dependence of the coefficient of friction under the micro PHL condition in sheet drawing. *CIRP Ann.* **1991**, *40–41*, 227–230. [CrossRef]

18. Mizuno, T.; Okamoto, M. Effects of lubricant viscosity at pressure and sliding velocity on lubricating conditions in the compression- friction test on sheet metals. *Trans. ASME J. Lubr. Technol.* **1982**, *104*, 53–59. [CrossRef]

19. Bay, N.; Azushima, A.; Groche, P.; Ishibashi, I.; Merklein, M.; Morishita, M.; Nakamura, T.; Schmid, S.; Yoshida, M. Environmentally benign tribo-systems for metal forming. *CIRP Ann.* **2010**, *59*, 760–780. [CrossRef]

20. Gachot, C.; Rosenkarnz, A.; Hsu, S.M.; Costa, H.L. A critical assessment of surface texturing for friction and wear improvement. *Wear* **2017**, *372–373*, 21–41. [CrossRef]

21. Shin-etsu Quartz, Co. Available online: https://www.sqp.co.jp/catalog/images/Technology_Guide1_j.pdf (accessed on 18 November 2017).

22. Tokyo Kohshin, Co. Available online: http://tokyokohshin.co.jp/wp/wp-content/themes/theme064/images/pdf/skd11_hi.pdf (accessed on 18 November 2017).

23. Sethian, J.A.; Smereka, P. Level set methods for fluid interface. *Ann. Rev. Fluid Mech.* **2003**, *35*, 341–372. [CrossRef]

24. Sussman, M.; Smereka, P.; Osher, S. A level set approach for computing solutions to incompressible two-phase flow. *J. Comput. Phys.* **1994**, *114*, 146–159. [CrossRef]

25. Wahl, R.; Schneider, J.; Gumbsch, P. In situ observation of cavitation in crossed microchannels. *Tribol. Int.* **2012**, *55*, 81–86. [CrossRef]

26. Qiu, Y.; Khonasari, M.M. Experimental investigation of tribological performance of laser textured stainless steel rings. *Tribol. Int.* **2011**, *44*, 635–644. [CrossRef]

27. Azushima, A. In lubro 3D measurement of oil film thickness at the interface between tool and workpiece in sheet drawing using a fluorescence microscope. *Tribol. Int.* **2005**, *38*, 105–112. [CrossRef]

28. Olver, A.V.; Fowell, M.T.; Spikes, H.A.; Pegg, I.G. 'Inlet suction', a load support mechanism in non-convergent, pocketed, hydrodynamic bearings. *Proc. Inst. Mech. Eng. Part J J. Eng. Tribol.* **2006**, *220*, 105–108. [CrossRef]

29. Karker, A.; Ostayen, R.A.J.; Beek, A.; Rixen, D.J. A multiscale method modeling surface texture effects. *J. Tribol.* **2007**, *129*, 221–230. [CrossRef]

Article

Formability Analysis and Oxidation Layer Effects in Dieless Drawing of Stainless Steel Wires

Yeong-Maw Hwang * and Han-Hsuan Liu

Department of Mechanical and Electro-Mechanical Engineering, National Sun Yat-sen University, No.70 Lianhai Rd., Gushan Dist., Kaohsiung City 804, Taiwan
* Correspondence: ymhwang@mail.nsysu.edu.tw; Tel.: +886-7-525-2000

Received: 14 June 2019; Accepted: 22 July 2019; Published: 25 July 2019

Abstract: In this study, dieless drawing experiments of stainless steel SUS304 wires of 1 mm in diameter were carried out using a self-developed dieless drawing machine. In order to prevent oxidation, argon gas was applied to a self-designed chamber during dieless drawing processes. The effects of the forming temperature and the oxide layer on the mechanical properties of the drawn SUS304 stainless wires obtained by tensile tests are discussed in this paper. A finite element model considering the high frequency induction heating mode in the finite element software DEFORM2D was developed to conduct the heat transfer analysis and the formability analysis of the drawn products in dieless drawing of stainless steel wires. The effects of the drawing speed and forming temperature on the maximal reachable area reductions are discussed. Through the comparisons of the maximal reachable area reduction between the finite element simulations and experiments, the finite element modelling for dieless drawing processes was validated.

Keywords: dieless drawing; SUS304 stainless steel wires; oxide layer; finite element simulation

1. Introduction

Weiss and Kot first proposed the idea of a dieless drawing process, which can reduce the cross-section of a wire without using a die, in 1969 [1]. There are no friction or lubrication problems between the workpiece and the die during the process. Compared with the traditional die drawing processes, dieless drawing has some advantages, such as no dies in the process, a larger area reduction at a single pass, no need for special surface treatment of the workpieces, a more flexible process, and so on. Recently, due to prominent development of micro-electro-mechanical-systems (MEMS) and biomedical technology, micro-metal forming has drawn more attention and demand for microtubes and microwires has increased dramatically [2]. Furushima and Manabe [3] have successfully manufactured micron-sized tubes of a diameter of 190 μm using dieless drawing processes.

Some research on analytical and experimental modelling in a dieless drawing process has been published. For example, Fortunier et al. [4] used a simple slice method and divided the wire into three zones to analyze the stability problems during dieless wire drawing processes. Wang et al. [5] investigated experimentally the wall thickness variation during drawing of stainless steel tube. Tubes with various inner and outer diameters were used and the effects of the area reduction were discussed. Wang et al. [6] also derived a velocity field and determined the drawing force using a power equivalent method during dieless tube drawing. The drawing force is affected by the width of deformation fields, sectional reduction ratios, deformation temperatures, and the drawing speed. Furushima et al. [7] proposed a drawing path control method during the drawing process to restrain the unstable deformation. The real deformation limit of the material was investigated by coupling the thermos-mechanical finite element analyses and experiments of the dieless drawing process. Furushima and Manabe [8] proposed an approach for fabrication of superplastic microtubes using a high-frequency induction heating apparatus and an air-cooling nozzle in a multi-pass dieless drawing process. After four-pass dieless drawing, a microtube with outer and inner diameters of 190 and 91 μm, respectively, was achieved successfully.

Furushima and Manabe [9] investigated the size effect on the deformation and heat transfer behavior of microtubes during a dieless drawing process using finite element analysis. They found that heat is not conducted from the outside surface to the inside of a macroscale tube in the case of very high-speed drawing. In contrast, due to the size effect at the microscale, the temperature of the outside surface and the inside of microtubes can be increased rapidly. Therefore, high-speed drawing can be realized for microtube dieless drawing. Furushima et al. [10] investigated the effects of surface roughening and oxidation on the drawing limit in dieless drawing of SUS304 stainless steel microtubes. They found that the limiting area reduction increases to a maximal value and then decreases with heating temperatures in both air and argon gas. The appropriate heating temperatures ranged from 950 to 1100 °C for the maximal drawing limit. Furushima et al. [11] used a laser irradiated from one direction of a metal microtube to propose a novel rotary laser dieless drawing process. They evaluated the rotating effects on outer diameter variations and outer radius errors in laser dieless drawing of stainless steel SUS304 microtubes. Milenin [12] used the special features of rheology to propose an approach to reduce the unevenness of the drawn product profile by dividing the drawing process into a few stages. The values of the strains in each stage correspond to the area of intensive hardening on the stress–strain curves of the processed material.

Some research about dieless drawing machines has been reported. For example, Twohig et al. [13] used a forward non-continuous dieless drawing machine to investigate dieless drawing of a 5 mm-diameter nickel-titanium rod. The effects of different forming parameters on the changes in microstructure were understood through hardness testing and microstructure observation. Naughton and Tiernan [14] pointed out that the reason that the dieless drawing method has not been fully accepted by the industry is that the forming system is unstable. Data collection and actuation control in the experimental process are the key components when designing and manufacturing the dieless drawing machine. A stability design for a continuous dieless drawing system has been proposed. Tiernan et al. [15] carried out dieless drawing experiments on AISI O1 steel wires. The experimental results showed that when the speed ratio is greater than 0.5, a large reduction and uniform diameter could be obtained. A load control system was developed to increase the formability of the product and reduce the product diameter tolerance to 0.01–0.05 mm. Supriadi and Manabe [16] used a visual sensor to monitor the minimum diameter of the deformation zone during dieless drawing to improve the diameter uniformity of the drawn product. The combination of the dieless pumping method and image recognition technology was first presented in their paper.

When the drawn metal is exposed to the atmosphere at a higher temperature, oxidation occurs and an oxide layer is generated on the product surfaces. The oxide layer protecting the inside material may break or crack due to uneven thermal or tensile stresses and lose its protective function. Eventually, the inside material will be oxidized more, which may result in poor surface quality. So far, few works discussed the formation of oxide layers and the effects of the oxide layers on the mechanical properties in dieless drawing of stainless steels. In this study, the formation of oxide layers and their effects on the mechanical properties of the dieless drawn SUS304 wires were investigated. Argon gas was used to surround the drawn wire to prevent the oxidation of the stainless steel during the dieless drawing processes. Tensile tests of stainless steel wires were conducted and the effects of the oxide layer on the mechanical properties are discussed in this paper. Also presented is a series of formability analyses of a wire dieless drawing process using DEFORM software coupled with a high frequency induction mode, which is different to the traditional approach of using a constant temperature zone moving with heating coil position during the drawing process.

2. Dieless Drawing Experiments

According to the movement direction of the heating coils, dieless drawing can be operated in three types of set-ups [17]: (a) continuous drawing type; (b) backward non-continuous drawing type; and (c) forward non-continuous drawing type, in which the heating coil is fixed, moves backward, and moves forward, respectively. A dieless drawing machine was developed by the present authors. The self-developed dieless drawing machine type is the same as that used by Weiss and Kot [1], a backward non-continuous drawing type, as shown in Figure 1.

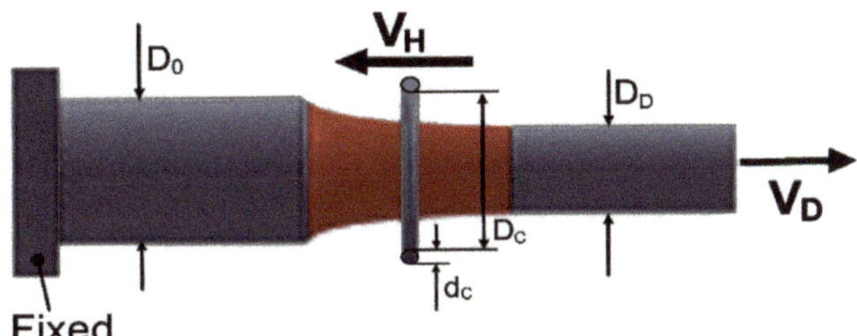

Figure 1. Schematic of a backward non-continuous dieless wire drawing process.

The area reduction in the dieless drawing process was determined according to the relationship between the drawing and heating coil speeds. From volume constancy, we get:

$$A_O V_H = A_D (V_D + V_H) \tag{1}$$

where V_D and V_H are the drawing and heating coil speeds, respectively. A_O and A_D are the cross-sectional areas of the wire before and after drawing, respectively. The area reduction R can be expressed as the function of V_H and V_D as given below [18].

$$R = 1 - \frac{A_D}{A_0} = 1 - \frac{V_H}{V_D + V_H} \tag{2}$$

A dieless drawing prototype machine was developed for tube or wire drawing processes. This self-developed dieless drawing machine was composed of three main parts: a supporting frame, a heating apparatus, two transmission systems and a control system [19]. The appearance of the completed prototype machine is shown in Figure 2. The heating apparatus, composed of a high frequency heater device and a heating coil, was designed to heat up the workpiece at a narrow heating zone. A cooling system was designed to cool down the workpiece at a cooling zone next to the heating zone. The temperature and drawing speed variations were recorded during the dieless drawing process. During the drawing process, one end of the wire workpiece was clamped at the fixing table, and the other end was clamped at the drawing table, which speed was controlled by a servomotor through a lead screw. The heating table and heating coils were moved by another servomotor in the opposite direction to that of the drawing table. The relative speed ratio of the heating coils to the drawing table was used to control the area reduction of the wire workpiece. The high-frequency induction heating apparatus was used to heat locally the wire and the infrared thermometer was used to measure the heating temperature. In order to avoid oxidation on the surface at the wire heating zone, argon gas was input into an argon gas chamber to enclose the whole wire surface at the heating zone. The argon gas chamber shown in Figure 3 was assembled by six acrylic plates and four small holes drilled on the plates were used for letting the wire, heating coil, and argon gas pass into the chamber.

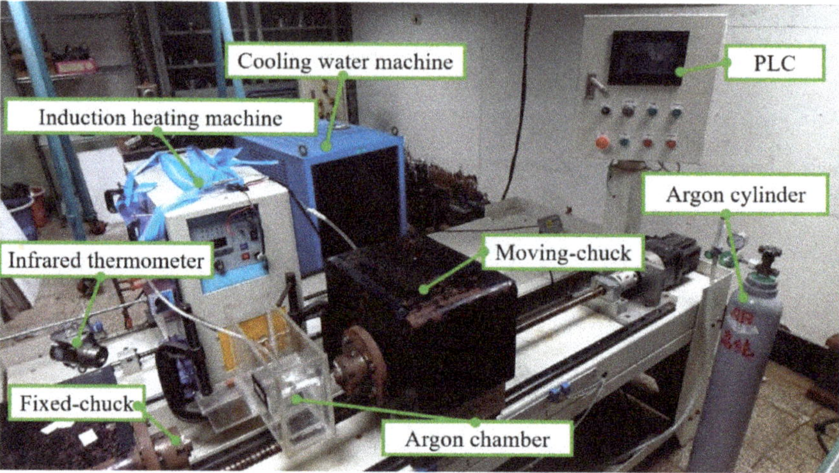

Figure 2. Layout of self-developed dieless drawing machine.

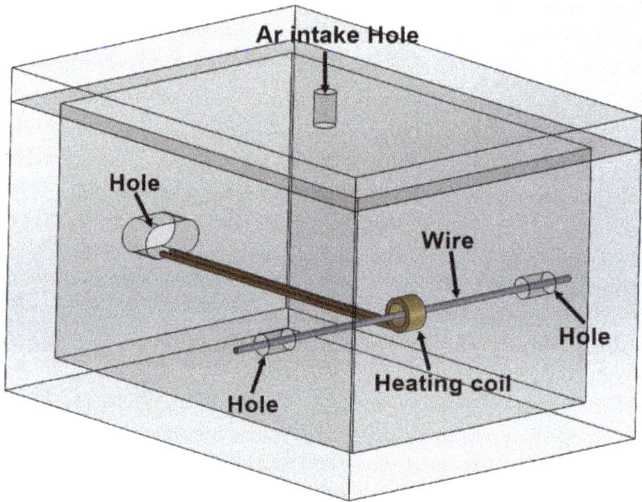

Figure 3. Schematic of self-designed argon gas chamber.

Before the dieless drawing experiments, the moving chuck and induction heating apparatus were calibrated to the set positions. Then, a stainless steel wire was moved to pass through the argon gas chamber and the center of the heating coil. The two ends of the wire were fixed at the fixed chuck and the moving drawing chuck. The measuring point of the infrared thermometer was adjusted to locate at the center of the wire and at the center of the heating coil to measure accurately the wire temperature during drawing. After that, an argon valve was opened to let the argon gas flow into the chamber continuously at a flow rate of 10 L/min for about 60 s to ensure no air existing inside the chamber. Finally, the drawing parameters of drawing speed, heating coil speed and heating temperature were input into the control panel, and then the dieless drawing experiment was started.

Stainless steel SUS304 wires with a diameter of 1 mm were used in the experiments. A series of experiments of dieless drawing processes were conducted under an atmosphere in air and argon gas. As the other parameters have relatively small effects on the oxide layers, only the forming temperature and the atmosphere were considered at several levels and the other parameters were fixed. The forming conditions for the experiments are shown in Table 1.

Table 1. Forming conditions for dieless wire drawing experiments.

Parameters	Values
Material	SUS304
Atmosphere	Air, Argon gas
Forming temperature, T [°C]	700, 800, 900, 1000, 1100
Area reduction, R [%]	30
Drawing speed, V_D [mm/s]	0.6
Heating coil speed, V_H [mm/s]	1.4
Workpiece wire diameter, D_0 [mm]	1
Heating coil loop inner diameter, D_C [mm]	5.5
Heating coil wire diameter, d_C [mm]	3

3. Experimental Results and Discussion

3.1. Product Surface Observation

The surface appearance of the stainless steel wires before and after drawing at different drawing temperatures and in different atmospheres is shown in Figure 4. It can be found that the surfaces of drawn wires exhibit different colors at different temperatures. As the forming temperature increases from 700 to 1100 °C, the wire surface becomes darker and the oxidation layer becomes thicker. The results indicate that even though the wire was drawn in argon gas, slight oxidation still occurred.

The oxides were very brittle and undergo fracture easily at the early stage of deformation, which usually resulted in the thickness variation of the oxide layer. According to Qin et al. [20], the structures of the oxide layer are mainly Fe_3O_4 and $FeCr_2O_4$. The diameter of the wire was so small that it is difficult to measure the surface roughness of the oxide layer. The thickness of the oxide layer exhibited some variation of about 1–7 μm, which is quite close to the results in Qin et al. [20]. Generally the thickness of the oxide layer increased with the forming temperature and the surface color of the product became slightly darker as the forming temperature increased. The surface colors of the products at 700 °C and 1100 °C exhibited more vivid contrast. Thus, only the appearance of the products at temperatures of 700 °C and 1100 °C is shown in Figure 4.

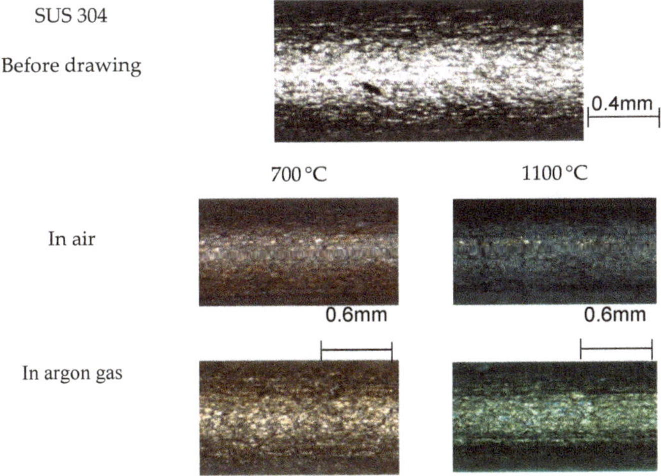

Figure 4. Appearance of stainless steel wire surfaces under different forming conditions.

3.2. Mechanical Properties of Drawn Wires

To investigate the effects of the oxide layer and forming temperature on the mechanical properties of the drawn wires, tensile tests were conducted at room temperature. The specimen gauge length was 30 mm and the tensile test speed was set as 10 mm/min. The engineering stress-strain curves of the drawn wires at different forming temperatures and in different atmospheres are shown in Figure 5. The yield stress, ultimate tensile strength (UTS) and elongation of the drawn products could be obtained from the stress-strain curves. Generally, the yield stress and ultimate tensile strength decreased with increases in the forming temperature in both air and argon gas.

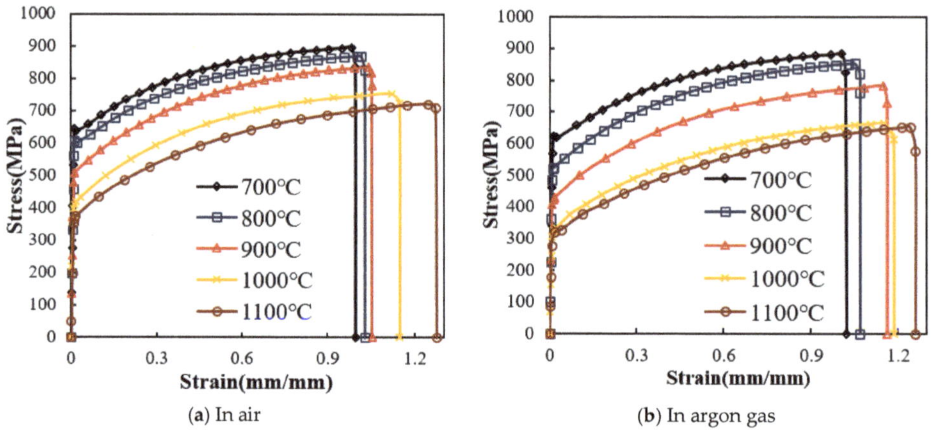

(a) In air   (b) In argon gas

Figure 5. Engineering stress-strain curves at different forming temperatures.

The yield stresses of the drawn wires were obtained from the engineering stress-strain curves using the 0.2% offset method. The yield stresses at different forming temperatures in air and argon gas are shown in Figure 6. From the figure it is knows that as the forming temperature increased, the yield stresses of the drawn wires decreased. The yield stresses in air were higher than those in argon gas.

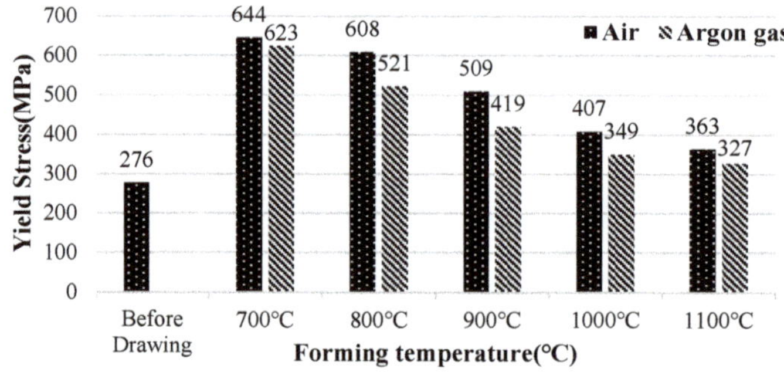

Figure 6. Comparisons of yield stresses of drawn wires in air and argon gas.

The ultimate tensile strengths (UTSs) at different forming temperatures in different atmospheres are shown in Figure 7. It is clear that the UTS decreased as the forming temperature increased, just as the case of yield stress. The UTSs of the drawn wires in air were slightly higher than those in argon gas. Because the strength of the oxide layer was generally larger than that of stainless steel SUS304.

The elongations at different forming temperatures in different atmospheres are shown in Figure 8. We can find that the elongation increased as the forming temperature increased. The elongations of the drawn wires in argon gas were slightly higher than those in air. The elongations of the drawn products were all smaller than the initial wire elongation.

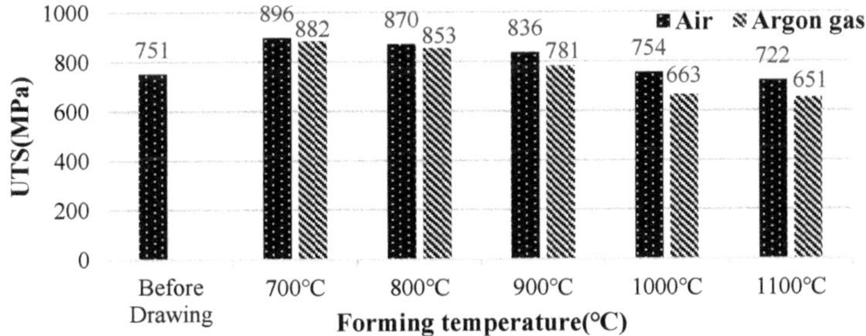

Figure 7. Comparisons of ultimate tensile strength of drawn wires in air and argon gas.

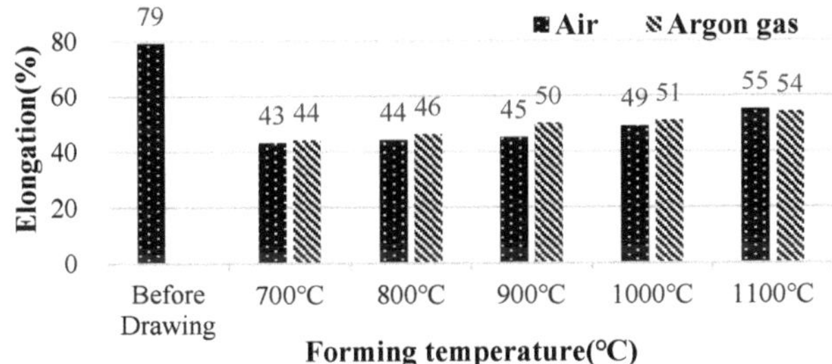

Figure 8. Comparisons of elongation of drawn wires in air and argon gas.

4. Finite Element Simulations

4.1. Finite Element Modelling with Induction Heating Mode

In this study, a high-frequency induction heating apparatus was used to heat the wire during dieless drawing. Therefore, a software DEFORM coupled with the induction heating mode was used, in which a boundary element method is used to analyze the heat transfer during the dieless drawing process. The flow stress curves of stainless steel SUS304 wires at different temperatures from the DEFORM database shown in Figure 9 were used. The strain rate variation in the dieless drawing process was not so big. The average strain rate was about 0.25. Therefore the strain rate effect was not taken into account. The electrical resistivity of the wire at 20 °C and 650 °C were set as 0.00072 and 0.00116 $\Omega \cdot$mm, respectively. The relative magnetic permeability was set as 1.02. After that, the current frequency and input power inside the induction coil were set. After element meshing in the wire was completed, the thermal conductivity of the wire at 100 °C and 500 °C were set as 17 and 22 W$\cdot$m$^{-1}\cdot$K^{-1}, respectively. The convection coefficient of the atmosphere was set as 70 W$\cdot$m$^{-2}\cdot$K^{-1}.

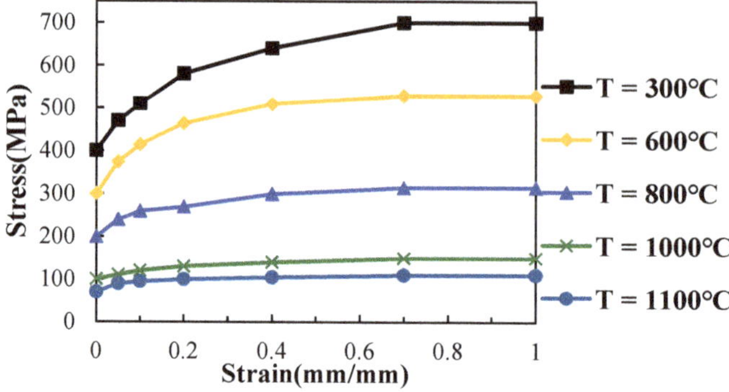

Figure 9. Flow stress curves of SUS304 at $\dot{\varepsilon} = 0.25$.

The configuration of the meshed wire and heating coil is shown in Figure 10a. The wire and the circular heating coil were axisymmetric, thus, DEFORM 2D was used to reduce the simulation time. The corresponding three dimensional perspective view of the wire and heating coil is shown in Figure 10b. Convergence analysis with different mesh numbers and different layers inside the wire were conducted. With 7 mesh layers in the radial direction and about 5500 mesh numbers, reasonably accurate simulation results within an error of 0.7% could be obtained. The accuracy of 0.7% in the simulation results of wire diameters is convergence accuracy with different element numbers inside the wire, which means even though many more element numbers were set, the variations of the simulation results were within 1%.

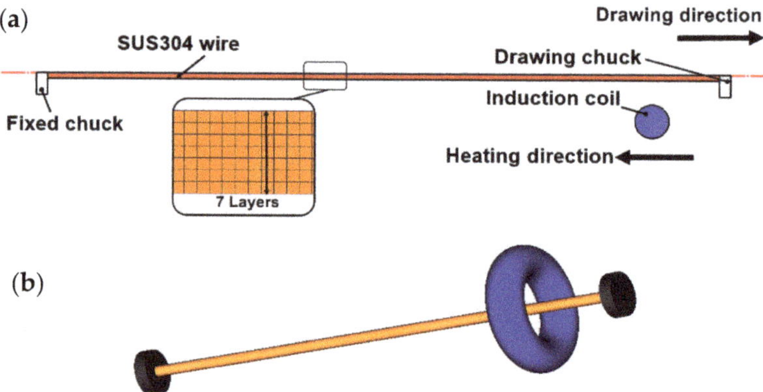

Figure 10. (**a**) Configuration of wire and heating coil used in DEFORM 2D, (**b**) 3D perspective view.

4.2. Formability Analysis

The drawing speed and the forming temperature are the key parameters that determine if a successful wireless drawing process can be achieved or not. Formability analysis with different drawing speeds on the maximal reachable area reductions was conducted. The parameters used in the formability analysis are shown in Table 2. The relationships between the forming temperature variation and the input power of 180 W are shown in Figure 11a. The relationships between the heating coil speed, drawing speed and forming temperature variation are shown in Figure 11b. The profiles of the heating coil speed and drawing speed are that same as those used in the self-developed dieless drawing machine. The values of input power, heating coil speed, drawing speed are input in the finite element (FE) simulations for formability analysis. After a series of formability analysis, formability experiments for the maximal reachable area reduction using the self-developed dieless drawing machine are also conducted.

Table 2. Parameters used in formability analysis.

Parameters	Values
Diameter of wire, D_0 [mm]	1
Length of wire, W_L [mm]	60
Wire mesh layers in radius, W_L	7
Wire mesh number, W_N	5500
Coil mesh number, C_N	4130
Coil loop inner diameter, D_C [mm]	5.5
Coil wire diameter, d_C [mm]	3
Current frequency, f [KHz]	1000

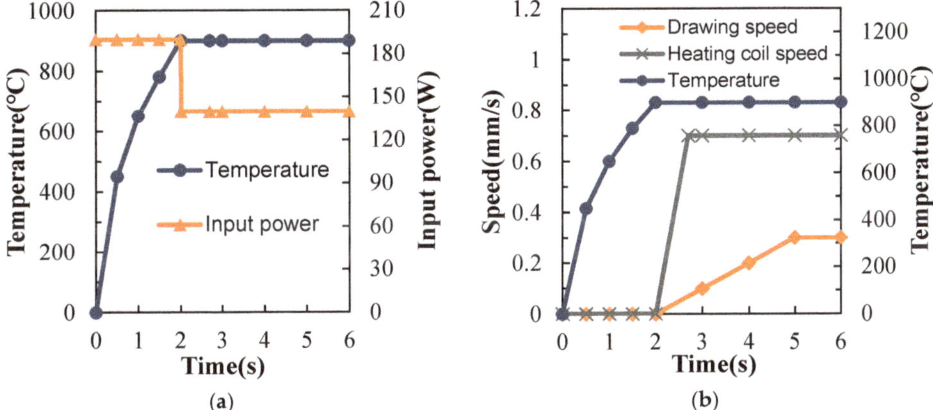

Figure 11. (a) Relationships between forming temperature variation and input power, (b) Relationships between heating speed, forming temperature and drawing speed.

5. Formability Simulation Results

FE simulations with forming temperature of 900 °C and drawing speed of 0.6 mm/s were conducted. The profiles of the drawn wires at reductions of 37% and 40% are shown in Figure 12a,b, respectively. The profile shown in Figure 12a is regarded as a successful case and that shown in Figure 12b is regarded as a failure case. The area reduction of 37% obtained in Figure 12a is called the maximal reachable area reduction under the conditions of forming temperature of 900 °C and drawing speed of 0.6 mm/s. The effects of the drawing speed and forming temperature on the maximal reachable area reduction are discussed below.

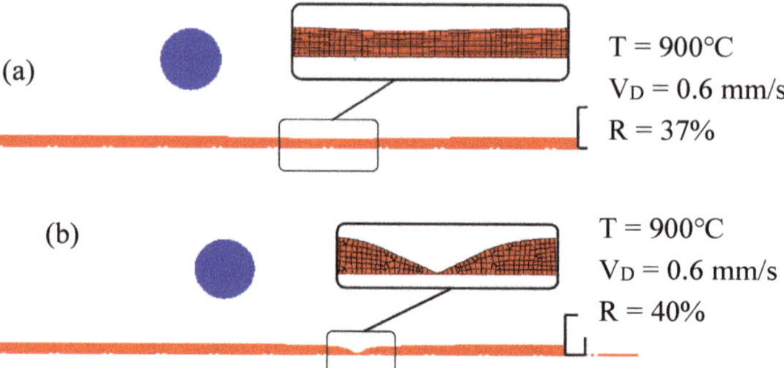

Figure 12. Profiles of drawn wires at different reductions. (**a**) Successful case, (**b**) Failure case.

The simulation results of maximal reachable area reductions for different drawing speeds and different forming temperatures are shown in Figure 13. Clearly the maximal reachable area reduction increased as the forming temperature increased, because stainless steel wires at higher temperatures have smaller flow stresses and better flowability. On the other hand, the maximal reachable area reduction decreased as the drawing speed increased. A larger drawing speed resulted in poor flowability and accordingly a smaller reachable area reduction was obtained.

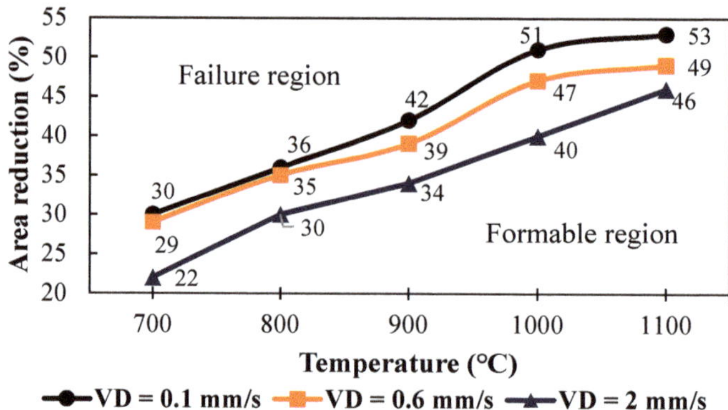

Figure 13. Simulative maximal reachable area reductions at different temperatures and drawing speeds.

The experimental results of the maximal reachable area reduction in air and argon gas are shown in Figure 14. As in the cases in Figure 13 for the simulation results, a larger maximal reachable area reduction could be obtained at a higher forming temperature and at a lower drawing speed. The maximal reachable area reductions during drawing in air are larger than those in argon gas. That is because continuously pouring of argon gas into the chamber cools down the wire and reduce the flowability of the deformation region.

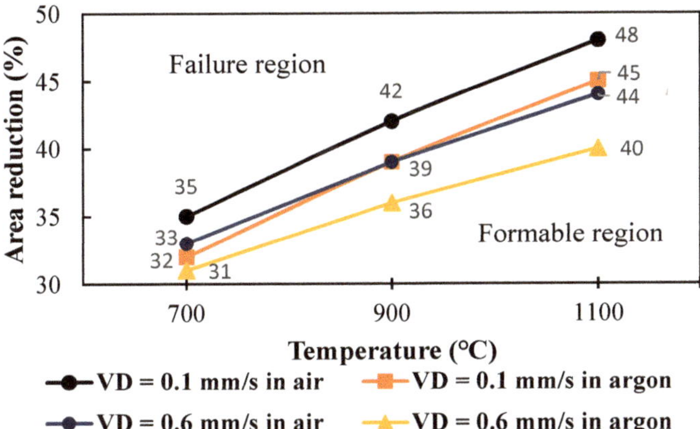

Figure 14. Experimental maximal reachable area reductions in air and argon gas.

The comparisons of maximal reachable area reductions at drawing speeds of 0.1 and 0.6 mm/s during drawing in argon between simulations and experiments are shown in Figure 15. Generally the tendency of simulative maximal reachable area reduction was the same as that from experimental results. A bigger error between the simulative and experimental values at a higher temperature of 1100 °C was found. That was probably because of some discrepancy in the material flow stresses used in the FE simulations and the real stainless steel wires.

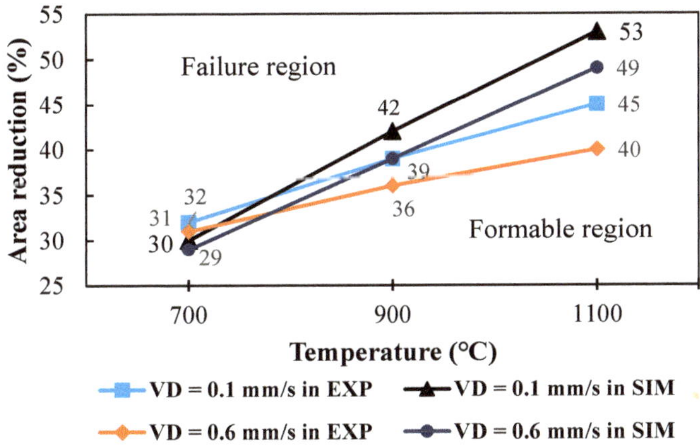

Figure 15. Comparisons of maximal reachable area reduction between simulations and experiments.

6. Conclusions

In this study, dieless drawing experiments on stainless steel SUS304 wires in air and argon gas were conducted using a self-developed dieless drawing machine. The effects of forming temperatures and the oxide layers on the mechanical properties of the drawn stainless steel SUS304 wires were discussed. The yield stress and UTS of the drawn wires decreased as the forming temperature increased. The yield stress and UTS of the drawn wires in air were slightly higher than those in argon gas, whereas the elongations of the drawn wires increased as the drawing temperature increased and the elongations in air were slightly smaller than those in argon gas. A finite element model considering the high frequency induction heating mode was used to conduct the formability analysis at different drawing speeds and forming temperatures during dieless drawing of stainless steel wires. A lower drawing speed and a higher forming temperature could raise the maximal reachable area reductions. The simulative maximal reachable area reductions at lower temperatures were quite coincident with the experimental results.

Author Contributions: Conceptualization, methodology, research supervision and writing—review and editing were conducted by Y.M.H. Simulation, experiments, validation and writing—original draft preparation were completed by H.H.L.

Funding: The authors would like to extend their thanks to the Ministry of Science and Technology of the Republic of China under Grant no. MOST 107-2221-E-110-038-MY3. The advice and financial support of MOST are gratefully acknowledged.

Acknowledgments: The authors would like to extend their thanks to the ministry of science and technology of the republic of china under grant no. most107-2221-e-110-038-my3. The advice and financial support of most are gratefully acknowledged.

Conflicts of Interest: The authors declare no conflict of interest.

References

1. Weiss, V.; Kot, R. Dieless wire drawing with transformation plasticity. *Wire J.* **1969**, *9*, 182–189.
2. Wicht, H.; Bouchaud, J. Nexus Market Analysis for MEMS and Microsystems III. *MST News* **2009**, *5*, 33.
3. Furushima, T.; Manabe, K. Experimental and numerical study on deformation behavior in dieless drawing process of superplastic microtubes. *J. Mater. Process. Technol.* **2007**, *191*, 59–63. [CrossRef]
4. Fortunier, R.; Sassoulas, H.; Momtheillet, F. A thermo-mechanical analysis of stability in dieless wire drawing. *Int. J. Mech. Sci.* **1997**, *39*, 615–627. [CrossRef]
5. Wang, Z.T.; Zhang, S.H.; Xu, Y.; Luan, G.F.; Bai, G.R. Experiment study on the variation of wall thickness during dieless drawing of stainless steel tube. *J. Mater. Process. Technol.* **2002**, *120*, 90–93. [CrossRef]
6. Wang, Z.T.; Luan, G.F.; Bai, G.R. Study of the deformation velocity field and drawing force during the dieless drawing of tube. *J. Mater. Process. Technol.* **1999**, *94*, 73–77. [CrossRef]
7. Furushima, T.; Hirose, S.; Manabe, K. Effective temperature distribution and drawing speed control for stable dieless drawing process metal tube. *J. Solid Mech. Mater. Eng.* **2009**, *3*, 236–246. [CrossRef]
8. Furushima, T.; Manabe, K. Experimental study on multi-pass dieless drawing process of superplastic Zn-22%Al alloy microtubes. *J. Mater. Process. Technol.* **2007**, *187–188*, 236–240. [CrossRef]
9. Furushima, T.; Manabe, K. FE analysis of size effect on deformation and heat transfer behavior in microtube dieless drawing. *J. Mater. Process. Technol.* **2008**, *201*, 123–127. [CrossRef]
10. Furushima, T.; Imagawa, Y.; Manabe, K.; Sakai, T. Effects of oxidation and surface roughening on drawing limit in dieless drawing process of SUS304 stainless steel microtubes. *J. Mater. Process. Technol.* **2015**, *223*, 186–192. [CrossRef]
11. Furushima, T.; Imagawa, Y.; Furushima, S.; Manabe, K. Deformation profile in rotary laser dieless drawing process for metal microtubes. *Procedia Eng.* **2014**, *81*, 700–705. [CrossRef]
12. Milenin, A. Rheology-based approach of design the dieless drawing processes. *Arch. Civ. Mech. Eng.* **2018**, *18*, 1309–1317. [CrossRef]
13. Twohig, E.; Tiernan, P.; Tofail, S.A.M. Experimental study on dieless drawing of Nickel-Titanium alloy. *J. Mech. Behav. Biomed. Mater.* **2012**, *8*, 8–20. [CrossRef] [PubMed]
14. Naughton, M.D.; Tiernan, P. Requirements of a dieless wire drawing system. *J. Mater. Process. Technol.* **2007**, *191*, 310–313. [CrossRef]
15. Tiernan, P.; Carolan, R.; Twohig, E.; Tofail, S.A.M. Design and development of a novel load-control dieless rod drawing system. *CIRP J. Manuf. Sci. Technol.* **2011**, *4*, 110–117. [CrossRef]
16. Supriadi, S.; Manabe, K. Enhancement of dimensional accuracy of dieless tube-drawing process with vision-based fuzzy control. *J. Mater. Process. Technol.* **2013**, *213*, 905–912. [CrossRef]
17. Tiernan, P. Dieless drawing of bars, rods, tubes, and wires. *Compre. Mater. Process.* **2014**, *3*, 149–157.
18. Sekiguchi, H.; Kobatake, K.; Osakada, K. A fundamental study on dieless drawing. In *Proceedings of the Fifteenth International Machine Tool Design and Research Conference*; Tobias, S.A., Koenigsberger, F., Eds.; Palgrave: London, UK, 1975; pp. 539–544.

19. Hwang, Y.M.; Kuo, T.Y. Dieless drawing of stainless steel tubes. *Int. J. Adv. Manuf. Technol.* **2013**, *68*, 1311–1316. [CrossRef]
20. Qin, F.; Liu, X.F.; Mao, H.E. The thickness distribution of oxidation film on tapered pipe surface in dieless drawing. *Inter. J. Corros.* **2011**, *2011*, 615197. [CrossRef]

 *metals*

Article

Material Flow in Ultrasonic Orbital Microforming

Wojciech Presz

Institute of Manufacturing Technologies, Warsaw University of Technology, Warsaw 02-524, Poland;
w.presz@wip.pw.edu.pl; Tel.: +48-696-844-998

Received: 18 March 2019; Accepted: 22 April 2019; Published: 24 April 2019

Abstract: Ultrasonic orbital microforming—UOM—uses the broadly understood idea of orbital forging but uses very different laws of physics. The only shaping force in this process is the inertia force resulting from the acceleration in the rotary motion of the workpiece. Micro specimen blanked from cold rolled aluminum sheet metal was used in the applied UOM process. Only the upper and lower part of the sample is deformed that gives about 70% of volume. The rest—the middle part—remains undeformed. The final shape of the product is influenced by the shape of the inside of the die in which the UMO process is carried out. However, this effect is not a direct one. The product shape does not repeat the shape of the interior of the die. The preliminary experiments with modular micro-die have been performed on the way of controlling the shape of deformed micro-objects. The microstructure analysis has been done as well as micro-hardness distribution.

Keywords: microforming; ultrasonic; orbital forming

1. Introduction

The progressive development of miniature systems is the driving force and challenge for miniaturization of all material technologies. The metal forming technology plays among them a significant role. Its miniaturization is connected with difficulties if object dimension is smaller than 1 mm. As a result, a new branch of metal forming which deals with the production of objects that meet the above criterion was created [1,2]. The microforming was introduced as a separate technology [3]. Deviations from the developed technology rules for years associated with miniaturization are called "scale effects". They concern, in principle, all elements of the technological process. Surface layer and contact phenomena [4–6], affecting the friction [7–9], lubrication [10], and galling [11], increasing the role of preparation methods for micro-billets surfaces [12,13]. They also refer to the internal structure [14–16], which affects the quality of the surface and cracking mechanisms [17,18]. Scale effect concerns even the construction of machines [19], tooling [20,21], and tools [22–26], as well as the design of the technological process plan [27,28]. Materials and friction tests using microsamples are also created because their results differ from those obtained in macrotests [29,30]. Reducing the size of manufactured parts on one hand is a serious challenge for traditional technologies, but on the other hand, mainly by removing the energy barrier, opens the opportunity of using other unconventional techniques. Electric power and magnetic wave assistance, laser treatment, and utilization of mechanical vibrations at various frequencies during forming can be mentioned here. In the last of these methods, ultrasonic vibrations are particularly in the sphere of greatest interest [31,32]. Ultrasonic-vibration reduce forces in the ECAP [33] and micro-extrusion processes by strong reduction of friction [34] and reduce the flow stress [35] by not only temperature increase, but also the softening effect. The use of ultrasonic vibrations can also cause phase changes [36] and specific macroscopic consequences in terms of cracking [37]. In studies on the influence of ultrasonic vibrations on the course of microforming process, often used is micro-upsetting under dry friction conditions [38,39] due to geometric simplicity and relative ease for modelling. In some conditions of upsetting with the use of ultrasonic vibrations, the formation of a specific shape of the lateral surface of the sample is reported [40–42]. The phenomenon leading to it was taken to call "anti-barrelling". The name is a reference to the phenomenon called "barrelling" [43–45], which concerns the formation of the convex side surface of the upset cylinder. It is created as a result of friction forces occurring on the contact surfaces with tools. Anti-barrelling is the formation of a concave lateral surface. This effect is not being observed in any upsetting conditions involving ultrasonic vibrations [46,47]. The causes and consequences of this phenomenon are not fully explained and are currently under investigation. One of the hypotheses is the temporary detachment of the surface of the punch from the surface of the sample [48,49]. The ultrasonic orbital microforming (UOM) process also leads to the creation of anti-barrelling shape however, it is based on completely different physical phenomena. UOM process was introduced in 2018 by Presz [50]. The punch is vibrating with the amplitude of magnitude large enough to obtain the so called "dynamic effect" that means detaching the front of the punch from the sample. At the same time, the end of the punch turns into small circles, which allows the shaped sample to be introduced into a rotary motion. This movement begins when the advancing punch touches the surface of a sample placed on the cylindrical die. The sample is lifted and stand on the edge and begins to roll around—similarly to the coin rolling on the table with the difference that the rotary motion of the sample accelerates because of actions of punch. The punch initially slides on the edge of specimen, temporary detaching from it. The edges begin to deform and their temperature increases, slippage decreases—this means that the rotational motion accelerates—which causes the "lifting" of the sample. The inertia force resulting from accelerated motion is the only shaping force, which makes this process unique. The process occurs cyclically and the workpiece is cyclically driven to rotary motion. It is a process in some way similar to the rotary forging, giving the possibility of manufacturing short objects with variable diameter. It is implemented in a micro-scale and has the potential for further miniaturization to an unpredictable degree. It is a completely new process with potential not yet recognized. Current work is a step towards understanding, controlling and possibly commercially exploiting the potential of this process.

2. Materials and Methods

UOM is a forming process, which, for technical reasons, can probably only be carried out as a microforming process. Since there are no known constructions of machines that would be capable of putting the macro-object in a free spin at such a high speed that it would stabilize it in space due to the gyroscopic effect. In the only so far presented version [38] UOM consists in inducing the rotational motion of the cylindrical workpiece with a flat punch vibrating with an ultrasonic frequency. The layout of the setup is shown in Figure 1.

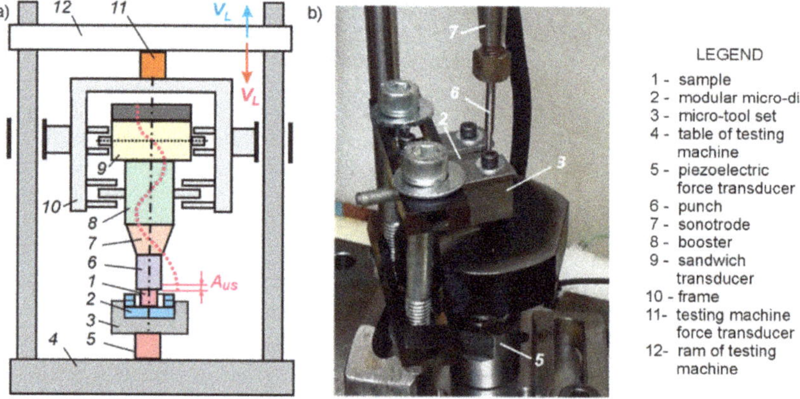

Figure 1. Experimental stand: (**a**) scheme of the test stand, and (**b**) close-up of working area.

The sample 1 is placed in the modular micro-die (2) in the Micro-tool set (3) standing on the force transducer (5) fixed on the table (4) of the testing machine. The sample is deformed using the punch (6), which is at the end of the sonotrode (7) of the ultrasonic system (8—booster, 9—sandwich transducer). This system is mounted in the frame (10), which through the force transducer (11) is connected to the ram (12) of the testing machine. An alternating voltage oscillating at ultrasonic frequency is applied by a power supply unit to the piezoelectric transducer. Booster and sonotrode work as half-wavelength resonators, vibrating lengthwise with standing waves at its resonant frequency. The frequency used is 20 kHz. Sonotrode acts as a displacement amplifier. On the basis of the laser displacement transducer, Keyence LK-H008, (Keyence, Osaka, Japan), the amplitude AUS of the ultrasonic vibrations on the surface of the booster and the face of the punch was determined, see Table 1.

Table 1. System parameters.

Amplitude			Ram Velocity	
At Surface of Booster	**At Surface of Punch Nose**	**Amplification of Sonotrode**	**Loading**	**Unloading**
(μm)	(μm)	(1)	(mm/min); (m/s)	(mm/min); (m/s)
2.5	16.0	6.3	$0.2; 3.3 \times 10^{-6}$	$0.02; 3.3 \times 10^{-7}$

Tools included in the test are shown in Figure 2. The micro-die used in the research has a modular structure. It consists of plates with precision holes drilled with the EDM, which can be combined in various configurations. Three versions of it were used, as shown in Figure 2: 0.2-mm thick plates designated as: C, F, and J were used. In each of presented experiments, two modules were used, matching them as follows: CC, FC, JC.

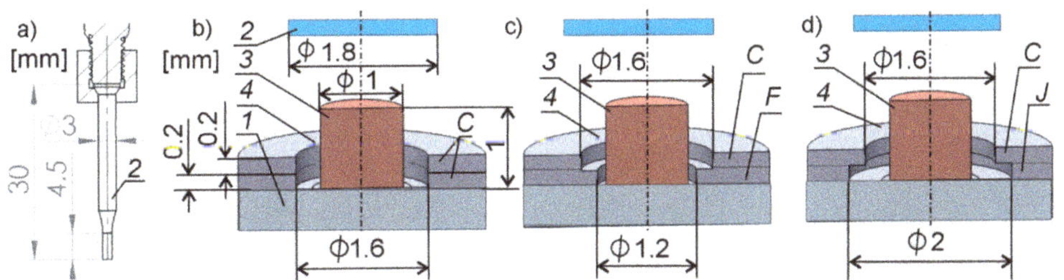

Figure 2. Tools included in the tests: (**a**) punch, (**b**) die CC: specimen "S3", (**c**) die FC: specimen "S4", and (**d**) die JC: specimen "S5".

The samples used for testing, see Figure 3, are blank with reduced clearance from cold rolled, 1-mm thick aluminum sheet. Diameter of samples $d = 1^{0}_{-0.02}$ mm. Samples after this process have a slightly convex upper surface, 1, and a small and very thin burr, 2, which before the process was manually removed. On some parts of the circumference, there is also a fracture zone, 3. The structure of the preforms is a modified structure after cold rolling process, see Figure 3b. Modification of the structure is associated with the zone of plastic deformation in the process of plastic blanking. In consequence there is a hardness distribution shown in Figure 11a. The plasticity characteristics of the material in the annealed state are represented by the stress strain curves given in Figure 3, the CR curve is drawn on the basis of experimentally determined equation of the two-parameters form (1). Experimental sites included only the range of $\varepsilon = (0, 0.3)$. The values of constants C and n are given in the Table 2.

$$\sigma_p = C \cdot \varepsilon^n \tag{1}$$

Table 2. Mechanical properties of specimen materials (cold rolled—*CR*).

E (GPa)	R_e (MPa)	ν (1)	*C* (MPa)	*N* (1)
70	135	0.32	170	0.05

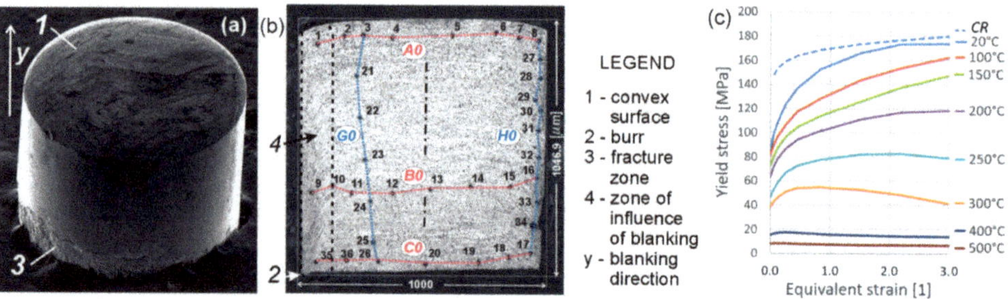

Figure 3. (**a**) SEM of specimen "S0": (**b**) Cross-section with marked micro-hardness, further described in text; and (**c**) stress strain curves of used material in the mild state and at the state of experiment, CR.

3. Results

During the processes, forces were recorded using a strain gauge dynamometer of a testing machine, Figures 1–11 and an additional piezoelectric dynamometer, Figures 1–5. The recorded runs of process forces are shown in Figure 4.

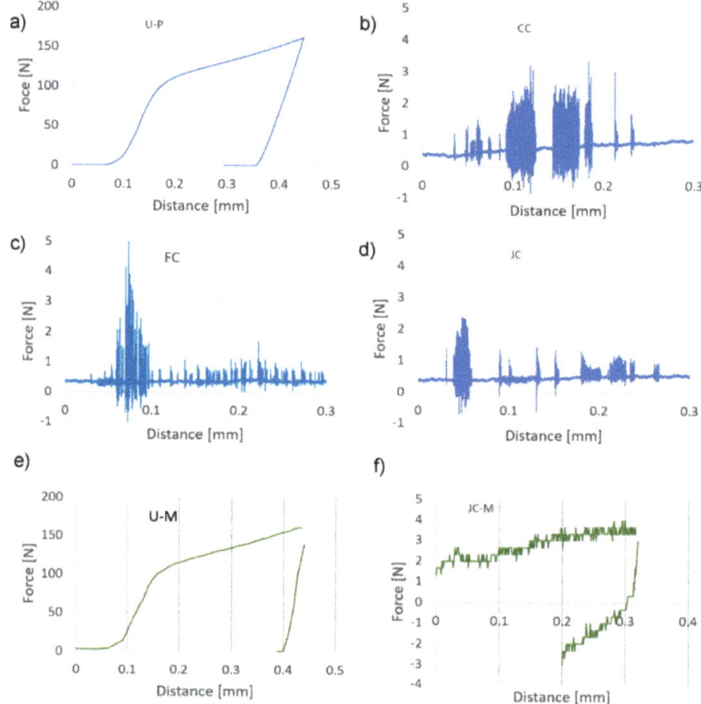

Figure 4. Process forces recorded by the auxiliary dynamometer (**a**)–(**d**) and by the testing machine dynamometer (**e**,**f**): (**a**) free upsetting—S1, (**b**) process CC—S3, (**c**) process FC—S4, (**d**) process JC—S5, (**e**) free upsetting—S1, and (**f**) process JC—S5.

In Figure 5 are reported images and outlines of the samples after upsetting and UOM. The outlines were obtained on the basis of microscopic photographs taken from the side.

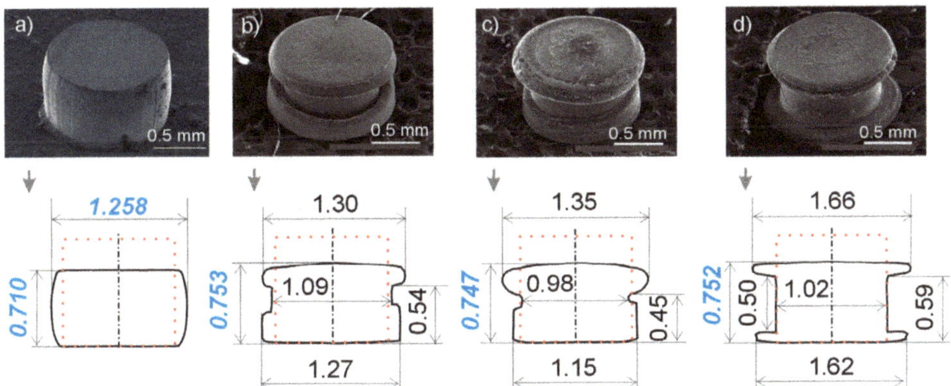

Figure 5. Microscopic (SEM) photos of deformed samples and their outlines drawn on the base of side views: (**a**) free expansion without vibrations—specimen S1, (**b**) UOM with a die CC—specimen S3, (**c**) UOM with a die FC—specimen S4, and (**d**) UOM with a die JC—specimen S5.

The metallographic cross-section of the samples are shown in Figure 6. On the basis of these, it was found that the samples in the middle part do not undergo deformation, and only their upper and lower parts are deformed. The cross-sections allowed to build 3D models and estimate the size of equivalent strains. In the model construction, the dimensions were determined based on the cross-sections from Figure 6 based on the height of the samples, which were measured with a micrometer.

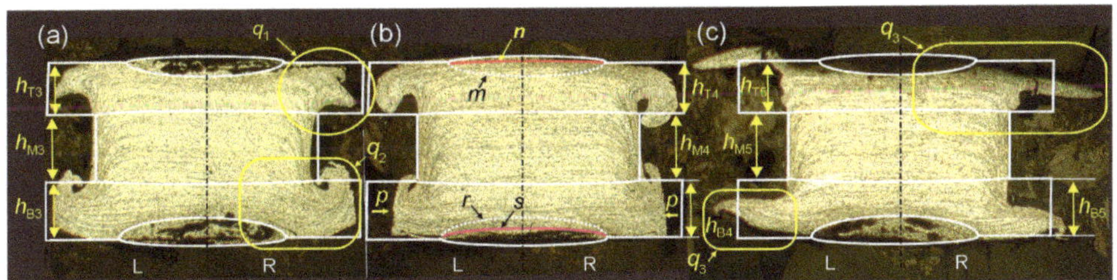

Figure 6. The metallographic cross-sections of the samples: (**a**) sample S3; (**b**) sample S4; and (**c**) sample S5.

The methodology for determining the average equivalent strain in deformed areas is shown in Figure 7a–d. At the beginning, the values $h^{(i)}_{T,M,B}$, $i = 3, 4, 5$ (top, middle, bottom) were determined and the half-section boundary was determined for each of the samples. Assuming the symmetry of shapes, 3D axially symmetric models of each of the specimens were prepared. Then the models were divided into three parts according to the determined $h^{(i)}_{T,M,B}$ and their volumes $V^{(i)}_{T,M,B}$ were calculated as well as the dimensions of cylinders $c^{(i)}_{T,B,n}$ and $c^{(i)}_{T,B,s}$. Assuming that their volumes are equal to the volumes $V^{(i)}_{T,M}$ and their dimensions meet (2) and (3) respectively, the values $h^{(i)}_{T,M,B,n}$ and $d^{(i)}_{T,M,B,n}$ were calculated according to (4) and (5).

$$d^{(i)}_{T,B,n} = d_0 \tag{2}$$

$$h^{(i)}_{T,B,s} = h^{(i)}_{T,B} \tag{3}$$

$$h^{(i)}_{T,B,n} = 4 \cdot \frac{V^{(i)}_{T,B}}{\pi \cdot d_0^2} \tag{4}$$

$$d^{(i)}_{T,B,n} = 2 \cdot \sqrt{\frac{V^{(i)}_{T,B,}}{\pi \cdot h^{(i)}_{T,B}}} \tag{5}$$

where all used quantities are defined in Figure 7.

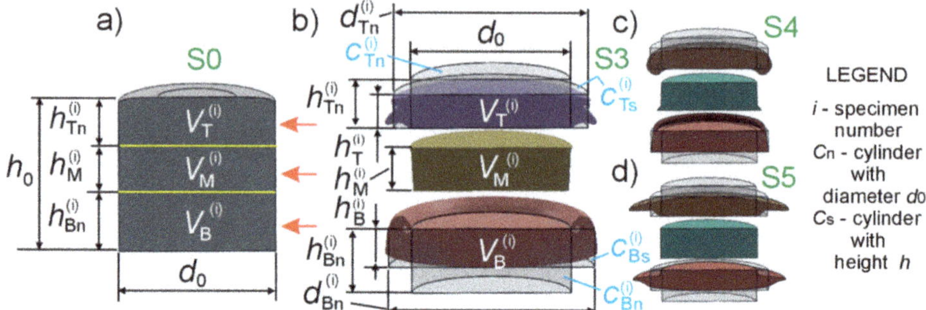

Figure 7. (**a**)–(**d**) The methodology for determining the average equivalent strain.

Mean equivalent strain in the upper and lower areas was calculated assuming that it is equal to the modulus of the axial deformation component according to Equation (6):

$$\bar{\varepsilon}^{(i)}_{T,B} = \left| 2 \cdot ln \frac{d_0}{d^{(i)}_{B,T,n}} \right| = \left| ln \frac{h^{(i)}_{T,B}}{h^{(i)}_{T,B}} \right| \tag{6}$$

In the sample S5 local strains were estimated according to Figure 8 and Equation (7):

$$\varepsilon^{(5)}_{T,B} = \left| ln \frac{h^{(5),\,j}_{T,B}}{h^{(5),\,j}_{T,B,n}} \right|, \; at \; r^{(5),\,j}_{T,B} \tag{7}$$

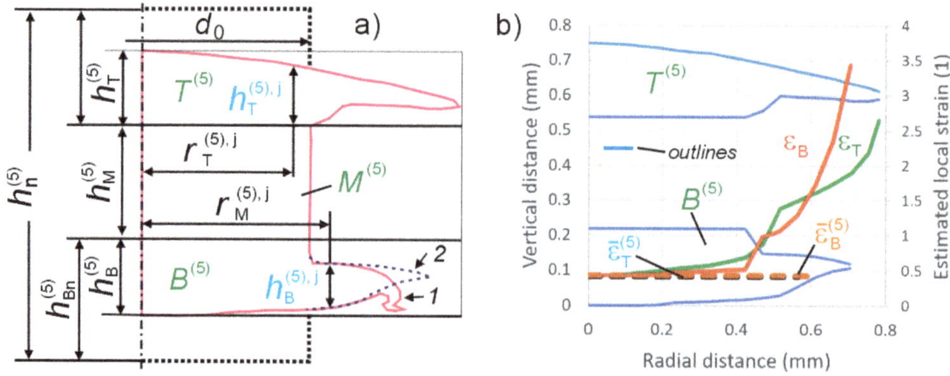

Figure 8. (**a**) The method of estimating local equivalent strain in specimen S5; and (**b**) mean and local equivalent strain distribution in the specimen S5.

The determined values for samples S3–S5 are collected in Tables 3 and 4. The strains distributions according to Equation (6) are shown in Figure 8b.

Table 3. Calculated values for V and h of specimens S3–S5.

Specimen	V (mm^3)				h (mm)			
	All	T	M	B	All	T	M	B
S3	0.783	0.25	0.206	0.327	0.752	0.22	0.283	0.25
S4	0.773	0.262	0.238	0.273	0.747	0.223	0.297	0.227
S5	0.781	0.253	0.262	0.266	0.752	0.212	0.32	0.22

Table 4. Calculated values for h_n, d_s and ε of specimens S3–S5.

Specimen	h_n (mm)			$d_{s.}$ (mm)			ε (1)		
	T	M	B	T	M	B	T	M	B
S3	0.318	0.263	0.417	1.202	0.964	1.291	0.368	0	0.511
S4	0.334	0.304	0.347	1.223	1.011	1.236	0.402	0	0.424
S5	0.322	0.334	0.339	1.232	1.022	1.241	0.417	0	0.432

The Vickers hardness under load of 0.09807 (N)–HV 0.01–(ISO 6507-1) was measured on the surfaces of metallographic specimens—Figure 9.

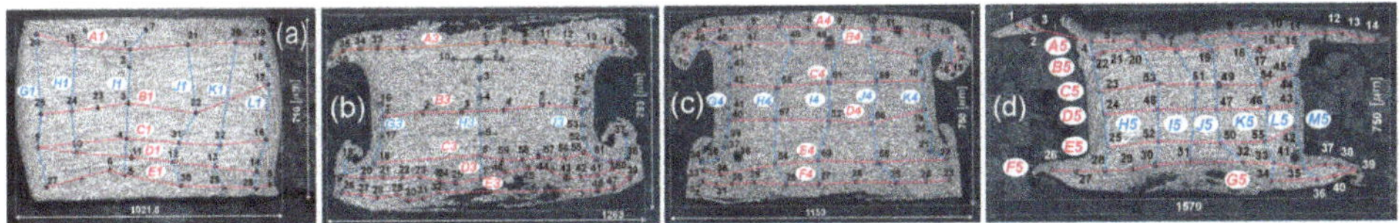

Figure 9. Metallographic cross-sections of samples: (**a**) Free upsetting; (**b**) UOM with a die CC; (**c**) UOM with a die FC; and (**d**) UOM with a die JC.

The hardness measures executed along the lines in Figure 9 are shown in Figure 10. The micro-part obtained in the UOM process is axially symmetrical. Starting from this assumption, the results of hardness measurements were expanded on the principle of mirror reflection in relation to the axis of the samples. The results in the form of micro-hardness distributions are shown in Figure 11.

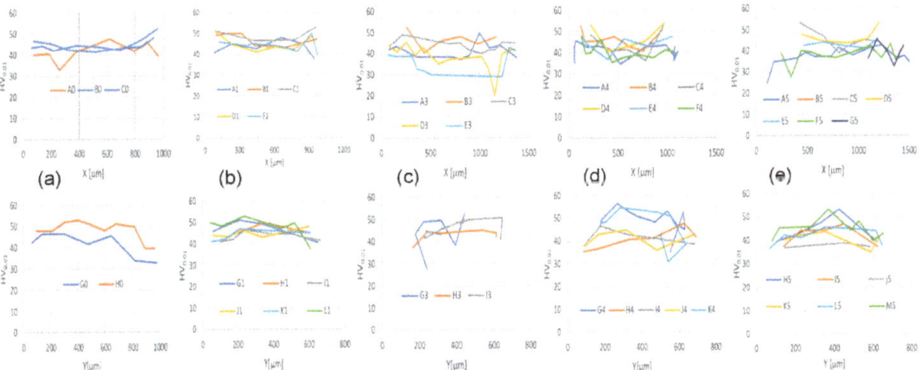

Figure 10. The results of micro-hardness measurements at the points marked in Figure 9 and in accordance with the courses of marked lines: (**a**) S0; (**b**) S1; (**c**) S3; (**d**) S4; (**e**) S5.

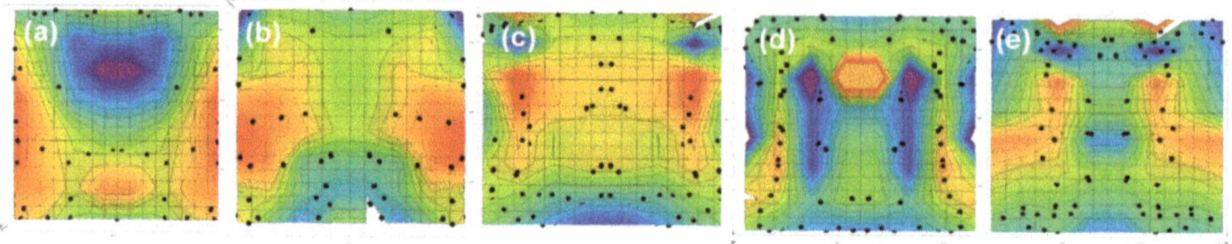

Figure 11. Micro-hardness distribution: (**a**) S0; (**b**) S1; (**c**) S3; (**d**) S4; and (**e**) S5.

4. Discussions

The deformation In UOM process starts at its periphery that is rolling on the surface of the bottom of the die and the face of the punch (Figure 12a,b). In the upper and lower parts of the samples, achieved ranges of deformations are impossible to be obtained in the process of free upsetting of this material [38]. Figure 12c shows the similar sample to these used in this work after the ultrasonic assisted upsetting process—the sample has cracked under deformation of the equivalent strain about 0.3. Achieving much larger deformations in the UOM process is associated with an increase in temperature resulting in increased deformability of the material.

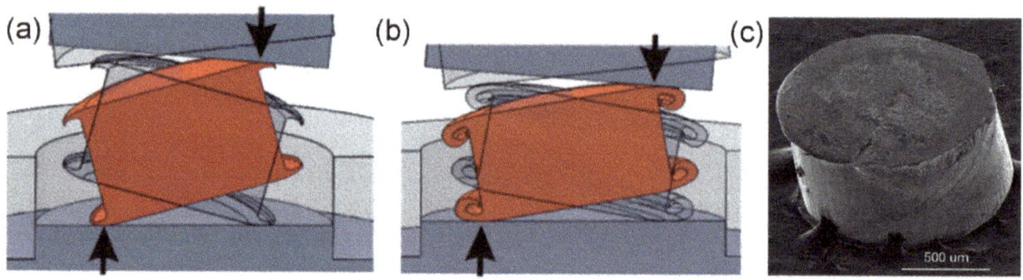

Figure 12. (**a,b**) The theoretical course of the process without the impact of die-cavity. (**c**) An example result of simple ultrasonic assisted micro-upsetting—previous investigation.

The UOM process creates a specific structure of the deformed sample. In the "classic" free upsetting, Figure 13a there are four zones: Z1 and Z2—the material is virtually unstrained; Z3—slight deformation and Z4—intense flow. These areas can be reflected in the microhardness distribution, Figure 11b. However, it should be remembered that the sample before deformation had the material structure after cold rolling, which was further changed in the blanking process.

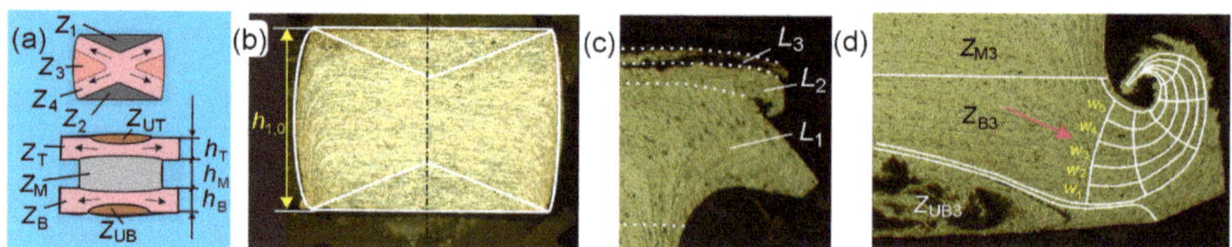

Figure 13. (**a**) Areas with diverse plastic features; (**b**) metallographic cross-section of S1 after simple upsetting; (**c**) close up q1of Figure 6a; close-up q2 of Figure 6a.

On the cross-section of samples deformed in the UOM process, five zones can be distinguished. They are shown in Figure 13d: Z_M-non-deformed material, Z_T and Z_B intense deformation, Z_{UT} and Z_{UB} material degradation zone under the influence of ultrasonic vibrations. In Z_M zones, the courses of the central hardness lines (B in Figure 10c, C and D in Figure 10d, C and D in Figure 10e) are consistent with the lines of the unstrained sample (B in Figure 10a). The Z_T and Z_B zones are zones of intense material flow. The material in these zones flows in layers with different flow speeds. The shape of the flow line (Figure 13) suggests that the highest speed is at the contact surfaces with the tools. This speed as it moves away from the surface falls, causing the rim to bend as shown in detail q_2 of Figure 6a. This part of specimen is closer shown in Figure 13d, where the drawn flow lines form spirals due to the gradual drop in the flow velocity in subsequent layers from w1 to w5. Such processes can follow each other creating "layers overlapping". This is shown by the detail q_1 from Figure 6a. This area is closer shown in Figure 13c. Differences in the flow velocity of the material in the layers: L_1, L_2, and L_3 are probably related to the temperature distribution, which decreases as it moves away from the surface of the contact between the tools and the sample. The appearance of elevated temperature is caused by intense plastic deformation and presumably slipping of the edge of the samples on the tool surfaces that may occur in the initial phase of the rotation/rolling motion. Two phenomena act in this zone oppositely: strain hardening and temperature based softening as dynamic recovery and recrystallization. The first causes an increase in the hardness of the material and a second decrease in this hardness.

For this reason, in layers with a lower temperature - further from the contact surface, the hardness is higher in comparison to layers which are closer to the contact surface: compare curve C3 with curve D3 in Figure 10c and curve E5 with curve F5 in Figure 10e. The ZUT and ZUB zones are characterized by the occurrence of rows and voids. At the border of these zones a decrease in hardness is observed, see curve E3 in Figure 10c, curve F4 in Figure 10d and curve A5 in Figure 10e. The above described zones can be recognized in the hardness distributions shown in Figure 11.

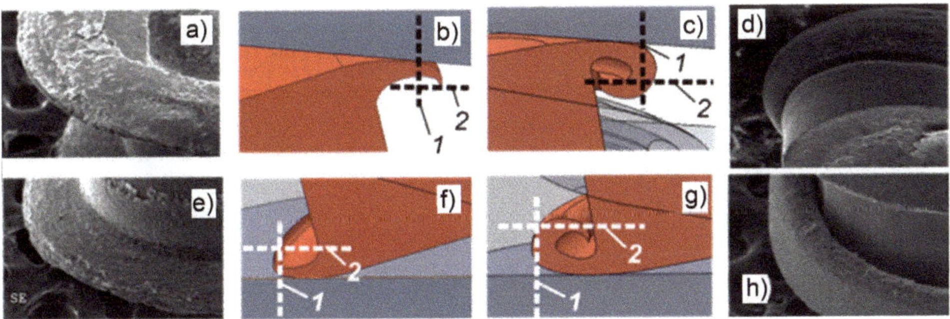

Figure 14. The course of the UOM process depending on the shape of die-cavity: (**a,b**) The theoretical course of the process without the impact of die-cavity; (**c**)–(**h**) The influence of vertical, 1, and horizontal, 2, limits.

The deformation proceeds according to the diagram shown in Figure 12a,b. The process runs in phases as evidenced by the forces in Figure 4b–d. Processes forces are very small, which is caused by a very long "distance of forming", small contact area and the so-called "dynamic effect". Forces are so small that their detection by the dynamometer of the testing machine is pointless. An exemplary record regarding the JC-S5 process, Figure 4f, is more a record of the resistances of the ultrasonic head guiding system than process force.

Along with the UOM process the phenomenon of folding of the deformed edge caused by the difference in temperature and flow velocity of the deformed layers is intensified by increasing the angle of inclination of the rotating sample with its deformation. The edge of the sample is successively "bended" in subsequent phases of the process (Figure 14c). The use of shaped dies (Figure 2) caused the material flow to be modified. The deformation is modified by two kinds of possible limitations: Vertical limits—Figure 14d, h and horizontal limits—Figure 14a,e. The shape of die cavity refers to the shape of the final sample only in the case of the FC process, see Figures 5c and 15b. In this case, the bottom diameter was calibrated, resulting in an increase in the force of the process (Figure 4c).

The probably presence of hydrostatic pressure in the bottom part of specimen results in the reduction of Z_{UT} and Z_{UB} zones by evolution of interface surfaces—white lines m and r in Figure 6b has been evolved into red lines n and s. In other cases, as shown in Figure 15 the final outline of the sample is different from the outline of the cavity of the die.

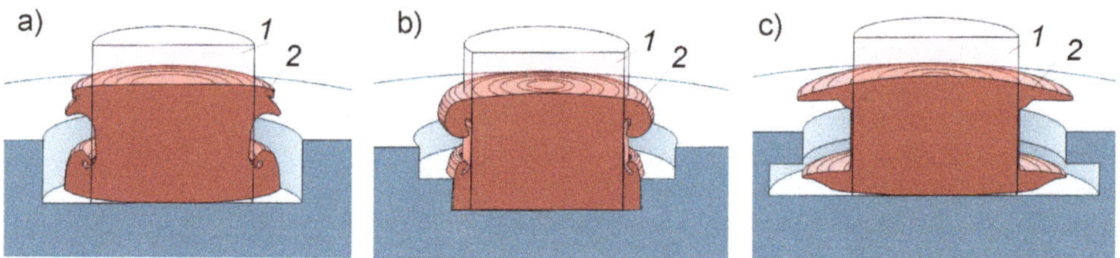

Figure 15. Initial, 1, and final, 2, shape of the workpiece in the die cavity in the processes: (**a**) CC; (**b**) FC; and (**c**) JF.

5. Conclusions and Future Work Suggestions

In this study, vibrations of 20 kHz frequency and 16 μm longitudinal amplitude were applied to deform an aluminum cylindrical micro-specimen inside three variants of modular-die in order to investigate the material flow during ultrasonic orbital microforming. Based on the experimental results of the testing stand with oscillatory stress measurement as well as the surface analysis by metallography, SEM, and micro-hardness tests, it can be concluded that:

- The OMF process causes the initial deformation of the edge material of the rotating workpiece. In further phases of the process, the deformation expands on both sides towards the center of the object.
- In the UOM process under investigation, only the upper and lower part of the workpiece is deformed, occupying approximately 70% of the volume. The middle part of the sample remains undeformed.
- During the UOM process, the material flows in layers of different flow velocity, which is most probably caused by the generation of heat at the interface of the workpiece-tool due to intense plastic deformation and friction resulting from the sliding edge of the rotating sample on the tool surface during the initial centrifugation phase.
- It was observed that the deformation takes place in phases between which breaks occur, during which the deformation does not take place.
- The progressive deformation can be modified by the internal shape of the micro-die in which the UOM process takes place.
- The internal shape of the die in which the UOM process takes place affects the final shape of the microproduct.
- The shape of the microproduct obtained during UOM process does not accurately reflect the shape of the matrix in which this process takes place.

UOM is a process in which dynamic phenomena related to the inertia of a rotating billet are used. In addition, ultrasonic vibrations that affect the contact phenomena and structural changes are involved. There is a high speed of deformation as well as rolling and sliding friction, which raise the local and global temperature during the process. The consequence is a whole range of thermally activated structural changes. The process is very complicated by the mutual interaction of many physical phenomena. The current state of knowledge on this subject should be considered as the initial.

Activities in many directions, such as: a description of traffic dynamics, numerical modeling, and modifications of the parameters as well as vibrating frequency and amplitude of the ultrasonic system as billet shape, material type, and structure are planned. The "statistical approach" to the research, and thus the development of results for more experiments is planned in the second phase of analysis. As the first, research is planned to lead to a better understanding of the occurring phenomena.

Funding: This research received no external funding.

Conflicts of Interest: The authors declare no conflict of interest.

References

1. Ma, J.; Gong, F.; Yang, Z.; Zeng, W. Microdeep drawing of C1100 microsquare cups using microforming technology. *Int. J. Adv. Manuf. Tech.* **2016**, *82*, 1363–1369. [CrossRef]
2. Olejnik, L.; Presz, W.; Rosochowski, A. Backward extrusion using micro-blanked aluminium sheet. *Int. J. Mater. Form.* **2009**, *2*, 617–620. [CrossRef]
3. Geiger, M.; Kleiner, M.; Eckstein, R.; Tiesler, N.; Engel, U. Microforming. *CIRP Ann.* **2001**, *50*, 445–462. [CrossRef]
4. Kocańda, A.; Presz, W.; Adamczyk, G.; Czyżewski, P.; Mazurek, W. Contact pressure distribution in upsetting of compound metals. *J. Mater. Process. Tech.* **1996**, *60*, 343–348. [CrossRef]
5. Presz, W. Contact phenomena in microblanking. *Int. J. Mater. Form.* **2008**, *1*, 471–474. [CrossRef]
6. Sałaciński, T.; Winiarski, M.; Chmielewski, T.; Świercz, R. Surface finishing using ceramic fiber brush tools. In Proceedings of the 26th Anniversary International Conference on Metallurgy and Materials (METAL), Brno, Czech Republic, 24–26 May 2017; pp. 1222–1226.
7. Tiesler, N. Microforming—Size effects in friction and their influence on extrusion process. *Wire* **2002**, *52*, 34–38.
8. Engel, U. Tribology in microforming. *Wear* **2006**, *260*, 265–273. [CrossRef]
9. Cheng, C.; Wan, M.; Meng, B. Size effect on the forming limit of sheet metal in micro-scaled plastic deformation considering free surface roughening. *Proc. Eng.* **2017**, *207*, 1010–1015. [CrossRef]
10. Wang, C.; Guo, B.; Shan, D.; Zhang, M.; Bai, X. Tribological behaviors in microforming considering microscopically trapped lubricant at contact interface. *Int. J. Adv. Manuf. Tech.* **2014**, *71*, 2083–2090. [CrossRef]
11. Presz, W. The method of determining the tendency to galling in vibration assisted microforming. In Proceedings of the 27th International Conference on Metallurgy and Materials, Brno, Czech Republic, 23–25 May 2018; pp. 450–456.
12. Muster, A.; Presz, W. Influence of initial surface roughness on galling behaviour of steel-steel couple. *Scand. J. Metall.* **1999**, *28*, 5–8.
13. Skowrońska, B.; Szulc, J.; Chmielewski, T.; Sałaciński, T.; Świercz, R. Properties and microstructure of hybride PLASMA+MAG welded joints of thermomechanically treated S700MC. In Proceedings of the 27th International Conference on Metallurgy and Materials, Brno, Czech Republic, 23–25 May 2018; pp. 849–854.
14. Ghassemali, E.; Tan, M.J.; Wah, C.B.; Jarfors, A.E.; Lim, S.C.V. Grain size and workpiece dimension effects on material flow in an open-die micro-forging/extrusion process. *Mater. Sci. Eng.* **2013**, *582*, 379–388. [CrossRef]
15. Presz, W.; Cacko, R. Analysis of the influence of of a rivet yield stress distribution on the micro-SPR joint—Initial approach. *Arch. Civ. Mech. Eng.* **2010**, *10*, 69–75. [CrossRef]
16. Pintu, K.; Sudhansu, S.P. A review on properties and microstructure of micro-extruded product using SPD and as-cast material. *Sadhana* **2018**, *13*, 77.
17. Ran, J.Q.; Fu, M.W.; Chan, W.L. The influence of size effect on the ductile fracture in micro-scaled plastic deformation. *Int. J. Plast.* **2013**, *41*, 65–81. [CrossRef]
18. Wang, J.L.; Fu, M.W.; Shi, S.Q. Influences of size effect and stress condition on ductile fracture behavior in micro-scaled plastic deformation. *Mater. Des.* **2017**, *131*, 69–80. [CrossRef]
19. Presz, W.; Andersen, B.; Wanheim, T. Piezoelectric driven micro-press for microforming. *J. Achiev. Mater. Manuf. Eng.* **2006**, *18*, 411–414.
20. Cannella, E.; Nielsen, E.; Stolfi, A. Designing a tool system for lowering friction during the ejection of in-die sintered micro gears. *Micromachines* **2017**, *8*, 214. [CrossRef]

21. Paldan, N.A.; Arentoft, M.; Eriksen, R.S.; Mangeot, C. Piezo driven prestressing of die-system for microforming of metal components. *Int. J. Mater Form. Suppl.* **2008**, *1*, 467–470. [CrossRef]

22. Presz, W. Scale effect in design of the pre-stressed micro-dies for microforming. *Comput. Methods Mater. Sci.* **2016**, *16*, 196–203.

23. Presz, W.; Rosochowski, M. Application of semi-physical modeling of interface surface roughness in design of pre-stressed microforming dies. Int. Conf. on the Technology of Plasticity, Cambridge, U.K. *Procedia Eng.* **2017**, *207*, 1004–1009. [CrossRef]

24. Presz, W.; Cacko, R. Application of complex micro-die for extrusion of micro-rivets for micro-joining. In Proceedings of the 26th International Conference on Metallurgy and Materials (METAL), Brno, Czech Republic, 24–26 May 2017; pp. 514–520.

25. Presz, W.; Cacko, R. Bimetallic micro-punches for micro-blanking processes. *Archiv. Metall. Mater.* **2018**, *1*, 29–34.

26. Piwnik, J.; Mogielnicki, K.; Gabrylewski, M.; Baranowski, P. The experimental tool for micro-extrusion of metals. *Archiv. Foundry Eng.* **2011**, *2*, 195–198.

27. Ghassemali, E.; Jarfors, A.E.W.; Tan, M.J.; Lim, S.C.V. On the microstructure of micro-pins manufactured by a novel progressive microforming process. *Int. J. Mater. Form.* **2013**, *6*, 65–74. [CrossRef]

28. Stellin, T.; van Tijum, R.; Engel, U. Modelling and experimental study of a microforging process from metal strip for the reduction of defects in mass production. *Prod. Eng. Re. Dev.* **2016**, *10*, 103–112. [CrossRef]

29. Gong, F.; Guo, B. Size effects on mechanical properties of copper thin sheet in uniaxial tensile tests. *Medziagotyra* **2014**, *20*, 509–512. [CrossRef]

30. Presz, W. The method of micro-upsetting in uneven temperature Distribution. In Proceedings of the 27th International Conference on Metallurgy and Materials, Brno, Czech Republic, 23–25 May 2018. s2.0-85056217326.

31. Liu, Y.; Wang, C.; Han, H.; Shan, D.; Guo, B. Investigation on effect of ultrasonic vibration on micro-blanking process of copper foil. *Int. J. Adv. Manuf. Tech.* **2017**, *93*, 2243–2249. [CrossRef]

32. Huang, G.; Gary, J.-C. Laser-induced high-strain-rate superplastic 3-D microforming of metallic thin films. *J. Microelectromech. Syst.* **2010**, *19*, 273–281.

33. Djavanroodi, F.; Ahmadian, H.; Koohkan, K.; Naseri, R. Ultrasonic assisted-ECAP. *Ultrasonics* **2013**, *53*, 1089–1096. [CrossRef]

34. Christina, B.; Gracious, N. Influence of ultrasonic vibration on micro-extrusion. *Ultrasonics* **2011**, *51*, 606–616.

35. Hung, J.-C.; Hung, C. The influence of ultrasonic-vibration on hot upsetting. *Ultrasonics* **2005**, *43*, 692–698. [CrossRef] [PubMed]

36. Presz, W.; Kulik, T. Ultrasonic vibrations as an impulse for glass transition in microforming of bulk metallic glass. *Archiv. Civ. Mech. Eng.* **2018**, *19*, 100–113. [CrossRef]

37. Abdul, A.S.; Lucas, M. The effect of ultrasonic excitation in metal forming tests. *Appl. Mech. Mater.* **2010**, *24*, 311–316. [CrossRef]

38. Presz, W.; Cacko, R. Ultrasonic assisted microforming. In Proceedings of the 26th International Conference on Metallurgy and Materials (METAL), Brno, Czech Republic, 24–26 May 2017; pp. 521–526.

39. Shimizu, T.; Kakegawa, T.; Yang, M. Micro-texturing of DLC thin film coatings and its tribological performance under dry sliding friction for microforming operation. *Procedia Eng.* **2014**, *81*, 1884–1889. [CrossRef]

40. Daud, Y.; Lucas, M.; Huang, Z. Superimposed ultrasonic oscillations in compression tests of aluminium. *Ultrasonics* **2006**, *44*, 511–515. [CrossRef]

41. Zhou, H.; Cui, H.; Qin, Q.H.; Wang, H.; Shen, Y. A comparative study of mechanical and microstructural characteristics of aluminium and titanium undergoing ultrasonic assisted compression testing. *Mater. Sci. Eng.* **2017**, *682*, 376–388. [CrossRef]

42. Hu, J.; Shimizu, T.; Yang, M. Investigation on ultrasonic volume effects: Stress superposition, acoustic softening and dynamic impact. *Ultrason. Sonochem.* **2018**, *48*, 240–248. [CrossRef]

43. Hung, C.; Tsai, Y.-C. Investigation of the effects of ultrasonic vibration-assistedmicro-upsetting on brass. *J. Mater. Sci. Eng.* **2013**, *580*, 125–132. [CrossRef]

44. Raja, R.; Lakshminarasimhan, S.; Murugesan, P. Investigation of barreling radius and top surface area for cold upsetting of aluminum specimens. *Int. J. Mod. Eng. Res.* **2013**, *3*, 3852–3862.

45. Zhou, H.; Cui, H.; Qin, Q.H. Influence of ultrasonic vibration on the plasticity of metals during compression process. *J. Mater. Process. Tech.* **2018**, *251*, 146–159. [CrossRef]

46. Yao, Z.; Kim, G.Y.; Faidley, L.; Zou, Q.; Mei, D.; Chen, Z. Effects of superimposed high-frequency vibration on deformation of aluminium in micro/meso-scale upsetting. *J. Mater. Process. Tech.* **2012**, *212*, 640–646. [CrossRef]
47. Liu, Y.; Suslov, S.; Han, Q.; Hua, L.; Xu, C. Comparison between ultrasonic vibration-assisted upsetting and conven-tional upsetting. *Metall. Mater. Trans.* **2013**, *44*, 3232–3244. [CrossRef]
48. Presz, W. Dynamic effect in ultrasonic assisted micro-upsetting. In Proceedings of the 21st International ESAFORM Conference on Material Forming, Palermo, Italy, 23–25 April 2018; p. 100012.
49. Hu, J.; Shimizu, T.; Yoshino, T.; Shiratori, T.; Yang, M. Ultrasonic dynamic impact effect on deformation of aluminum during micro-compression tests. *J. Mater. Process. Tech.* **2018**, *258*, 144–154. [CrossRef]
50. Presz, W. Ultrasonic orbital microforming—A new possibility in the forming of microparts. *Metals* **2018**, *8*, 889. [CrossRef]

Article

A New Compression Test for Determining Free Surface Roughness Evolution in Thin Sheet Metals

Tsuyoshi Furushima [1],*, Kohei Aoto [2] and Sergei Alexandrov [3]

[1]   Institute of Industrial Science, The University of Tokyo, Tokyo 153-8505, Japan
[2]   Department of Mechanical Engineering, Tokyo Metropolitan University, Tokyo 192-0397, Japan; kohei.aoto@gmail.com
[3]   Ishlinsky Institute for Problems in Mechanics, Russian Academy of Sciences, Moscow 119526, Russia; sergei_alexandrov@spartak.ru
*   Correspondence: tsuyoful@iis.u-tokyo.ac.jp; Tel.: +81-3-5452-6809

Received: 22 March 2019; Accepted: 16 April 2019; Published: 17 April 2019

Abstract: In sheet microforming processes, in-surface principal strain rates may be compressive such that the thickness of the sheet increases in the process of deformation. In general, the evolution of free surface roughness depends on the sense of the principal strain normal to the free surface. Therefore, in order to predict the evolution of free surface roughness in processes in which this normal principal strain is positive by means of empirical equations, it is necessary to carry out experiments in which the thickness of the sheet increases. Conventional experiments, such as the Marciniak test, do not provide such strain paths. In general, it is rather difficult to induce a sufficiently uniform state of strain in thin sheets of increasing thickness throughout the process of deformation because instability occurs at the very beginning of the process. The present paper proposes a compression test for thin sheets. Teflon sheets are placed between support jigs and the metallic sheet tested to prevent the occurrence of instability and significantly reduce the effect of the support jigs on the evolution of surface roughness. The test is used to determine the evolution of surface roughness in thin sheets made of C1220-O under three strain paths.

Keywords: free surface roughness evolution; compression; thin sheet metal; micro metal forming

---

1. Introduction

Micro sheet metal forming for manufacturing thin structural parts has been applied in various fields of medical, information/communication and electrical devices. Especially, the further miniaturization of copper alloy parts widely used in electronic device parts such as micro connecter pins and terminals has been required with growing interest in the next-generation communication of 5G. Micro sheet metal forming is an attractive manufacturing method capable of mass production at high productivity. However, it is difficult to miniaturize the size of products using conventional sheet metal forming techniques, because various size effects [1,2] such as material properties, forming condition, heat transfer [3], friction [4], formability [5] and machine functions seriously influence the quality of products. As one of the size effects in micro sheet metal forming, the ratio of grain size [6] and surface roughness [7,8] to thickness becomes large relative to the scaling down of product size. The increasing ratio of surface roughness to thickness of material at the micro-scale causes decreasing ductility [7]. The effect of the number of grains through the thickness affects the formability of microforming [8]. In addition, it is well known that the surface roughness evolution on the free surface with the non-contact surface of the material occurs with increasing plastic deformation [9–11]. Thus, surface roughness evolution in micro sheet metal forming strongly effects not only surface appearance, but also tribological behavior [12] and formability [13].

In micro sheet metal forming processes, in-surface principal strain rates may be compressive such that the thickness of the sheet increases in the process of deformation. This mode of deformation occurs, for example, in flanges and side walls of some stamping processes [14,15]. Using conventional empirical equations for free surface roughness evolution, for example [16–18], it is in general possible to predict the evolution of free surface roughness in such processes without having experimental results in the domain of compressive in-surface principal strain rates, since these empirical equations do not account for the effect of the strain path on free surface roughness evolution. In particular, experimental data found by means of tests in which the thickness of the sheet decreases can be used to identify the empirical free surface roughness evolution equation. However, the accuracy of such predictions is questionable. In particular, in most cases the sense of the principal strain normal to the free surface, ε_n, has a significant effect on the evolution of free surface roughness. It has been shown in [19,20] using a great number of independent experimental data. In this paper, a qualitative dependence of the roughness parameters, Ra and Rz, on ε_n and the equivalent strain, ε_{eq}, has been proposed. This dependence is confirmed by numerical results reported in [21]. In this work, a model of crystallographic plasticity has been adopted to predict the evolution of free surface roughness. Several tests are available to identify the empirical equation for free surface roughness evolution proposed in [19,20] in the range $\varepsilon_n < 0$ (for example, uniaxial tension, equibiaxial tension, Marciniak test [22] and others). However, to the best of the authors' knowledge, no test has been proposed to determine the evolution of free surface roughness in the range of $\varepsilon_n > 0$. The present paper concerns itself with such a test. The test is used to find the evolution of free surface roughness in thin sheets made of C1220-O for three values of the ratio $\varepsilon_n/\varepsilon_{eq}$.

2. Compression Test

A number of compression tests for thin sheets have been proposed in the literature [23,24]. A distinguished feature of the test described in this section is that it is designed to prevent instability and significantly reduce the effect of support jigs on the evolution of surface roughness. A schematic diagram of the experimental setup is shown in Figure 1. A metallic sheet to be tested and two teflon sheets are placed between two support jigs to prevent the occurrence of instability. The teflon sheets are used to significantly reduce the effect of the support jigs on the evolution of surface roughness in the sample. The use of soft material for this purpose has been justified in [25]. It has been shown in this work that the evolution of roughness at the contact surface between the material tested and the rubber is practically the same as the evolution of free surface roughness at large plastic strains. Furthermore, it has been shown in [26] that a hydrostatic pressure of 405 MPa does not affect the evolution of surface roughness. Thus, it is believed that the teflon sheets and the contact pressure between these sheets and the material tested do not affect the evolution of surface roughness. Pressure (*p*) is applied to the support jigs to prevent wrinkling at early stages of the process. The metallic and teflon sheets are compressed by two punches. The speed of each punch is *V*. Grid lines should be applied on the surfaces of the metallic sheet to measure in-surface strains. Then, the through thickness strain can be found from the equation of incompressibility. The test may be interrupted several times to determine the dependence of roughness parameters on the equivalent strain and through thickness strain.

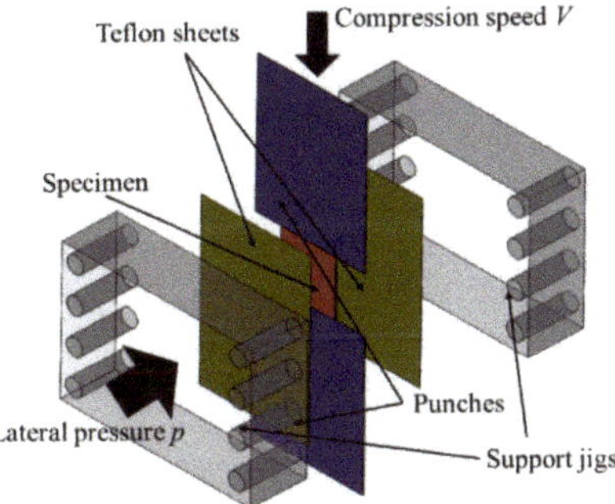

Figure 1. Schematic diagram of the experimental setup of compression test for thin sheet metals.

3. Evolution of Surface Roughness in C1220-O Sheets in the Range $\varepsilon_n > 0$

Samples for testing were cut from commercial phosphorous-deoxidized copper C1220-O (99.90 wt% or more of copper and 0.015–0.040 wt% Phosphorus) sheets of thickness $t = 0.5$ mm. For microstructure observation, the samples were mounted with resin and polished. After polishing, the surface treatment by ion milling SVM-721 (SANYU ELECTRON Co., Tokyo, Japan) was conducted. The microstructure was observed using a HITACHI S-3700N Scanning Electron Microscope (SEM) (HITACHI Ltd., Tokyo, Japan) with a Nordlys NL 04-2201-03 Electron Back-Scatter Diffraction (EBSD) system (Oxford Instruments plc., Abingdon, UK) The average grain size $\alpha_g = 40$ µm was found by Hyde's method (Figure 2). In order to create different strain paths, three types of rectangular samples were used (type 1: 40 mm long × 15 mm wide; type 2: 30 mm long × 15 mm wide; type 3: 25 mm long × 15 mm wide). Three nominally identical samples of each type were cut using a milling machine. Samples of types 1 and 3 were compressed along the long side of the rectangular and samples of type 2 along the short side (Figure 3). The compression direction and the rolling direction are matched. The teflon sheets (Nitto Denko Corporation, NITOFLON No. 900UL) of thickness 0.05 mm, width 40 mm and length 50 mm were used in all tests. Fasteners were used to join two support jigs together (Figure 4). The torque applied to turn the fasteners was measured by using a torque wrench and this torque controlled the lateral pressure p (Figure 1). For each type of sample, the magnitude of the torque was chosen by trial-and-error to prevent wrinkling (1.5 Nm for type 1, 1.5 Nm for type 2, 4 Nm for type 3). The test jig of width 60 mm, height 50 mm and thickness 25 mm was made of carbon steel S45C. The test jig fitted in the test pieces is shown in Figure 5. The compression test was conducted using a SHIMAZU AUTOGRAPH AG-50kN universal testing machine with maximum load of 50 kN (SHIMAZU Co., Kyoto, Japan). The test pieces were each tested to $\varepsilon_1 = -0.24$ under compression speed $V = 1$ mm/min (Figure 1). In order to measure in-surface principal strains, ε_1 and ε_2, grid lines were applied by a scriber on the free surface on each sample. To calculate ε_1 and ε_2, the dimensions of the initial 5 mm × 5 mm grids after deformation were measured using a KEYENCE VHX900 digital microscope (KEYENCE Co., Osaka, Japan). Figure 6 shows the typical shape of grids before and after deformation using a sample of type 1. The through thickness strain, $\varepsilon_3 \equiv \varepsilon_n$, was calculated by means of the equation of incompressibility The surface roughness in the area surrounded by grid lines was measured using KEYENCE VK-100 confocal laser microscope (KEYENCE Co., Osaka, Japan). The arithmetic average roughness Ra along the compression direction was evaluated. Assuming that strain paths are proportional, the von Mises equivalent strain was found as:

$$\varepsilon_{eq} = \sqrt{\frac{2}{3}\left(\varepsilon_1{}^2 + \varepsilon_2{}^2 + \varepsilon_n{}^2\right)}. \tag{1}$$

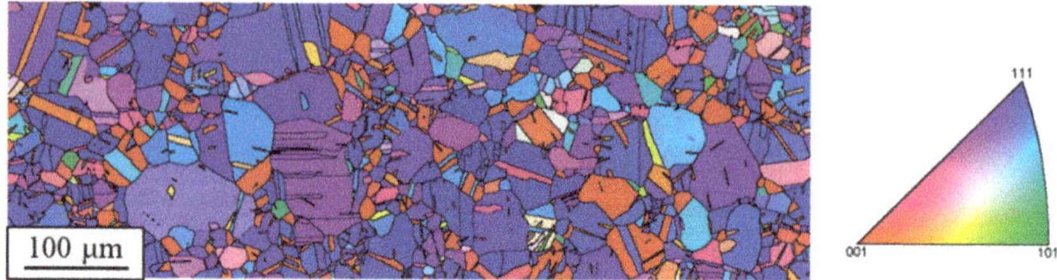

Figure 2. Microstructure of pure copper C1220-O.

Figure 3. Specimen size used in compression experiments.

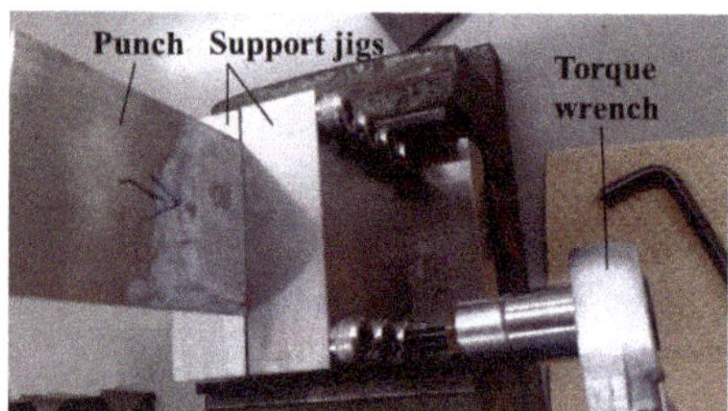

Figure 4. Torque controlled the lateral pressure using turning fasteners.

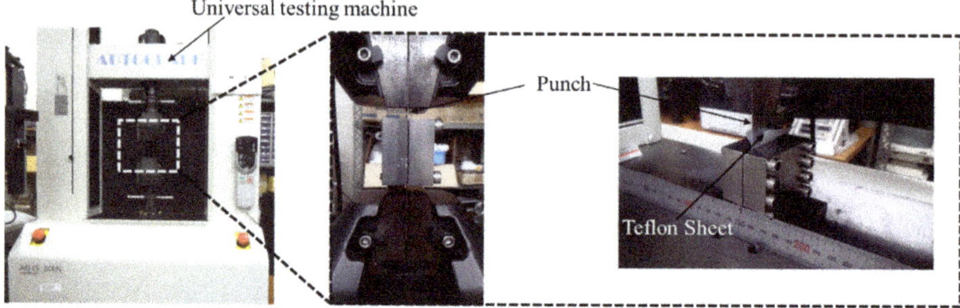

Figure 5. Photo of compression test set up for thin sheet metal.

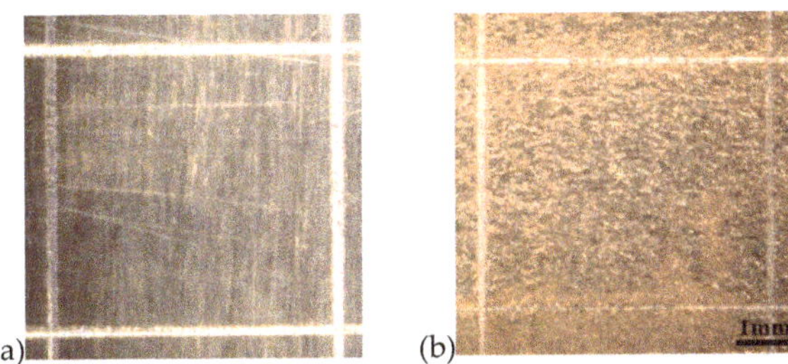

Figure 6. Change in square grid (**a**) before and (**b**) after deformation for sample of type 1.

The shape of samples of type 1 at several values of the equivalent strain is shown in Figure 7. It is seen from this figure that deformation is practically uniform up to $\varepsilon_{eq} = 0.199$. The shape of samples of type 2 at several values of the equivalent strain is shown in Figure 8. It is seen from this figure that deformation is practically uniform up to $\varepsilon_{eq} = 0.146$. The shape of samples of type 3 at several values of the equivalent strain is shown in Figure 9. It is seen from this figure that deformation is practically uniform up to $\varepsilon_{eq} = 0.382$. The experimental dependence of ε_2 on ε_1 for all three types of samples is depicted in Figure 10. These experimental data were fitted to polynomials of the first degree. As a result,

$$\varepsilon_2 = \beta_i \varepsilon_1, \tag{2}$$

where $1 \leq i \leq 3$ and

$$\beta_1 = -0.49, \ \beta_2 = -0.34, \ \beta_3 = -0.7. \tag{3}$$

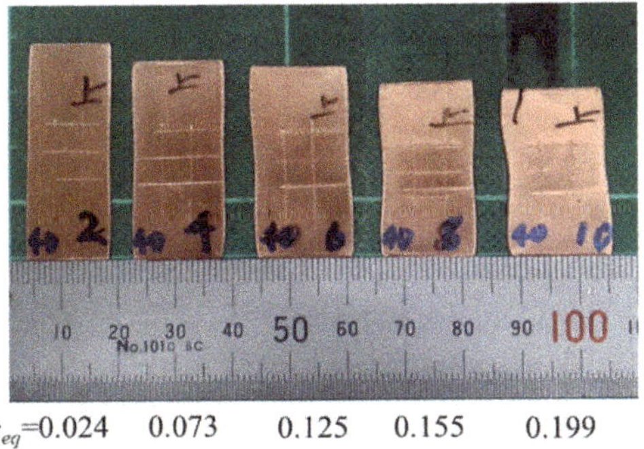

Figure 7. Shape of samples of type 1 at several values of the equivalent strain.

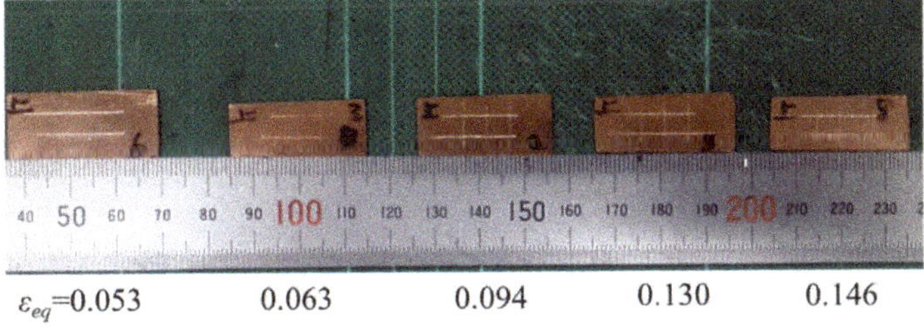

Figure 8. Shape of samples of type 2 at several values of the equivalent strain.

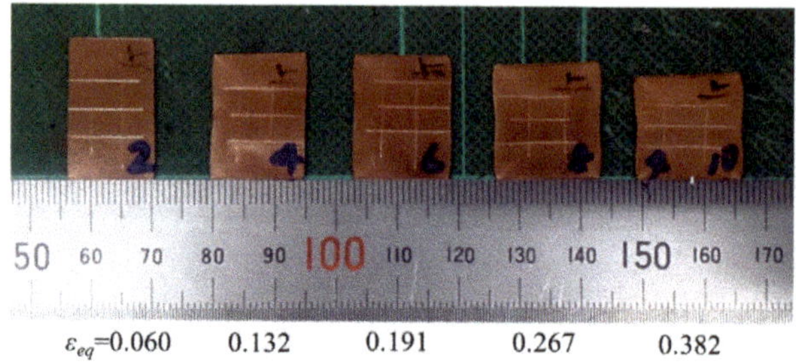

Figure 9. Shape of samples of type 3 at several values of the equivalent strain.

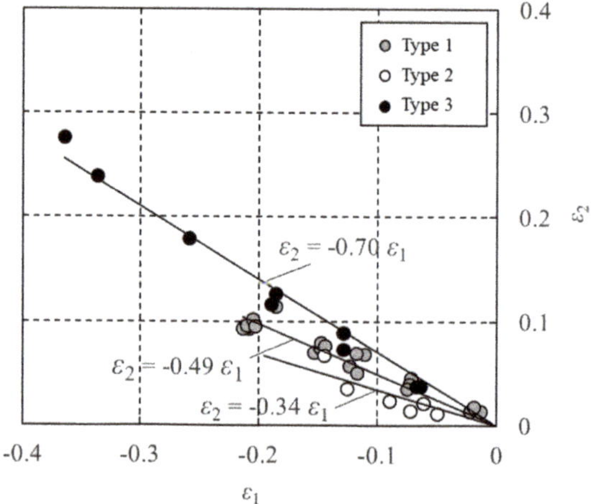

Figure 10. Experimental dependence of of ε_2 on ε_1 for all three types of samples.

Here β_1 corresponds to samples of type 1, β_2 to samples of type 2, and β_3 to samples of type 3. Equation (2) is illustrated in Figure 10. It is seen from this figure that the assumption that the strain paths are proportional is satisfied with a high degree of accuracy. Using Equation (2) and the equation of incompressibility, $\varepsilon_1 + \varepsilon_2 + \varepsilon_3 = 0$, it is possible to transform Equation (1) to:

$$\frac{\varepsilon_n}{\varepsilon_{eq}} = \frac{\sqrt{3}(1+\beta_i)}{2\sqrt{1+\beta_i+\beta_i^2}}. \tag{4}$$

4. Results

One of the main outputs of the test proposed is that the empirical equation for free surface roughness evolution proposed in [4] can be identified. In particular, this equation reads:

$$\frac{R - R_0}{\alpha_g} = \frac{\Delta R}{\alpha_g} = \varepsilon_{eq}\Omega\left(\frac{\varepsilon_n}{\varepsilon_{eq}}\right). \tag{5}$$

where R is the surface roughness parameter (Ra or Rz), R_0 is its initial magnitude, and $\Omega\left(\varepsilon_n/\varepsilon_{eq}\right)$ is a function of its argument. This function should be determined from experiment. Figure 11 shows the three-dimensional behavior of free surface roughness evolution with increasing ε_{eq} for the sample of type 2. It is seen that the surface roughens with ε_{eq}. Thus, the effectivity of the proposed compression method for thin sheets to evaluate free surface roughness evolution is verified. The variation of $\Delta R \equiv Ra - R_0$ with ε_{eq} found from the experiment is depicted in Figure 12. It is seen from Equation (4) that the ratio $\varepsilon_n / \varepsilon_{eq}$ is constant along each strain path. Therefore, Equation (5) predicts that ΔR is a linear function of ε_{eq} for each strain path. Figure 12 confirms this prediction. The experimental data shown in Figure 12 were fitted to polynomials of the first degree to give:

$$\frac{\Delta R}{\alpha_g} = \begin{cases} 0.141 \times \varepsilon_{eq} \text{ for samples of type 1} \\ 0.122 \times \varepsilon_{eq} \text{ for samples of type 2} \\ 0.092 \times \varepsilon_{eq} \text{ for samples of type 3} \end{cases} \qquad (6)$$

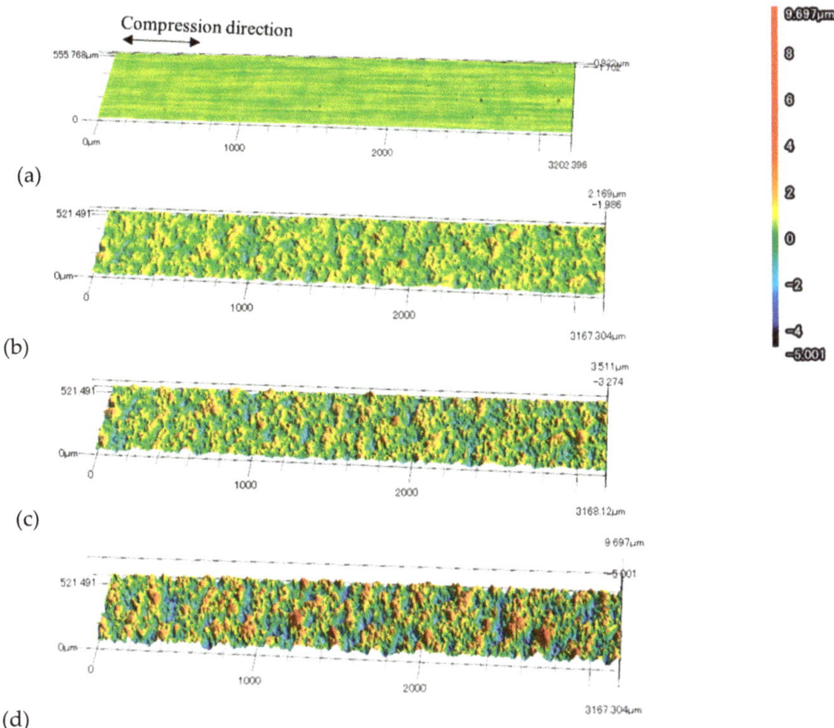

Figure 11. Free surface roughness evolution with increasing ε_{eq} for the sample of type 2: (**a**) $\varepsilon_{eq} = 0$, (**b**) $\varepsilon_{eq} = 0.053$, (**c**) $\varepsilon_{eq} = 0.063$, (**d**) $\varepsilon_{eq} = 0.094$.

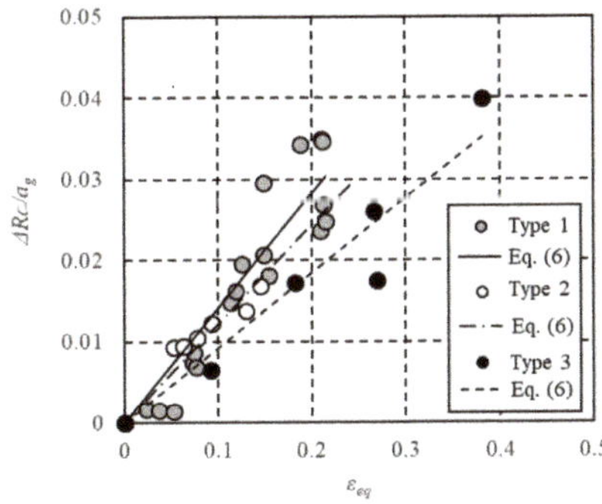

Figure 12. Comparison of surface roughness evolution between predicted equation and experiments.

Substituting Equation (3) into Equation (4) one can get $\varepsilon_n/\varepsilon_{eq} = 0.65$ for samples of type 1, $\varepsilon_n/\varepsilon_{eq} = 0.51$ for samples of type 2 and $\varepsilon_n/\varepsilon_{eq} = 0.29$ for samples of type 3. Then, it follows from Equations (5) and (6) that

$$\Omega(0.51) = 0.122, \quad \Omega(0.65) = 0.141, \quad \Omega(0.29) = 0.092 \tag{7}$$

The domain of the function $\Omega\left(\varepsilon_n/\varepsilon_{eq}\right)$ is $-1 \leq \varepsilon_n/\varepsilon_{eq} \leq 1$. The Marciniak test (see, for example, [19,20]) can be adopted to produce various strain paths in the range $-1 \leq \varepsilon_n/\varepsilon_{eq} \leq 0$. Then, the procedure for determining the function $\Omega\left(\varepsilon_n/\varepsilon_{eq}\right)$ at given values of $\varepsilon_n/\varepsilon_{eq}$ is the same as that used to arrive at Equation (7). In our previous study, the Marciniak test was carried out on C1220-O sheets for three strain paths [20]. As a result,

$$\Omega(-0.57) = 0.181, \quad \Omega(-0.86) = 0.207, \quad \Omega(-1) = 0.141. \tag{8}$$

The experimental data given in Equations (7) and (8) are illustrated in Figure 13. The approximating curve is determined by the equation:

$$\Omega = 0.0595\left(\frac{\varepsilon_n}{\varepsilon_{eq}}\right)^5 - 0.4586\left(\frac{\varepsilon_n}{\varepsilon_{eq}}\right)^4 - 0.1529\left(\frac{\varepsilon_n}{\varepsilon_{eq}}\right)^3 + 0.4549\left(\frac{\varepsilon_n}{\varepsilon_{eq}}\right)^2 + 0.0092\left(\frac{\varepsilon_n}{\varepsilon_{eq}}\right) + 0.0605 \tag{9}$$

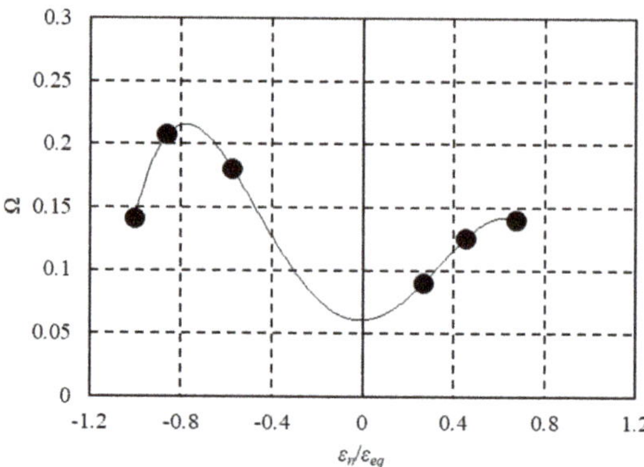

Figure 13. Discrete and continuous functions Ω for C1220P-O sheets.

Both the experimental data and approximating curve suggest that the function $\Omega\left(\varepsilon_n/\varepsilon_{eq}\right)$ has a local minimum in the vicinity of $\varepsilon_n/\varepsilon_{eq} = 0$. The same conclusion has been drawn in [4] based on analysis of many independent experimental results. Thus, the surface roughness evolution is affected by the strain path and minimum surface roughness evolution is obtained under the strain path in which the ratio of $\varepsilon_n/\varepsilon_{eq}$ is close to zero. Therefore, a crucial issue for determining an accurate approximation of the function $\Omega\left(\varepsilon_n/\varepsilon_{eq}\right)$. over its entire domain is to propose and carry out a test in which the ratio $\varepsilon_n/\varepsilon_{eq}$ is close to zero.

5. Conclusions

A new experiment has been proposed to reveal the effect of the strain path on the evolution of surface roughness. The experiment consists of a compression test under lateral pressure. The latter is used to verify that the effect of the lateral pressure on the evolution of surface roughness is small. Therefore, the evolution of surface roughness in the compression test can be regarded as free surface roughness evolution. The compression test can be adopted for identifying the empirical equation for free surface roughness evolution proposed in [19,20] in the range $\varepsilon_n/\varepsilon_{eq} > 0$. Both compression and tensile tests have been carried out on C1220-O sheets. As a result, the value of the function $\Omega\left(\varepsilon_n/\varepsilon_{eq}\right)$ involved in the empirical equation has been found at three points (see Equation (7)). In addition, the Marciniak test has been conducted to determine the value of the function $\Omega\left(\varepsilon_n/\varepsilon_{eq}\right)$ at three points in the range $\varepsilon_n/\varepsilon_{eq} < 0$ (see Equation (8)). It has been emphasized that an accurate approximation of the function $\Omega\left(\varepsilon_n/\varepsilon_{eq}\right)$ over its entire domain requires a test in which $\varepsilon_n/\varepsilon_{eq} \approx 0$. This will be the subject of a subsequent investigation. The conventional study on the evolution of surface roughness has been discussed mainly in the tension state ($\varepsilon_n/\varepsilon_{eq} < 0$). In this study, the proposed test can evaluate the evolution of free surface roughness under the compression strain path ($\varepsilon_n/\varepsilon_{eq} > 0$). The testing results will contribute to predict the evolution of surface roughness in flanges and side walls of some micro sheet metal forming processes with compressive modes such that the thickness of the sheet increases in the process of deformation. Therefore, it is concluded that the prediction range of the evolution of surface roughness can be expanded through a series of these studies.

Author Contributions: T.F. and S.A. made the experimental conceptual design and the theoretical framework, wrote the paper. K.A. carried out the experiments.

Funding: The main part of this work was done while S.A. was with Tokyo Metropolitan University as a visiting professor under the short-term program sponsored by Japan Society for the Promotion of Science (grant S16056). S.A. also acknowledges support from the Russian Ministry of Science and Higher Education (Project AAAA-A17-117021310373-3).

Conflicts of Interest: The authors declare no conflict of interest.

References

1. Geiger, M.; Kleiner, M.; Eckstein, R.; Tiesler, N.; Engel, U. Microforming. *CIRP Annals* **2001**, *50*, 445–462. [CrossRef]
2. Vollertsen, F.; Biermann, D.; Hansen, H.N.; Jawahir, I.S.; Kuzman, K. Size effects in manufacturing of metallic components. *CIRP Annals* **2009**, *58*, 566–587. [CrossRef]
3. Furushima, T.; Manabe, K. FE analysis of size effect on deformation and heat transfer behavior in microtube dieless drawing. *J. Mater. Process. Technol.* **2008**, *201*, 123–127. [CrossRef]
4. Engel, U. Tribology in microforming. *Wear* **2006**, *260*, 265–273. [CrossRef]
5. Liu, H.; Zhang, W.; Gau, J.-T.; Shen, Z.; Ma, Y.; Zhang, G.; Wang, X. Feature Size Effect on Formability of Multilayer Metal Composite Sheets under Microscale Laser Flexible Forming. *Metals* **2017**, *7*, 275. [CrossRef]
6. Kals, T.A.; Eckstein, R. Miniaturization in sheet metal working. *J. Mater. Process. Technol.* **2000**, *103*, 95–101. [CrossRef]
7. Fu, M.W.; Chan, W.L. Geometry and grain size effects on the fracture behavior of sheet metal in micro-scale plastic deformation. *Mater. Des.* **2011**, *32*, 4738–4746. [CrossRef]
8. Furushima, T.; Tsunezaki, H.; Manabe, K.I.; Alexsandrov, S. Ductile fracture and free surface roughening behaviors of pure copper foils for micro/meso-scale forming. *Int. J. Mach. Tools Manuf.* **2014**, *76*, 34–48. [CrossRef]
9. Osakada, K.; Oyane, M. On the Roughening of Free Surface in Deformatio Process. *Bull. JSME* **1971**, *14*, 171–177. [CrossRef]
10. Yamaguchi, K.; Mellor, P.B. Thickness and Grain Size Dependence of Limit Strains in Sheet Metal Stretching. *Int. J. Mech. Sci.* **1976**, *18*, 85–90. [CrossRef]
11. Furushima, T.; Masuda, T.; Manabe, K.; Alexandrov, S. Inhomogeneous 3D Finite Element Model for Prediction of Free Surface Roughening. *Adv. Mater. Res.* **2012**, *418–420*, 1040–1043. [CrossRef]
12. Shimizu, T.; Yang, M.; Manabe, K. Classification of mesoscopic tribological properties under dry sliding friction for microforming operation. *Wear* **2015**, *330–331*, 49–58. [CrossRef]
13. Luo, L.; Jiang, Z.; Wei, D.; Manabe, K.I.; Zhao, X.; Wu, D.; Furushima, T. Effects of surface roughness on micro deep drawing of circular cups with consideration of size effects. *Finite Elem. Anal. Des.* **2016**, *111*, 46–55. [CrossRef]
14. Hu, P.; Li, D.Y.; Li, Y.X. Analytical models of stretch and shrink flanging. *Int. J. Machine Tools and Manuf.* **2003**, *43*, 1367–1373. [CrossRef]

15. Shimizu, T.; Ogawa, M.; Yang, M.; Manabe, K. Plastic anisotropy of ultra-thin rolled phosphor bronze foils and its thickness strain evolution in micro-deep drawing. *Mater. Design* **2014**, *56*, 604–612. [CrossRef]

16. Klimczak, T.; Dautzenberg, J.H.; Kals, J.A.G. On the Roughening of a Free Surface During Sheet Metal Forming. *CIRP Annals* **1988**, *37*, 267–270. [CrossRef]

17. Parmar, A.; Mellor, P.B.; Chakrabarty, J. A New Model for the Prediction of Instability and Limit Strains in Thin Sheet Metal. *Int. J. Mech. Sci.* **1977**, *19*, 389–398. [CrossRef]

18. Jain, M.; Lloyd, D.J.; Macewen, S.R. Hardening Laws, Surface Roughness and Biaxial Tensile Limit Strains of Sheet Aluminium Alloys. *Int. J. Mech. Sci.* **1996**, *38*, 219–232. [CrossRef]

19. Alexandrov, S.; Manabe, K.; Furushima, T. A New Empirical Relation for Free Surface Roughening. *Trans. ASME J. Manufact. Sci. Eng.* **2011**, *133*, 014503. [CrossRef]

20. Furushima, T.; Sato, H.; Manabe, K.; Alexandrov, S. Identification of an Empirical Equation for Predicting Free Surface Roughness Evolution in Thin Sheets of Aluminum Alloy and Pure Copper. *J. Manuf. Sci. Eng. Trans. ASME* **2018**, *140*, 034501. [CrossRef]

21. Becker, R. Effects of Strain Localization on Surface Roughening During Sheet Forming. *Acta Mater.* **1998**, *46*, 1385–1401. [CrossRef]

22. Marciniak, Z.; Kuczynski, K. Limit strains in the processes of stretch-forming sheet metal. *Int. J. Mech. Sci.* **1967**, *9*, 609–612. [CrossRef]

23. Bae, G.H.; Huh, H. Tension/compression test of auto-body steel sheets with the variation of the pre-strain and the strain rate. *WIT Trans. Eng. Sci.* **2011**, *72*, 213–225.

24. Piao, K.; Lee, J.K.; Kim, J.H.; Kim, H.Y.; Chung, K.; Barlat, F.; Wagoner, R.H. A sheet tension/compression test for elevated temperature. *Int. J. Plast.* **2012**, *38*, 27–46. [CrossRef]

25. Fukuda, M.; Yamaguchi, K.; Takakura, N.; Sakano, Y. Roughening Phenomenon on Free Surface of Products in Sheet Metal Forming. *J. Japan Soc. Technol. Plast.* **1974**, *15*, 994–1002.

26. Azushima, A.; Miyagawa, M. Effect of Working Factors and Metallurgical Factors on Roughening Phenomenon on Free Surface of Sheets—Investigation on Free Surface of Sheet Deformed I. *J. Japan Soc. Technol. Plast.* **1986**, *27*, 1261–1267.

Article

Development of a Novel Resistance Heating System for Microforming Using Surface-Modified Dies and Evaluation of Its Heating Property

Ming Yang * and Tetsuhide Shimizu

Tokyo Metropolitan University, Tokyo 191-0065, Japan; simizu-tetsuhide@tmu.ac.jp
* Correspondence: yang@tmu.ac.jp; Tel.: +81-42-585-8440

Received: 20 March 2019; Accepted: 11 April 2019; Published: 15 April 2019

Abstract: For this study, a novel resistance heating system for microforming was developed using surfaces of forming dies as heating resources. The electrical resistance of the die surfaces was designed and the hard-coating material AlCrSiN was selected to coat the die surfaces for heating. To clarify the effects of the thickness and modified surfaces on heating efficiency, the temperature and stress reduction were evaluated in a micro-compression test using dies coated with 0.5 and 1 μm AlCrSiN films. Furthermore, the formability was also demonstrated using 1 μm thick AlCrSiN-coated tools in a microforging test. By applying surface-modified dies to the forming processes, we found that not only was the heating efficiency improved, but also the dependence of heating on the product's shape and the material's electrical properties was reduced.

Keywords: resistance heating system; surface modification; microforming

---

1. Introduction

In the fields of medical, precision, and electronic information equipment, the demand for fine metal components is increasing, and microforming processing technology has attracted much attention. However, there are several issues due to the scaling-down of material, such as reduction in ductility and uneven deformation of materials; therefore, the development of materials and processes suitable for microforming is required [1–3]. Heat-assisted microforming is a promising approach since the effects of heating, such as reduction in flow stress and uneven deformation, are expected to contribute to improved formability in microforming processes [4,5]. Several localized heating systems including laser heating, ultrasonic heating, and electrical heating have been investigated. It has been established that heat assistance could reduce the flow stress of a material. In laser heating, efficient heating can be realized by using high-power or pulsed lasers, which have been applied in flexible sheet bending [6], dieless micro-tube forming [7], and stamping processes with transparent dies [8]. However, the use of transparent dies could be a limitation in industrial applications. In electrical heating, the heating system can be embedded in dies for efficient heating [9]. Moreover, the reduction in flow stress in the deep drawing of a metal foil material, the reduction in the springback in foil bending [10–12], and further surface smoothing [9] in microforging have been confirmed. However, some issues in the electrical heating system should be overcome as shown below. Since the current flow in a workpiece shows a dependence on the shape of the products, it is necessary to optimize the electrode allocation for various product shapes to prevent a non-uniform temperature distribution due to the current density gradient in the workpiece [13]. Since the heating depends on the electrical resistance of the workpiece, a high electric energy is required for materials with lower electrical resistivity. Furthermore, since the resistance of the workpiece varies owing to deformation, the temperature will also change accordingly with the progress of the deformation during a process.

In this study, we aimed to develop a new electric heating system that does not depend on the material properties and shape to overcome the issues encountered in previous heating systems. We designed and fabricated this new electrical heating system for generating heat on the die surface by forming a modified layer with high electrical resistivity on the surface of the dies. The heating system was subjected to a compression test to evaluate the effects of materials with various electrical resistivities. The heating system was also applied to microforging to evaluate its effects on formability and accuracy in microforging.

2. Design and Surface Treatment of Electrode Dies

Figure 1 shows a schematic diagram of the proposed method in this study. Surface treatment was applied to modify the die surfaces to design their electrical resistance as a heat source for heating. It is a novel idea that the surfaces of dies can function as a heat source by modifying not only their mechanical properties, namely, heat resistance, and seizure resistance, but also their electrical properties. As a result, a workpiece can be heated through thermal conduction from itself of a die in contact with the workpiece. Finite element (FE) analysis was carried out to design the electrical resistivity of the die surface for an electric heating system. Then, the surface coating material was selected based on simulation results and deposited on the die surface by a commercial physical vapor deposition (PVD) process. Finally, the heating characteristics were evaluated by microforming tests using the coated dies.

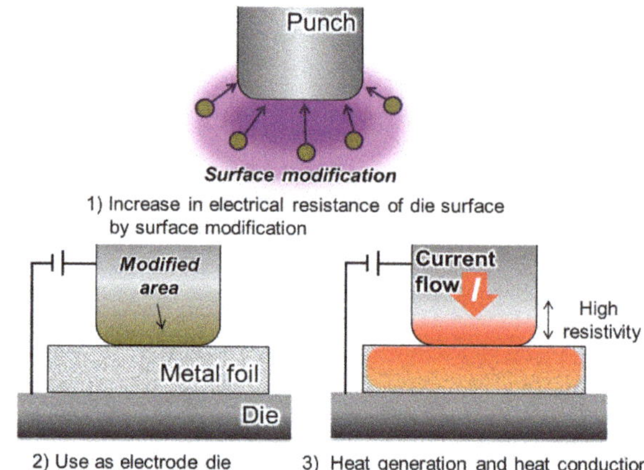

Figure 1. Concept of heating by surface-modified dies.

2.1. Simulation for Design of Die Surface

The effects of the thickness and electrical resistivity of the surface layer of dies on the heating properties were evaluated by FE analysis in order to determinate the conditions of the surface layer required for heating. A coupled thermal-electric analysis was performed using the commercial code ABAQUS 6.17 (Dassault Systemes, Vélizy-Villacoublay, France, 2017) and the conditions of FE analysis are shown in Table 1. Figure 2 shows the boundary conditions for the analysis. The materials used for surface coating are shown in Table 2. A step input of a direct current of 40 A was applied from the top of the punch through the lower die which had a potential of 0 V. The initial and ambient temperatures of the surfaces of the punch, workpiece, and die were all 25 °C.

Figure 3 shows the FE analysis results of the temperature distribution 60 s after heating. It was confirmed that local heating is achieved by thermal conduction from the punch surface to the workpiece. Electric heating was found to be much less dependent on the resistance property of the workpiece in the case of surface modification of the die by coating a material with a much higher electrical resistivity on the die surface. A higher temperature was achieved by coating a material with a greater thickness and higher electrical resistivity at the same input energy. The maximum temperature in the workpiece was 780 °C in the case of the 100 μm thick $CrN_{0.6}$ layer. Based on the FE analysis results, the relationship between the achievable temperature and the resistivity of the coated material was estimated. Figure 4 shows the prediction results. The prediction curve fitted the simulation results well, indicating that an electrical resistivity of 2.8×10^6 μΩcm is necessary for the coating film with a thickness of 50 μm and that of 6.7×10^7 μΩcm is necessary for thickness of 5 μm in the case of a target temperature of 800 °C. In our previous study, a high-density plasma nitriding was performed to modify a surface of tool steel in order to increase the electrical resistivity of the die surface [14]. The surface layer with a thickness of 50 μm was modified, but still not enough to achieve the desired temperature. It is necessary to choose materials with higher resistivities for the surface modification.

Table 1. Conditions in the FE analysis.

Punch			SUS420J2
Die			SKD-11
Workpiece	Material		SUS304
	Thickness	(mm)	0.1
Surface modification	Thickness	(m)	5, 50, 100
	Shape		Hollow
	Outer/ Inner	(mm)	5 / 3.53
Initial temperature		(°C)	25
Atmosphere		(°C)	25
Current		(A)	40
FEM code			Abaqus6.17

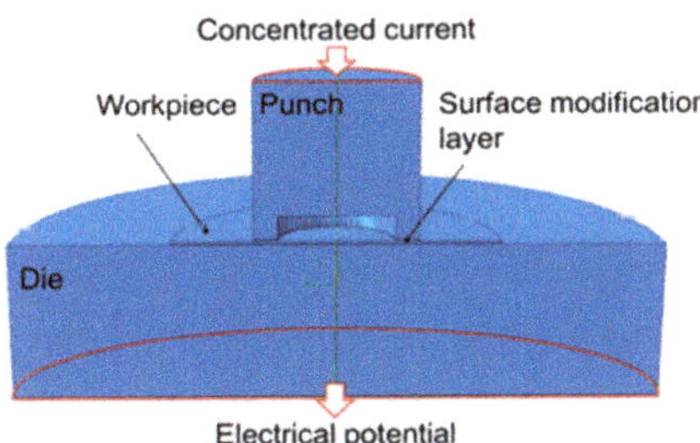

Figure 2. Schematic illustration of the finite element (FE) modeling of the resistance heating.

Table 2. Resistivities of various materials used for surface modification.

Material	Fe_4N	ZnO	$CrN_{0.6}$
Resistivity(cm)	162	2.2×10^4	1.16×10^6

Figure 3. Analytical result of temperature distribution in 100 μm thick $CrN_{0.6}$ layer heated for 60 s.

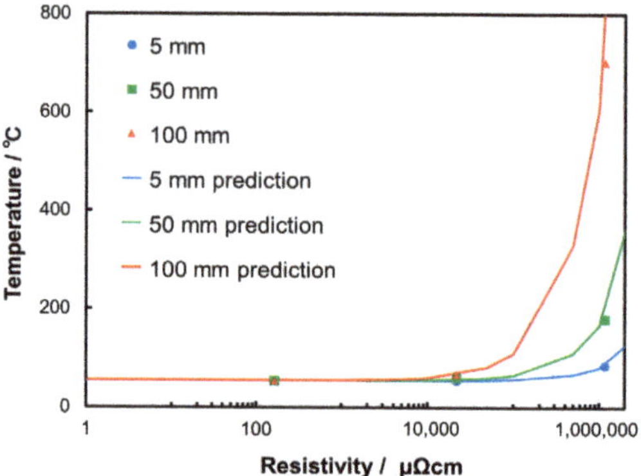

Figure 4. Predicted relationship between temperature and electrical resistivity.

2.2. Selection of Material and Die Coating

Three kinds of commercial ceramic coating materials used for cutting tools were investigated from the viewpoint of heat resistance and seizure resistance at more than 800 °C. The ceramics coated to a thickness of 1 μm on stainless steel (JIS: SUS420J2) by PVD process and their electrical resistance were measured using a digital multimeter (VOAC7522H, IWATSU ELECTRIC CO., LTD., Tokyo Japan). The results are shown in Table 3. All three coated materials show higher electrical resistivities than the uncoated material. Since the AlCrSiN coating had the highest heat resistance and a maximum electrical resistivity of 2×10^{12} μΩcm, it was coated on the die for the forming test.

Table 3. Through-thickness resistances of ceramic-coated stainless steel and resistivities of ceramics.

	Original	TiAlN	TiSiN	AlCrSiN
Resistance (Ω)	0.2	0.6–0.8	0.42	300
Resistivity (μΩcm)	-	4×10^9	1.7×10^9	2×10^{12}

3. Evaluation of Heating System

The configuration of the microforming system with surface-modified dies is shown in Figure 5. The surface-modified dies are attached to a tabletop servo press machine (HJ50, Micro Fabrication Lab, Tokyo, Japan) and the dies are insulated in the press machine with a ceramic plate. The dies and workpiece are heated by connecting a power supplier to the dies during the forming process and the temperature is measured during heating using a radiation thermometer (FTK-R160R-5R21, TMC9-MPN). A compression test using the dies coated with AlCrSiN with thicknesses of 0.5 and 1 μm was carried out in comparison with uncoated dies. To evaluate the heating dependence on the material properties, pure titanium, pure copper, and stainless steel (JIS: SUS304) with the same cylindrical shape were used as test specimens. Changes in the temperature during the compression test were measured to evaluate the temperature dependence on the deformation. After the initial loading at 50 N, a DC current in the range of 10 to 60 A was applied, and the temperature of the specimen surface in the steady state was recorded. In the compression test, the strain rate was 0.005 s^{-1}, and the stroke for the compression was 0.8 mm. In order to suppress the effect of friction during processing, graphite powder was sprayed on the upper and lower surfaces of the specimen.

Stress–strain curves for pure titanium with a uniform initial temperature of 600 °C are shown in Figure 6. The flow stress increases gradually as the deformation progresses in the uncoated dies, whereas it is relatively constant in the AlCrSiN-coated dies. Since the flow stress for Ti decreases significantly with the increase of temperature [15], this result indicates that the changes in temperature during the process may affect the flow stress of the workpiece. The changes in temperature are shown in Figure 7. Under all conditions, the temperature tends to decrease with increasing strain, but the rate of temperature decrease was suppressed with the increase in the thickness of the die coating. The tendency of the temperature decrease is caused by the reduction in the joule heat of the material caused by the increase in the compressive strain in the axial direction of the material; as a result, the current density decreases owing to the decrease in the length and increase in the diameter of the specimen. In the coated dies, the temperature drop due to material deformation was suppressed by heat conduction from the die surface and this suppression effect was greater in the coating with higher resistivity. As a result, although the initial temperatures were the same under the three conditions, the flow stress markedly increased with the progression of the process in the uncoated dies compared with the coated dies.

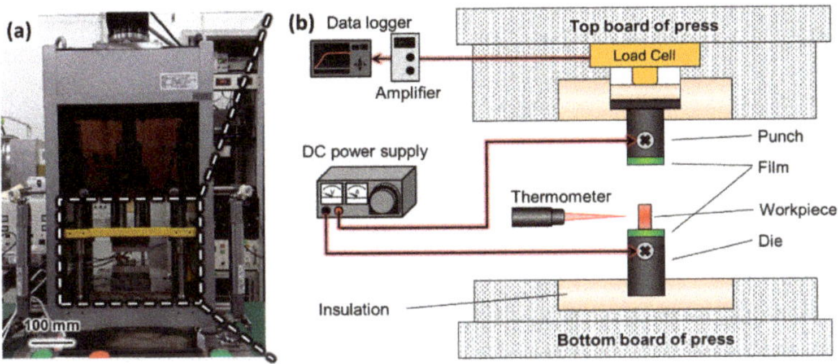

Figure 5. (**a**) Image of the miniature servo press machine; (**b**) schematic illustration of the resistance heating system.

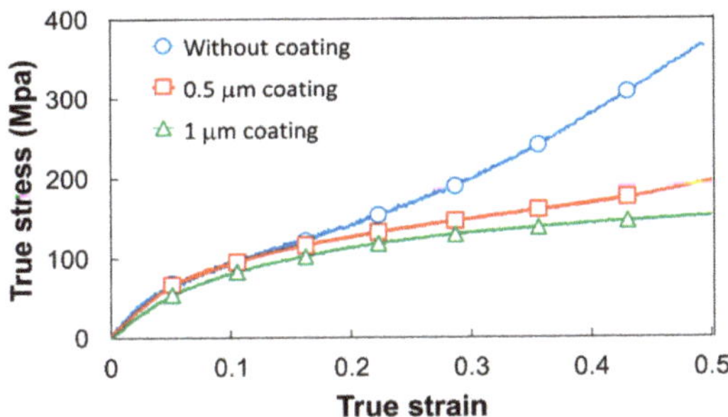

Figure 6. Stress–strain curves for pure titanium at 600 °C.

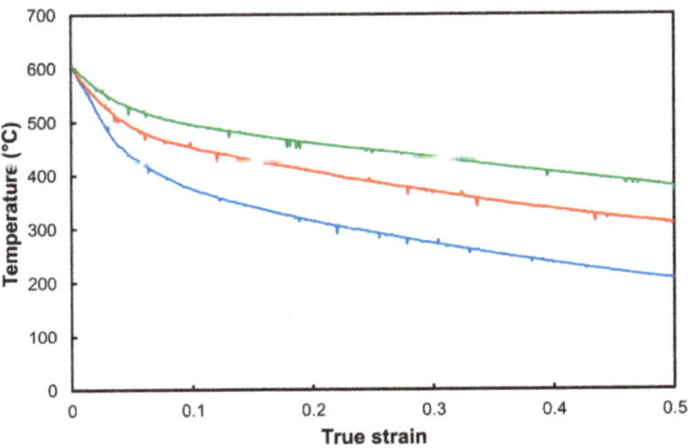

Figure 7. Variation of temperature as a function of strain.

Figure 8 shows heating temperatures at different current densities for the three materials. Higher heating temperatures were achieved for the coated dies than for the uncoated dies, and higher temperatures were confirmed for the coated dies with a thicker coating. In the case of processing the pure titanium with the coating thickness of 1 µm, a temperature of 780 °C was achieved at a current density of 51 A/mm^2, which is 350 °C higher than the average temperature achieved at the same current density for the uncoated dies. In order to quantitatively evaluate the increase in temperature for individual materials, the rate of temperature increase at a current density of 38 A/mm^2 was evaluated, and the results are shown in Figure 9. The rate of temperature increase was defined as follows:

$$T_{increase\ rate} = (T_{coated} - T_{uncoated})/T_{uncoated} \times 100\%. \tag{1}$$

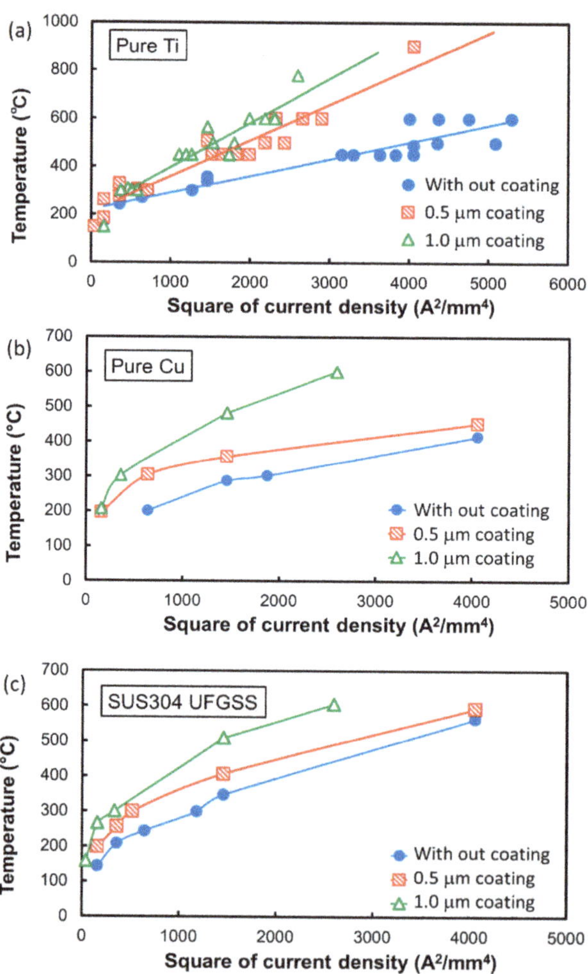

Figure 8. Maximum temperatures plotted against square of current density for (**a**) pure titanium, (**b**) pure copper, and (**c**) SUS304 UFGSS.

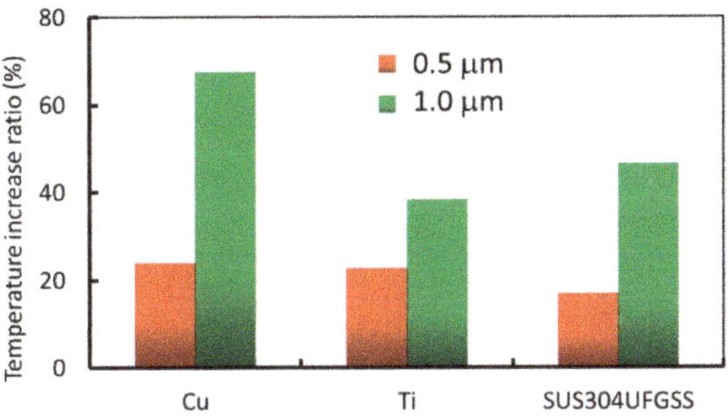

Figure 9. Rate of temperature increase for different work materials.

The electrical resistivities of the three materials used are 1.7 μΩcm for Cu, 54 μΩcm for Ti, and 72 μΩcm for SUS304. As shown in Figure 9, a higher rate of temperature increase was obtained for all materials with thicker coated dies. The maximum rate of temperature increase was confirmed for pure copper, which has the lowest electrical resistivity and is difficult to heat using the conventional heating system. This means that the new heating system can reduce the temperature dependence of the electrical resistivity of the workpiece.

In addition, a compression test was performed to verify the effect of temperature decrease suppression on the dependence of deformation. As an example, the rate of temperature decrease before and after processing in the compression test of pure titanium material is shown in Figure 10. At relatively high temperatures of 450 and 600 °C, the rate of temperature decrease was suppressed in the coated dies. Furthermore, the dies with thicker coatings generated a larger volume of heat, which transferred to the workpiece from the surface of the die and contributed to the higher ratio of this heat to the joule heat generated by the workpiece.

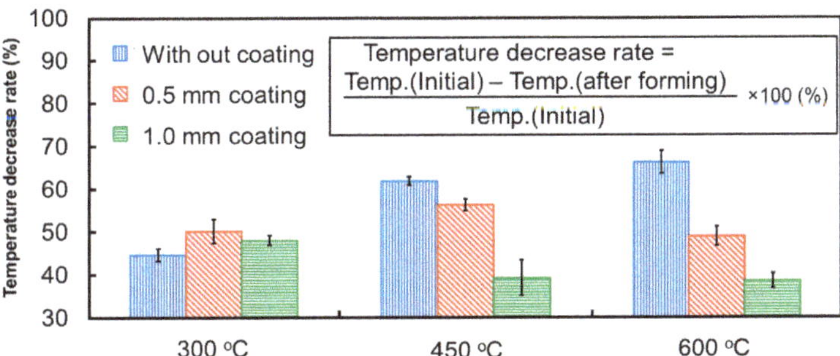

Figure 10. Rate of temperature decrease for pure titanium.

On the other hand, Figure 11 shows the change in the electrical resistivity of AlCrSiN coated to a thickness of 1 μm with the temperature when heating the pure titanium material in comparison with that for the untreated die. It was found that the electrical resistivity of the AlCrSiN coating decreases with increasing temperature. Since the electrical resistivity of ceramic materials shows a temperature dependence, it is necessary to select the appropriate coating material with high electrical resistivity within the appropriate desired heating temperature range.

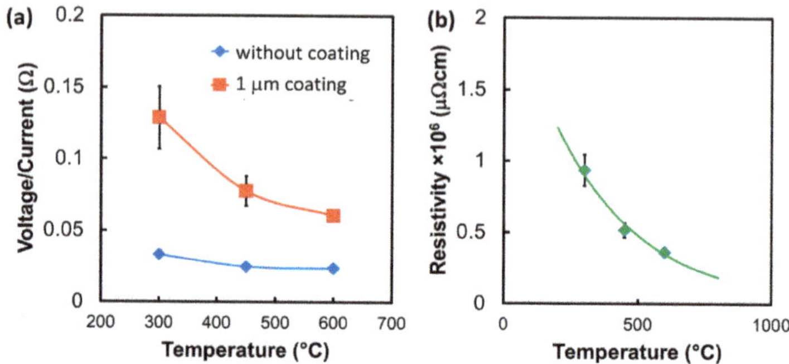

Figure 11. (**a**) Electrical resistivity as a function of temperature in resistance heating of pure titanium, (**b**) temperature dependence of electrical resistivity of 1 µm AlCrSiN.

4. Application of Coated Die for Microforging

The developed electric heating system was applied to the microforging of a metallic foil material and its effects on the formability were evaluated. As shown in Figure 12, a punch of φ 1 mm with an AlCrSiN coating of 1 µm was used for the process. In order to verify the dependence on the shape of the workpiece, the workability and shape-freezing property were evaluated using a fine-grained stainless steel foil with a thickness of 0.2 mm, which is difficult to heat using a conventional heating system. The strain rate was 0.005 s^{-1} and the punch stroke was 80 µm. Graphite spray was applied to the upper and lower surfaces of the workpiece as a lubricant. In order to prevent the displacement effect due to thermal expansion of the dies, the punch position was adjusted to 50 N after a given current was applied. Figure 13 shows the punch load-stroke results of the microforging. It was confirmed that the load was reduced by applying heating during the process, and the processing load was reduced by more than 20% in comparison with the process at room temperature. Resultant depths forged with various conditions were measured using an interference microscope (CCI, AMETEK, Berwyn, PA, USA). Figure 14 shows the images of the measurement and Figure 15 shows the values for the forging depth. It was found that the forging depth increased with heating during the process, and the forging depth further increased in coated dies in comparison with the uncoated dies. The temperature near the top of the punch was measured and it was found that the temperatures were 400 °C for the uncoated punch with a current density of 50.9 A/mm^2 and 600 °C for the coated punch. These findings indicate that coated dies heated the material more efficiently than uncoated dies. As a result, the flow stress of the material was reduced, and the forging depth was increased.

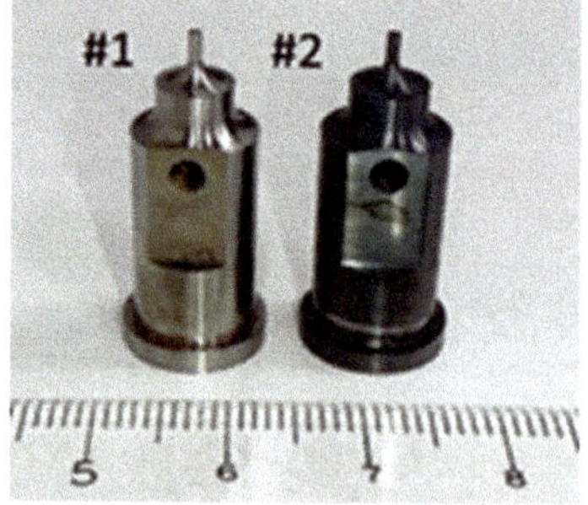

Figure 12. Forging punches: #1 uncoated, #2 AlCrSiN-coated.

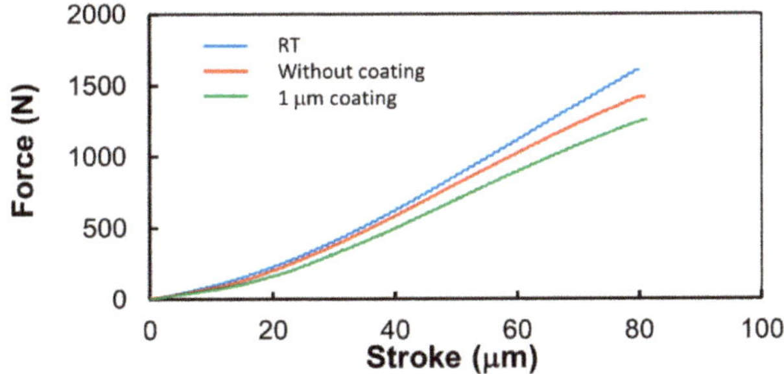

Figure 13. Punch force during forging process for UFG stainless steel foil of 0.2 mm thickness.

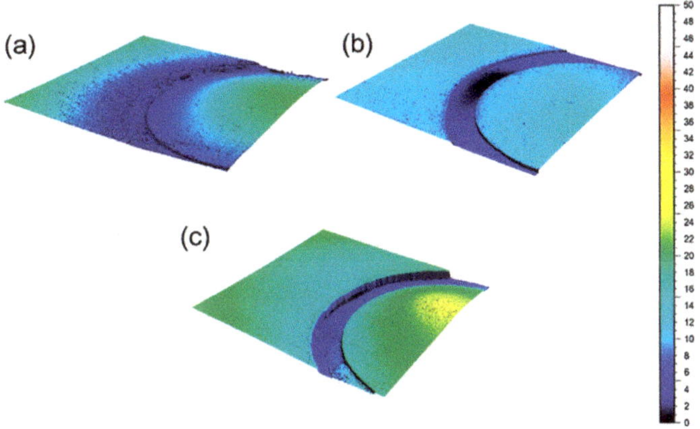

Figure 14. Three-dimensional profiles of ring-shape forging. (**a**) RT; (**b**) without coating with 50.9 A/mm^2; (**c**) 1 mm coating with 50.9 A/mm^2.

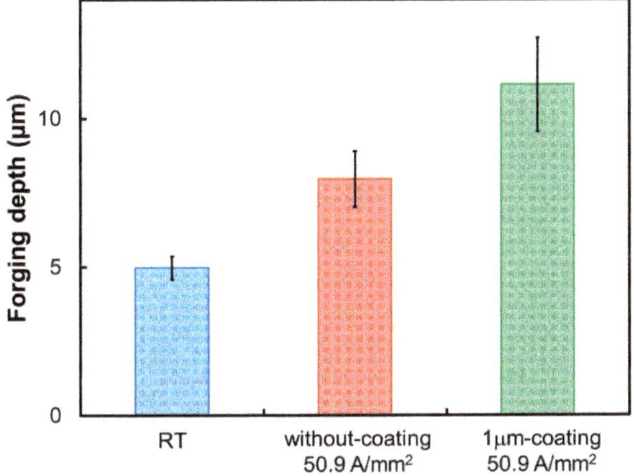

Figure 15. Forging depth of UFG stainless steel foil.

5. Conclusions

For this study, a novel electric heating system with a ceramic material coated on the dies was developed for microforming processes. The ceramic coating, which has been used to improve tribological properties and heat resistance, was applied as a heat source by utilizing its high electrical resistivity. The heating system depends much less on the shape, physical properties, and deformation of the workpiece than previous ones. The effect of the thickness and electrical resistivity of the surface coating layer on the electric heating efficiency was verified by a micro-compression test, and the applicability to an actual process was verified by a microforging test. The following findings were obtained.

(1) A layer with high electrical resistivity coated on the surface of the dies for the heating system was designed based on the prediction from FE analysis results. The desired temperature was achieved by selecting the appropriate thickness and electrical resistivity of the surface modification layer.

(2) The superior heating efficiency of all the processed materials, namely, pure titanium, pure copper, and stainless steel, was confirmed by the compression test using the die coated with AlCrSiN. The heat generated from the surface layer of the die suppressed the temperature decrease of the workpiece during the compression deformation.

(3) It was confirmed that the forging depth was increased by the surface modification in the electric heating while microforging a fine-grain stainless steel foil material with a thickness of 0.2 mm, which is difficult to heat using a conventional heating system.

Author Contributions: M.Y. designed and executed experiments. T.S. executed FE simulation and design of surface coating system.

Funding: This work was funded by JST foundation Grant-in-Aid for Scientific Research (B).

Acknowledgments: The authors thank Ryota Ogura for his contribution to the experimental implementation, Kaoru Ikenaga of Japan Electronics Co., Ltd., for cooperation with the coating of the dies using various ceramic materials, and Tomomi Shiratori for cooperation with the fabrication of microforged dies in the development of this electric heating system.

Conflicts of Interest: The authors declare no conflict of interest.

References

1. Geiger, M.; Kleiner, M.; Eckstein, R.; Tiesler, N.; Engel, U. Microforming. *CIRP Ann.* **2001**, *50*, 445–462. [CrossRef]
2. Vollertsen, F.; Biermann, D.; Hansen, H.N.; Jawahir, I.S.; Kuzman, K. Size effects in manufacturing of metallic components. *CIRP Ann.* **2009**, *58*, 566–587. [CrossRef]
3. Engel, U.; Eckstein, R. Microforming—from basic research to its realization. *J. Mater. Process. Technol.* **2002**, *125*, 35–44. [CrossRef]
4. Egerer, E.; Engel, U. Process characterization and material flow in microforming at elevated temperatures. *J. Manuf. Process.* **2004**, *6*, 1–6. [CrossRef]
5. Peng, X.; Qin, Y.; Balendra, R. A numerical investigation to the strategies of the localised heating for micro-part stamping. *Int. J. Mech. Sci.* **2007**, *49*, 379–391. [CrossRef]
6. Magee, J.; Watkins, K.G.; Steen, W.M. Advances in laser forming. *J. Laser Appl.* **1998**, *10*, 235. [CrossRef]
7. Milenin, A.; Kustra, P.; Furushima, T.; Du, P.; Němeček, J. Design of the laser dieless drawing process of tubes from magnesium alloy using FEM model. *J. Mater. Process. Technol.* **2018**, *262*, 65–74. [CrossRef]
8. Gillner, A.; Holtkamp, J.; Hartmann, C.; Olowinsky, A.; Gedicke, J.; Klages, K.; Bosse, L.; Bayer, A. Laser applications in microtechnology. *J. Mater. Process. Technol.* **2005**, *167*, 494–498. [CrossRef]
9. Tanabe, H.; Yang, M. Design and evaluation of heat assisted microforming system. *Steel Res.* **2011**, *Special Edition*, 1020–1024.
10. Aoyama, T.; Shimizu, T.; Zheng, Q.; Yang, M. Effect of heating on springback in heat assisted micro bending. *Adv. Mater. Res.* **2014**, *939*, 409–414. [CrossRef]
11. Zheng, Q.; Aoyama, T.; Shimizu, T.; Yang, M. Experimental and numerical analysis of springback behavior under elevated temperatures in micro bending assisted by resistance heating. *Procedia Eng.* **2014**, *81*, 1481–1486. [CrossRef]
12. Zheng, Q.; Shimizu, T.; Yang, M. Scale effect on springback behavior of pure titanium foils in micro bending at elevated temperature. *J. Mater. Process. Technol.* **2016**, *230*, 233–243. [CrossRef]

13. Zheng, Q.; Shimizu, T.; Yang, M. Numerical analysis of temperature distribution and its optimization for thin foils in micro deep drawing assisted by resistance heating. *Steel Res.* **2015**. [CrossRef]
14. Ogura, R.; Shimizu, T.; Shiratori, T.; Yang, M. Development of a novel resistance heating system for microforming by modifying surface of dies. *Procedia Eng.* **2017**, *207*, 1016–1021. [CrossRef]
15. Zheng, Q.; Shimizu, T.; Yang, M. Effect of heat on tensile properties of thin pure titanium foils. *Manuf. Rev.* **2015**, *2*, 3. [CrossRef]

Article

Plasma Printing of an AISI316 Micro-Meshing Punch Array for Micro-Embossing onto Copper Plates

Tomomi Shiratori [1,*], Tatsuhiko Aizawa [2], Yasuo Saito [3] and Kenji Wasa [4]

[1]   Komatsu-Seiki Kosakusho, Co., Ltd., R&D section, Suwa, Nagano 392-0012, Japan
[2]   Surface Engineering Design Laboratory, Shibaura Institute of Technology, Ohta, Tokyo 144-0045, Japan;
      taizawa@sic.shibaura-it.ac.jp
[3]   Chuo Denshi Kogyo, Co., Ltd., Ujyou, Kumamoto 869-0512, Japan; yasuo.saito@cdk.co.jp
[4]   MICROTEX Inc., Ohta, Tokyo 144-0051, Japan; wasa@mui.biglobe.ne.jp
*   Correspondence: shiratori@komatsuseiki.co.jp; Tel.: +81-266-52-6100

Received: 23 February 2019; Accepted: 27 March 2019; Published: 30 March 2019

Abstract: Packaging using thermoplastic molding for hollowed GaN chips were requested for a leak-proof micro-joining between plastic molds and copper-based substrates. The design and engineering of micro-textures is a key technology for putting leak-proof packaging into practice. In the present paper, a micro-meshing punch array was prepared using plasma-nitriding-assisted printing. Two-dimensional original patterns were screen-printed onto an AISI316 die substrate and plasma nitrided at 673 K for 14.4 ks (or 4 h). The unprinted surfaces were selectively nitrogen super-saturated to have more nitrogen content than 5 mass% and a higher hardness than 1200 HV. The printed surfaces were selectively sand blasted to fabricate the micro-meshing punch array for micro-embossing. A computer numerically controlled stamping system was utilized to describe the micro-embossing behavior onto copper substrates and to investigate how the micro-textures on the array was transcribed onto the copper. Reduction of takt time as well as flexibility in the micro-grooving were discussed with reference to the picosecond laser machining and mechanical milling processes.

Keywords: plasma printing; micro-texturing; screen printing; low-temperature plasma nitriding; selective anisotropic nitrogen embedding; selective hardening; sand blasting; AISI316; micro-meshing punch array; copper plates

---

1. Introduction

Welding and chemical adhesion processes have been utilized as a reliable means for joining of metallic components and parts. In recent years, mechanical clinching has grown as a key manufacturing step to join dissimilar products [1] and to stack the constituent sheets into an integrated unit [2]. In particular, mechanical micro-joining is needed to make innovative packaging for power transistors and computer-processing units [3]. In those micro-joining processes, the surface to be joined must have micro-textures to increase the joining strength between two dissimilar parts, e.g., a plastic mold which must be micro-joined to a metallic substrate [4]. How to make suitable micro-textures into metallic substrates has become an urgent issue in engineering these packaging processes.

In the literature, there has been many micro-texturing methods reported, e.g., the micro-milling method [5], the micro-EDM (electrical discharging machining) method [6], and the laser micro-texturing method [7]. In the former two approaches, their micro-texturing process into a die substrate is very much dependent on the durability of thin tools and wires during the long cutting time. Although the pico-second laser machining is free from this difficulty, the total takt time for micro-texturing significantly increases in mass production. The authors [8] proposed a non-traditional method to accommodate the designed micro-textures into the molds and dies for their transcription onto the metallic and polymer sheets. In addition, an original two-dimensional micro-pattern was CAD (Computer Aided Design)-designed. Then, it was directly printed onto the DLC (Diamond Like Carbon) coatings on the die and mold surface. Plasma oxidation was utilized to remove the unprinted DLC films and to fabricate the DLC-punch array on the DLC-coated dies and molds [9,10]. Plasma nitriding was also employed to selectively nitride the unprinted surfaces and to build up the nitride-punch array by mechanical and chemical etching processes [11]. These micro-textures of the punch array were transcribed into metallic and polymer sheets via precise rolling and stamping. This selective-nitrogen embedding into stainless steel substrates came from the microstructure evolution only below the unprinted substrate surfaces [12]. Without this anisotropic microstructure evolution, this plasma-assisted printing often fails in making nitrided micro-textures into the substrate [13].

In the present paper, this plasma printing method is further advanced to fabricate a nitrided micro-meshing punch array and to make micro-embossing for transcription of this texture into the copper substrate. Screen printing was utilized to print the negative pattern onto the micro-meshing texture. These unprinted sections were selectively nitrogen-supersaturated and hardened after plasma nitriding process. The un-nitrided die parts were removed by a sandblasting process to apper the micro-meshing punch head. This micro-meshing punch was fixed onto a cassette die and was utilized to conduct micro-embossing onto a copper plate. The micro-grooves were formed onto the copper plate after micro-embossing process.

2. Experimental Procedure

An AISI316 die substrate with the size of 24 mm × 12 mm × 5 mm was utilized as a substrate material. Its surface was mirror-polished for plasma printing. The average roughness (*Ra*) of the AISI316 die substrate was 0.010 μm. The plasma printing procedure was stated with comments on the screen printing, the low-temperature plasma nitriding, the blasting, and the CNC (Computer Numerical Control)-micro-embossing. This present plasma printing process consists of three steps as illustrated in Figure 1. First, the negative micro-pattern of arrayed punch heads was printed as a two-dimensional mask onto the mirror-polished AISI316 substrate in Figure 1a. Figure 1b depicts the screen-printed mask pattern. Second, this printed substrate was plasma-nitrided at 673 K for 14.4 ks to make nitrogen super-saturation selectively to the unprinted substrate surfaces in Figure 1c. The printed surfaces were not nitrided and they maintained the same hardness as the matrix so that they could be mechanically removed with ease from the substrate to form the multi-punch array as shown in Figure 1d.

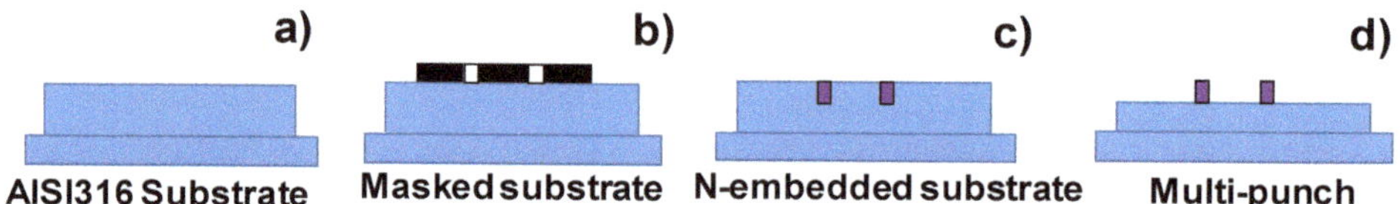

Figure 1. Plasma printing procedure in the present study. (**a**) The starting AISI316 substrate, (**b**) masked substrate by the screen printing, (**c**) nitrogen-embedded substrate by plasma nitriding, and (**d**) multi-punch formed by the blasting.

2.1. Screen Printing onto Die Substrate

The screen-printing system (NEWLONG, Co., Ltd., Shinagawa, Tokyo, Japan) was employed to print the designed micro-pattern onto the substrate's surface after the CAD data. The mask-pattern on the screen corresponded to the arrayed multi-punch alignment. The square cell units with 200 μm × 200 μm were directly printed onto the AISI316 die surface, as shown in Figure 2. In the following experiments, a screen with a micro-meshing texture was employed to print its negative pattern onto the surface of the AISI316 die. An ink for screen-printing must be optimally selected among several candidates to have sufficient thermal resistance during the plasma nitriding at 673 K. A polymer-based ink has a risk of diminishing itself during plasma nitriding at 673 K. These inks after nitriding have to be easily removed from the processed substrate surface. In a previous study [11], a $CaCO_3$-based ink was invented and used in trials. In the present experiment, TiO_2-based ink with sufficient thermal resistance at 673 K was selected among several candidates to improve the spatial resolution in micro-patterning. In practical operations, specially formulated TiO_2 ink (Teikoku Printing Inks Mfg. Co., Ltd., Arakawa, Tokyo, Japan) without the use of thinning agents was used for directly screen printing onto the die surface and dried at 673 K for 600 s in the nitrogen atmosphere.

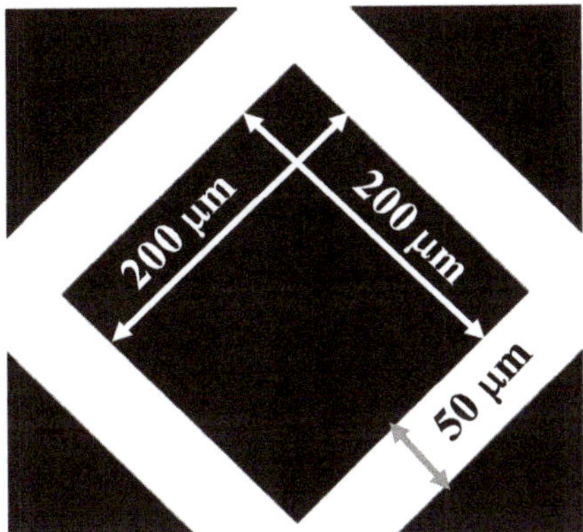

Figure 2. Geometry of a square unit cell on the screen.

2.2. Low-Temperature Plasma Nitriding

High-density RF (radio frequency)/DC (direct current) plasma nitriding system (YS-Electric Industry, Co., Ltd., Koufu, Yamanashi, Japan) was utilized to make selective nitrogen super-saturation to the unprinted substrate surfaces at 673 K for 14.4 ks by 70 Pa. After evacuation down to 0.1 Pa, the nitrogen gas was introduced to pre-sputter the printed die surface for 1 ks under the DC-bias of −600 V. After re-evacuation, the specimen was heated up to 673 K under the nitrogen atmosphere by 250 Pa. Then, the nitrogen hydrogen mixture gas was introduced with a flow rate of 160 mL/min for nitrogen and 30 mL/min for hydrogen, respectively. After plasma nitriding, the specimen was cooled down in the chamber under the nitrogen atmosphere. The micro-printed AISI316 substrate's surface was fully covered by the plasma sheath with high nitrogen ion and NH-radical densities, enough to drive the nitrogen super-saturation even at lower temperatures [12]. This selective anisotropic nitrogen-embedding process results in the selective hardening and selective nitrogen concentration. The printed surface remains as a matrix hardness while the unprinted surfaces were selectively hardened up to 1400 HV for the AISI316 substrates as reported in Reference [14]. This hardness distribution by selective anisotropic nitrogen embedding drives to selectively remove the screen-printed parts from substrate.

2.3. Mechanical Blasting Process

Mechanical blasting equipment (Fuji Manufacturing Co., Ltd., Edogawa, Tokyo, Japan) was also employed to selectively remove the un-nitrided parts and masked ink from the substrate. Owing to the hardness distribution, the nitrided areas were left as a punch head while the un-nitrided areas were completely removed by this processing. Figure 3 depicts the blasting apparatus for manual operation. The blasting rate was controlled by the shooting speed of the blasting media. The punch height also varied by the duration time. In the following blasting, fine silica particles with an average diameter of 5 μm were utilized as a blasting medium. The shooting rate was constant by 2 m/s and the shooting angle by 60°. As depicted in Figure 3b, the specimen was fixed into a jig on the stage for continuous shooting operation. The duration time was selected to be 300 s in the following experiments.

After Reference [11], the punch height reached 60 μm by blasting the printed parts of substrate for 300 s.

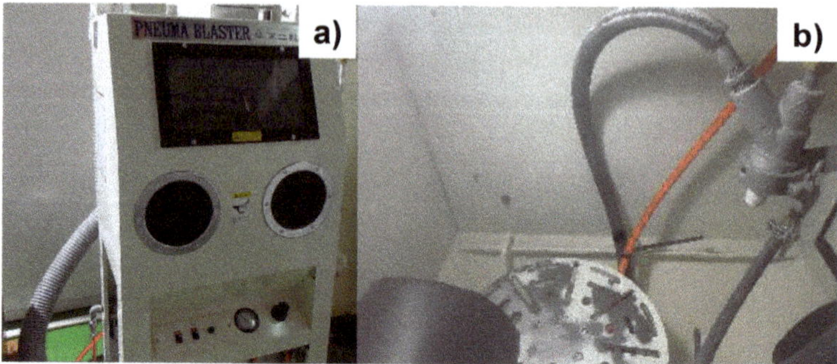

Figure 3. Mechanical blasting equipment. (**a**) Overall image of blasting equipment and (**b**) the shooting stage of substrates in manual operation.

2.4. Micro-Embossing Process

The CNC stamper (Hoden Seimitsu-kako Kenkyusho, Co., Ltd., Atsugi, Kanagawa, Japan) was utilized for micro-embossing the micro-meshing punch array onto the copper specimen in Figure 4a. The multi-punch array was fixed into a cassette die in Figure 4b. This cassette die set was placed and fixed between the upper and lower bolsters in Figure 4a. The incremental loading sequence was used to control the punch stroke every 20 μm until the maximum load reached 30 kN. The total duration was 10 s for a single operation including the set-up and pick-up time of the work. An oxygen-free copper plate 20 mm × 10 mm × 1 mm was employed as a work material in the micro-embossing. The average roughness of the oxygen-free copper plate was 0.093 μm.

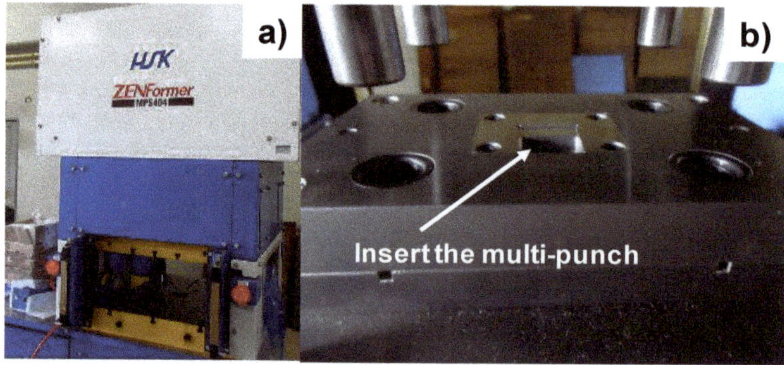

Figure 4. CNC-controlled stamper for micro-embossing onto the copper plates. (**a**) Overview of the CNC-stamper and (**b**) the cassette die set.

2.5. Evaluation Method for Micro-Embossed AISI316 Die Substrate

The AISI316 die substrate in each step of the plasma printing was evaluated by optical microscopy (STZ-168; Shimazu, Co., Ltd., Kyoto, Japan), as well as the scanning electron microscopy (SEM; JSDM-IT300LV, JEOL Ltd., Akishima, Tokyo, Japan). Energy dispersive X-ray spectroscopy (EDX; Pegasus, EDAX, Inc., Minato-ku, Tokyo, Japan) was utilized for fine element mapping. The surface roughness was measured by a three-dimensional measurement machine (Infinite-Focus; Alicona Imajing GmbH., Raaba, Graz, Austria). The hardness was measured by a micro Vickers hardness testing machine (HM-210C; Mitsutoyo Co., Ltd., Kawasaki, Kanagawa, Japan) with an applied load of 49.03 mN and for a holding time of 10 s.

3. Experimental Results

The plasma printing procedure in Figure 1 was put into practice to shape the micro-punch array with the micro-meshing heads.

3.1. Two-Dimensional Micro-Patterning onto AISI316 Die Materials

The screen was prepared to have a micro-meshing pattern with the spatial resolution of 0.2 µm as depicted in Figure 5a. The square cell units of 200 µm × 200 µm were directly printed onto the AISI316 die surface. Figure 5b shows the screen-printed micro-pattern onto the die surface. The square cell units were printed in white by using the TiO_2-based ink to form the micro-meshing pattern with the line width of 50 µm and the pitch of 250 µm. Their deviation in width and pitch was within ±0.3 µm in practice.

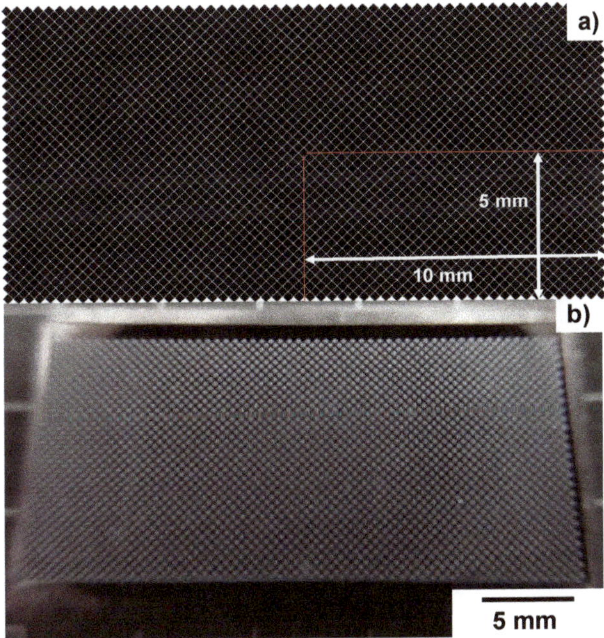

Figure 5. Micro-patterned AISI316 die substrate by the screen printing. (**a**) Micro-pattern on the screen corresponding to the CAD data and (**b**) the printed micropattern on the AISI316 die substrate.

3.2. Plasma Printing with Use of High-Density Nitrogen–Hydrogen Plasmas

Figure 6 depicts the plasma nitrided AISI316 die substrate after cleansing the masking inks. The unprinted micro-meshing lines in Figure 5b were selectively plasma nitrided to form the micro-meshing network with nitrogen super-saturated microstructure on the whole substrate surface. After References [11,12,14], the nitrided layer thickness after nitriding by the same conditions reached to 60 µm on average. Average surface hardness was measured to be 1200 HV with the standard deviation of hardness within only 30 HV.

Scanning electric microscopy as well as EDX were utilized to analyze the microstructure of the nitrided AISI316 die surface and to map nitrogen across the square cell units. As shown in Figure 7a, the micro-meshed lines were regularly aligned by the pitch of 250 µm. These lines also had higher nitrogen content than 5 mass% selectively in their widths, as analyzed in Figure 7b. Although the residuals of masking inks were left as a thin film, no height difference was seen between the masked and nitrided areas on the surface.

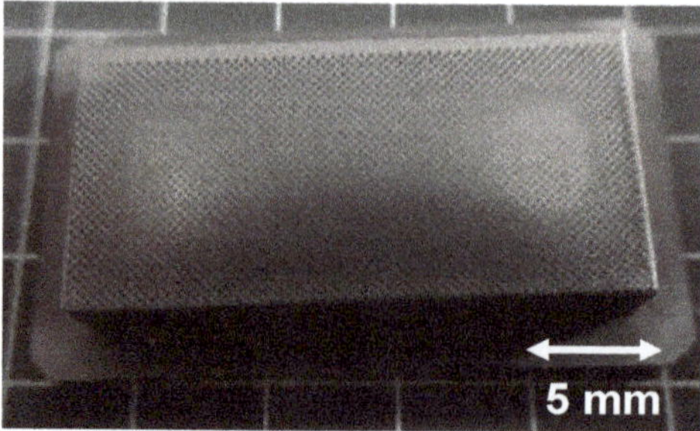

Figure 6. Selectively plasma-nitrided AISI316 die substrate at 673 K for 14.4 ks.

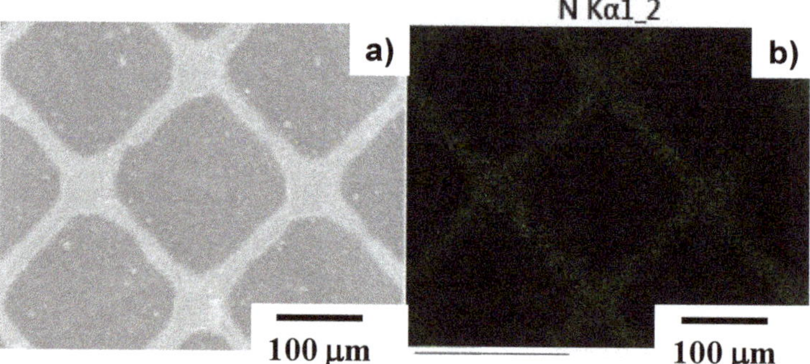

Figure 7. Microstructure of plasma-nitrided AISI316 die surface at 673 K for 14.4 ks. (**a**) SEM image and (**b**) nitrogen mapping by EDX.

3.3. Formation of a Micro-Meshing Punch Array by Blasting Process

The blasting process was employed to remove the printed parts from the nitrided AISI316 die substrate in Figures 6 and 7. Figure 8 depicts the blasted die surface. The whole surface became a multi-punch head with meshing line network. Figure 9 shows its optical microscopic image and SEM image where a square micro-cavity with a size of 180 μm × 180 μm and a depth of 75 μm was formed in regular alignment. In particular, the SEM image in a narrow range in Figure 9b proves that the un-nitrided square cells in Figure 7 were selectively dug in depth by this sand blasting to form the micro-meshed punch head array with the line width of 70 μm.

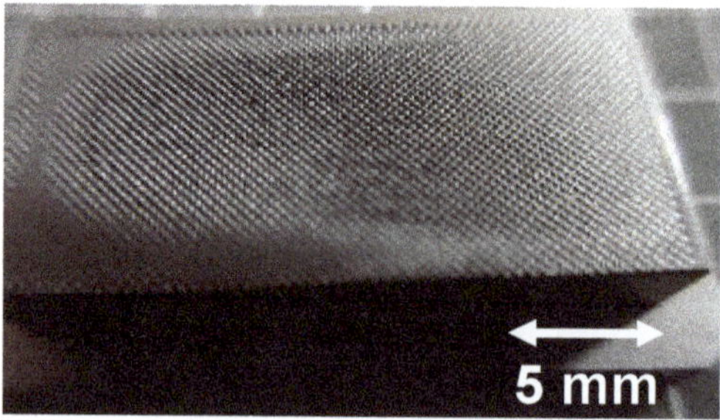

Figure 8. Multi-punch array formed on the surface of AISI316 die by sand blasting for 300 s.

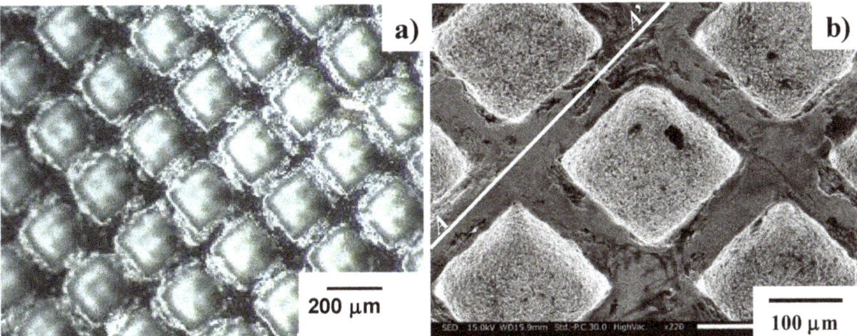

Figure 9. Microscopic view of the multi-punch array. (**a**) Optical microscopic image in wide range of the micro-meshed punch array and (**b**) the SEM image in narrow.

3.4. Micro-Embossing onto Copper Plates

This multi-punch die was inserted into a cassette die in Figure 4 for micro-embossing. Figure 10 shows the optical and SEM images of the micro-embossed copper plate. The multi-punch head was pushed down onto the copper plate to coin the meshed micro-grooves with the width of 70 μm. The widely scanned image in Figure 10a reveals that this coining took place uniformly on the copper surface. Each square cell with the size of 180 μm × 180 μm was backwards extruded to form a quadratic prism pillar with a maximum height of 10 μm. Metallic scratches, which were coined onto the copper plate by cold stamping, were seen on these prism head surfaces.

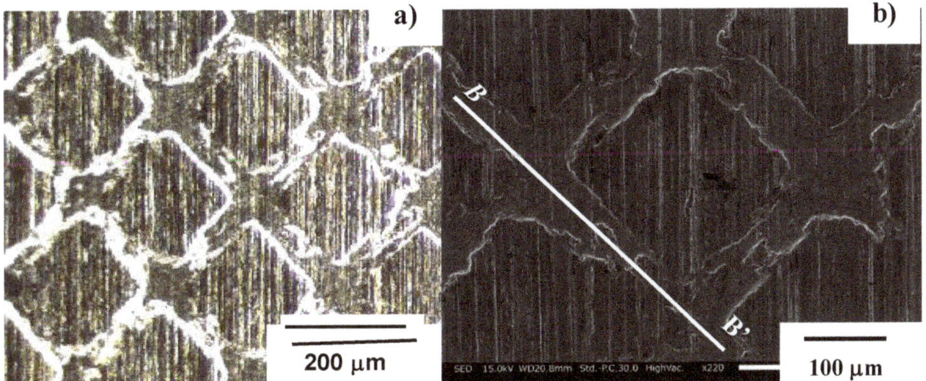

Figure 10. Micro-embossed copper plate by incremental loading every 20 μm until the maximum load reached 30 kN. (**a**) Optical microscopic image in wide range and (**b**) the SEM image in narrow range.

3.5. Transcription Processes from Micropattern on Screen to Copper Micro-Textures

A two-dimensional micropattern on the screen was transcribed into three-dimensional micro-textures on the copper plate by the present plasma printing. The micro-meshing pattern on the screen corresponds to the CAD data in Figure 11a; the white square unit cell with the edge length of 200 μm was directly printed onto the AISI316 die surface. There was little error in dimension at this step of the screen printing. Figure 11b shows the square unit cell on the nitrided AISI316 substrate at 673 K for 14.4 ks after cleansing the surface. Although thin TiO_2 ink residuals overlapped at the vicinity of the mask edges and corners, the un-nitrided unit cell geometry was preserved to be a square in Figure 11b. Let us compare the unit cell geometries before and after sand blasting in Figure 11b,c. In correspondence to the original square pattern in Figure 11a, the square micro-cavities, each with an edge length of 180 μm and a depth of 75 μm, were shaped into the AISI316 die. The four edges and corners of each micro-cavity had sufficient sharpness to be working as a micro-embossing punch.

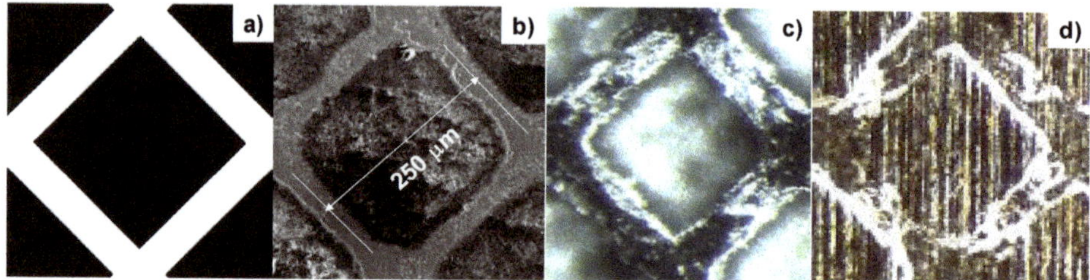

Figure 11. Comparison of the geometric configurations among (**a**) the micropattern on the screen, (**b**) the SEM image of the nitrided AISI316 die, (**c**) the three-dimensional profile of the blasted AISI316 die, and (**d**) the micro-embossed copper plate.

Figure 11c,d describes the micro-embossing behavior to transcribe the square unit cell. The copper was backward extruded into a square micro-cavity in the multi-punch array in Figure 9 to form the quadratic prism array with an edge length of 180 μm and a maximum height of 10 μm.

The dimensional accuracy in this plasma printing was mainly determined by the spatial resolution in screen printing. The white linear segments in Figure 11a transcribed into the micro-punch head in Figure 11c, with an average line width of 70 μm. This head width became broader than the line width of 50 μm in the original screen film because the TiO_2 ink subdues and bleeds at the masking edges. This deviation of punch width was reduced by increasing the viscosity and adhesiveness of TiO_2-based inks. In particular, higher viscosity ink was worked with non-subduing and non-bleeding at the square cell edges in Figure 11b. Each linear segment of the punch heads had sharp edges, even at the vicinity of cross-points as seen in Figure 11c. This edge sharpness was improved by the reduction of the edge curvature at the micro-punch shoulder via the ion beam treatment [15].

In the micro-embossing process, the copper was backward extruded to flow into each square micro-cavity of the punch by incrementally stamping the micro-meshed punch array into the copper plate as seen in Figures 10 and 11d. The micro-punch heads in Figure 11c penetrate into the copper plate and formed the micro-grooves in Figure 11d. At the same time, the copper quadratic prism array was shaped on the micro-embossed copper plate by the backward extrusion. The burrs and debris were seen around the prism edges; these were also diminished by sharpening the micro-punch shoulder edges.

3.6. Roughness Profiles on the Punch Array Heads and the Meshed Cavities in Copper Substrate

The surface roughness on the punch array heads influenced the dimensional accuracy of the meshed copper substrate cavities. A three-dimensional profilometer was employed to measure the surface roughness profile along the line *A–A'* in Figure 9b. As shown in Figure 12a, *Ra* of the meshed punch head surface was 1 μm except for the crossed line-heads where deep dimples were seen heterogeneously. This punch head roughness reflects on the micro-cavity quality of the copper products. Figure 12b shows the surface roughness along the line *B–B'* in Figure 10b on the micro-cavities, which were cut into the copper substrate by micro-embossing. Three convex bumps were seen in correspondence to the roughness along the line *A–A'* in Figure 12a. Since the dimensional deviation between two surface profiles in Figure 12 was less than 1 μm, the original surface profile on the arrayed punch heads was accurately transcribed onto the copper substrate by the present micro-embossing with use of a multi-punch array.

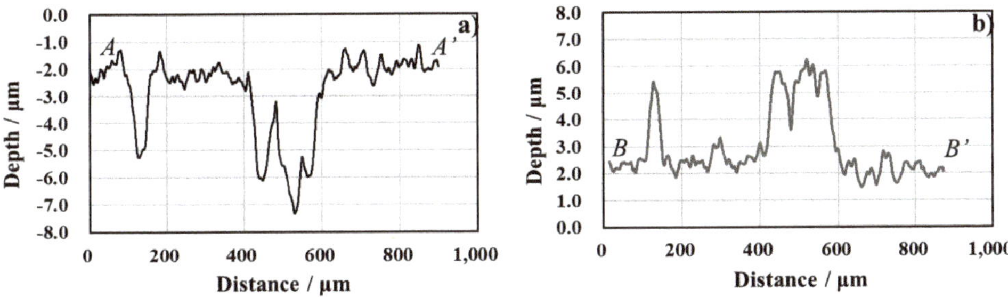

Figure 12. Measured surface roughness profiles by three-dimensional laser profilometer. (**a**) Surface roughness along *A–A'* in Figure 9b, and (**b**) surface roughness along *B–B'* in Figure 10b.

3.7. Evaluation of the Dimensional Accuracy of the Punch Array

A three-dimensional profilometer was also employed to measure the surface geometry of the multi-punch array. Figure 13 shows the surface geometry across four micro-punches along the line *C–C'*. The average height and pitch of micro-punches were 75.7 µm and 251 µm, respectively. The deviation of height was +1 µm and −2 µm among the four micro-punches. This deviation was corresponding to the surface roughness on the punch head. Each micro-punch in Figure 13 had dull shoulder edges; the borders between masked and unmasked zones might have had a lower hardness than the nitrided zone at the center of the punch head. The average head height was higher by 15 µm than the previous data of 60 µm in Reference [4]. This might be because AISI316 has a softer matrix hardness of 200 HV than the martensitic stainless steel AISI420 in Reference [4].

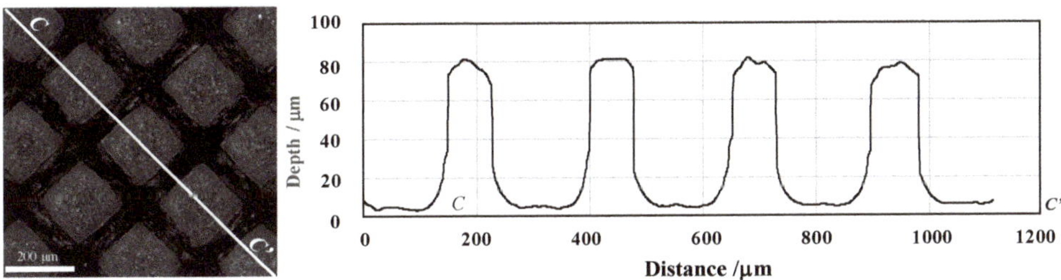

Figure 13. Height distribution of the multi-punch array measured along the line *C–C'*. The average punch height was 76 µm.

4. Discussion

A micro-milling process was employed to compare the processing time for the fabrication of the same micro-meshed AISI316 punch array as made by the present plasma printing. A milling tool with the diameter of 10 µm was prepared to achieve the fine corner curvatures of the micro-cavities in Figures 9 and 11c. The average machining speed, cutting depth, as well as cutting distance of a single cavity layer were assumed to be 1 mm/s, 5 µm, and 2.25 mm, respectively, without fracture of the thin milling tools. Two milling paths were needed to attain the same roughness, as seen in Figure 12a. The milling time to remove a single micro-cavity 200 µm × 200 µm × 75 µm reached 120 s. Since the whole surface of a AISI316 die with the dimensions of 20 mm × 10 mm had 5000 micro-cavities (Figure 8), the total milling time became 167 h. Including the takt time to prepare the CAM (Computer Aided Machining) data for this micro-milling, the practical takt time was nearly doubled to become 320 h. The present plasma printing requires 10 min at most for screen printing, 5 h for plasma nitriding including heating and cooling, and 5 min for set-up and blasting. No CAM data were necessary since the CAD data was reflected on the screen. This comparison proves the superiority of plasma printing for fabrication of micro-punch arrays for precision mechanical milling processes.

Pico-second laser machining, as well as fiber-laser machining, have been utilized for the formation of micro-groove textures into each copper plate [3,4]. Let us evaluate the takt time for the production of the micro-textured copper plate by this plasma printing with comparison to the picosecond laser machining to make a micro-grooved copper plate. Each micro-groove has the width, depth, and pitch of 70 μm, 10 μm, and 250 μm, respectively. Then the number of micro-grooves on a copper plate with a size of 20 mm × 10 mm is counted to be 400. When using two path laser-machining, the number of paths reaches 800. When utilizing the picosecond laser with a higher repetitive frequency than 10 MHz [7], the duration time per single path is 1 s, including the on/off operation and beam positioning control. Total takt time per copper plate is 800 s. In the present approach, this takt time is reduced down to 10 s, even including the setting and stamping durations.

Microgroove textures for joining must be tailored enough to have geometric compatibility to the substrate size and chip allocation on the substrate. When using laser machining, more takt time is necessary to prepare for CAM data and to actual machining operations. The present plasma printing has intrinsic flexibility to transcribe the tailored micropattern to the multi-punch array on the die unit for micro-embossing the microgroove textures onto the copper substrate without increasing the takt time in production. Furthermore, this nitrided multi-punch array has sufficient hardness to prolong the die life in practical micro-embossing operations.

5. Conclusions

A plasma printing method with the assistance of low-temperature plasma nitriding was proposed to fabricate the micro-meshing punch array for micro-embossing the micro-groove textures onto copper plates for plastic mold packaging. A two-dimensional micro-meshing pattern with a line width of 50 μm and a pitch of 250 μm on the screen was utilized to print its negative image directly onto the AISI316 substrate's surface. This print worked as a mask to make selective nitrogen super-saturation onto the unprinted areas. A punch array with micro-meshing textures was fabricated by blasting the printed substrate parts. This regular network of meshed punches had a meshing line width of 70 μm, height of 75 μm, and pitch of 250 μm. This array was fixed into a cassette die for micro-embossing. The takt time to fabricate this multi-punch array was significantly reduced from 320 h by precise milling process down to 6 h by the present plasma printing process. High qualification in the design of inks for screen printing is needed to improve the dimensional accuracy of punch heads.

In the micro-embossing process, copper was backwards extruded to flow into the micro-cavity array in the plasma-printed micro-punch. A quadratic prism array 180 μm × 180 μm was formed onto the copper plate. This micro-embossing process, with the use of the plasma-printed micro-punch, was useful for fabricating copper-based devices and sensors with fine alignment of prism pillars.

Author Contributions: T.S. and T.A. made the conceptual design, and planned and executed the series of experiments, with Y.S. and K.W., and wrote this paper.

Funding: This study was financially supported in part by the METI-Program for Supporting Industries in 2018.

Acknowledgments: The authors would like to express their gratitude to S. Kurozumi and H. Morita (Surface Engineering Design Laboratory, llc.), S. Hashimoto (TecDia. Co., Ltd.) and T. Yamaguchi (Sanko-Light Kogyo, Co., Ltd.), for their help in experiments.

References

1. Balden, A. Adhesively bonded joints and repairs in metallic alloy, polymers and composite materials: Adhesives, adhesion theories and surface pretreatment. *J. Mater. Sci.* **2004**, *39*, 1–49. [CrossRef]
2. Satoh, T.; Aizawa, T.; Shiratori, T.; Sugita, Y.; Anzai, M. Micro-joining of multi stainless steel sheets into mechanical element by low temperature diffusion process. *Procedia Manuf.* **2018**, *15*, 1475–1480. [CrossRef]
3. Doughetty, D.; Mahalingam, M.; Viswanathan, V.; Zimmerman, M. Multi-lead organic air-cavity package for high power high frequency RFICs. In Proceedings of the 2009 IEEE MTT-S International Microwave Symposium Digest, Boston, MA, USA, 7–12 June 2009; pp. 473–476.

4.   Saito, Y.; Aizawa, T.; Wasa, K.; Nogami, Y. Leak-proof packaging for GaN chip with controlled thermal spreading and transients. In Proceedings of the 2018 IEEE BiCMOS and Compound Semiconductor Integrated Circuits and Technology Symposium, San Diego, CA, USA, 14–17 October 2018. in press.

5.   Denkena, B.; Koehler, J.; Laestner, J. Efficient machining of micro-dimples for friction reduction. In Proceedings of the 7th ICOMM, Chicago, IL, USA, 12 March 2012; pp. 85–89.

6.   Jiang, Y.; Zhao, W.S.; Kang, X.M.; Gu, L. Adaptive control for micro-hole EDM process with wavelet transform detecting method. In Proceedings of the 6th ICOMM, Tokyo, Japan, 7 September 2011; pp. 207–211.

7.   Aizawa, T.; Inohara, T. *Pico- and Femtosecond Laser Micromachining for Surface Texturing*; IntechOpen: London, UK, 2019.

8.   Aizawa, T.; Yamaguchi, T. Plasma nitriding assisted micro-texturing into martensitic stainless steel molds for injection molding. In Proceedings of the 9th 4M/ICOMM-2015, Milan, Italy, 26 March 2015; pp. 114–118.

9.   Aizawa, T. Micro-texturing onto amorphous carbon materials as a mold die for micro-forming. *Appl. Mech. Mater.* **2014**, *289*, 23–37. [CrossRef]

10.   Aizawa, T.; Wasa, K.; Tamagaki, H. A DLC-punch array to fabricate the micro-textured aluminum sheet for boiling heat transfer control. *Micromechanics* **2018**, *9*, 147. [CrossRef]

11.   Aizawa, T.; Saito, Y.; Yoshihara, S.-I. Micro-embossing into copper plates for mold-packaging into power semiconducting devices. In Proceedings of the 44th Autumnal Meeting of JSTP, Kumamoto, Japan, 27 October 2018; pp. 3–4.

12.   Duriagina, Z. (Ed.) *Low Temperature Plasma Nitriding of Austenitic Stainless Steels*; InTechOpen: London, UK, 2019; Chapter 3; pp. 31–50.

13.   Marcos, G.; Guilet, S.; Clymand, F.; Thiriet, T.; Czerwuec, T. Stainless steel patterning by combination of micro-patterning and driven strain produced by plasma assisted nitriding. *Surf. Coat. Technol.* **2011**, *205*, 5275–5279. [CrossRef]

14.   Aizawa, T.; Yoshino, T. Plastic straining for microstructural refinement in stainless steels by low temperature plasma nitriding. In Proceedings of the 12nd SEATUC Conference, YagYakarta, Indonesia, 8 March 2018; pp. 121–126.

15.   Shiratori, T.; Yoshino, T.; Suzuki, Y.; Nakano, S.; Yang, M. Effects of ion processing of cutting edge of tools on tool wear and punching characteristics in micropunching process. *J. JSTP* **2019**, *60*, 58–63. [CrossRef]

Article

Feedstocks of Aluminum and 316L Stainless Steel Powders for Micro Hot Embossing

Omid Emadinia [1,2,*], Maria Teresa Vieira [2] and Manuel Fernando Vieira [1,3]

1   CEMMPRE, Centre for Mechanical Engineering, Materials and Processes, Department of Metallurgical and Materials Engineering, University of Porto, Rua Dr. Roberto Frias, 4200-465 Porto, Portugal; mvieira@fe.up.pt
2   CEMMPRE, Centre for Mechanical Engineering, Materials and Processes, Department of Mechanical Engineering, University of Coimbra, Rua Luís Reis Santos, 3030-788 Coimbra, Portugal; teresa.vieira@dem.uc.pt
3   INEGI—Institute of Science and Innovation in Mechanical and Industrial Engineering, Rua. Dr. Roberto Frias, 4200-465 Porto, Portugal
*   Correspondence: omid.emadinia@fe.up.pt; Tel.: +35-192-434-2143

Received: 15 October 2018; Accepted: 23 November 2018; Published: 28 November 2018

Abstract: In metal powder, shaping the preparation and characterization of the feedstock is an aspect commonly recognized as fundamental. An optimized composition is required to ensure the successful shaping of the feedstock. In this study, a commercial binder system, pure aluminum and 316L austenitic stainless-steel powders were used for micro hot embossing. The optimization process revealed that powder characteristics such as shape and the stability of the torque mixing, were important parameters. Manipulating the feedstock composition by adding multi-walled carbon nanotubes or stearic acid or using a higher powder concentration considerably influenced the torque mixing values. The steady state of torque mixing was achieved for all feedstocks. This torque behavior indicates a homogeneous feedstock, which was also confirmed by microscopic observations. Nevertheless, an extruding process was required for greater homogeneity of the aluminum feedstocks. The presence of the carbon nanotubes increased the homogeneity of green parts and reduced microcrack formation. The roughness was essentially dependent on the feedstock composition and on the plastic deformation of the elastomer die. Shaping the prepared feedstocks (with or without carbon nanotube) was achieved by the optimized powder concentrations and it did not increase by the stearic acid addition.

Keywords: carbon nanotubes; feedstock; homogeneity; metallic powders; micro hot embossing; shaping

---

1. Introduction

Micro hot embossing is a well-known replicating technique for the production of polymeric micro components through which large scale production of substrates with micropatterns (and even in nanoscales) is feasible, being reported aspect ratios from 0.06 up to 2.00. For industrial applications three manufacturing methods are applied: plate to plate, roll to plate and roll to roll. The shaping is carried out by pressing with a determined force the mould into a substrate at a temperature at which the material (thermoplastic) behaves as a viscous flow [1]. Micro hot embossing is interesting since the application of laboratory machines helps to reduce the interval between product development and production [2]. It has also been adapted as a metal powder processing technique [3], which involves feedstock (powder-binder mixture) preparation, shaping to obtain a so-called green part, debinding and sintering to attain final product. Major differences between micro hot embossing applied to polymers and metal feedstocks include high series production for polymers and low series for metals, in addition, shaping is achieved using a hard mould to promote replication on the surface of the polymeric substrate, and an elastomer die for the metallic feedstock. Related studies reported the shaping of 316L stainless steel, Cu, Fe-Ni, and WC-Co feedstocks [3–7]. Some feedstocks were prepared based on the critical powder volume concentration (CPVC) having a value slightly smaller than CPVC, so-called optimized feedstocks. Determination of this value was reported by using torque mixer equipment and by plotting the torque values of the softened binder and powder mixture against the powder concentrations. The CPVC is the volume at which the slope of the torque growth increases suddenly [8,9].

This study evaluated the Al and AISI 316L feedstock preparation based on the CPVC criteria and verified the optimized compositions by shaping green micro-components. Multi-walled carbon nanotubes (MWCNTs) were also used for some feedstocks, this carbon allotrope was used because it is strong and stiff, widely used for strengthening metal matrices [10]. This being the case, shaping metal matrix nanocomposites through micro hot embossing will be evaluated using large amount of feedstock that needs the dispersion of MWCNTs and metallic powders in large scale, in a similar manner to what was done in micro powder injection moulding (µPIM) [11,12].

A torque rheometer was used for the feedstock preparation. It is expected that the feedstocks will be homogeneous when the torque reaches a steady state [13]. The homogeneity of the feedstocks was also evaluated by scanning electron microscopy (SEM) observations. The torque value represents the energy required for mixing a certain volume of materials under specified processing conditions, and it can be used to evaluate the behaviour of the feedstock during mixing. It can further provide information on viscosity, in terms of powder concentration (the higher the torque value, the more viscous the mixture [13]), or feedstock composition manipulation (the addition of MWCNTs will increase feedstock viscosity due to the high surface area of the nanotubes). The SEM and infinite focus microscopy (IFM) were used to evaluate the shaping of green micro hot embossed parts.

2. Materials and Methods

AISI 316L stainless steel powder (0.012% C, 0.57% Si, 2.40% Mo, 17.10% Cr, 11.10% Ni and 1.17% Mn) and pure Al powders (99.80% purity), a commercial binder material (called M1) composed of polyolefin waxes, and MWCNT were used (Table 1).

Table 1. Suppliers and characteristics of input materials.

Material	Supplier	D_{50} (µm)	Density (kg/m^3)	Specific Surface Area (m^2/kg) [1]	Aspect Ratio
AISI 316L	Sandvik Osprey Ltd [2]	3.5	7937	2120	~1
Al	Alfa Aesar [3]	10.0	2670	695	>1
MWCNT	Fibermax Composites [4]	-	2010	-	
M1	Atect® Corporation [5]	-	970	-	

[1] This value was given along with the results of particle size analyses made by Malvern equipment. We report these values, although they are not as accurate as those obtained by the gas absorption method. [2] Sandvik Osprey Ltd., West Glamorgan, UK. [3] Thermo Fisher (Kandel) GmbH, Karlsruhe, Germany. [4] Fibermax Ltd. Agria Volou, Greece. [5] Atect Corporation, Shiga, Japan.

The density, morphology (Figure 1a,b) and particle size distribution (PSD) of the powders (Figure 1c) were analyzed by Helium pycnometer (Micromeritics AccuPyc 1330, Micromeritics Instrument Corporation, Norcross, GA, USA), SEM, backscattered electron (BSE) and secondary electron (SE) modes, (FEI-Quanta 400 FEG equipment, FEI Company, Hillsboro, OR, USA), and laser diffraction analyzer (Mastersizer 3000, Malvern Instruments Limited, Worcestershire, UK) equipment, respectively. Thermal properties of the binder material (Figure 1d) were analyzed by a thermal analyzer (Setaram SetSys equipment, Setarem Instrumentation, Caluire, France), it was carried out in Argon (99.999% purity) atmosphere at heating-cooling rates of 10 °C/min. Feedstock processing was performed on a Brabender Plastograph (Brabender®GmbH & Co. KG, Duisburg, Germany), with a chamber volume of 42 cm^3 and a torque rheometer with twin Z-blades. A single screw extruder Brabender equipment was used to eliminate the porosities left inside feedstocks prepared by the

Plastograph, if necessary. The Alicona IFM G4 equipment (Alicona Imaging GmbH, Graz, Austria) determined two surface roughness parameters: Sa which is the arithmetical average of the absolute values of the roughness profile, and Sz as the difference between the highest peak and the deepest valley in the analyzed area.

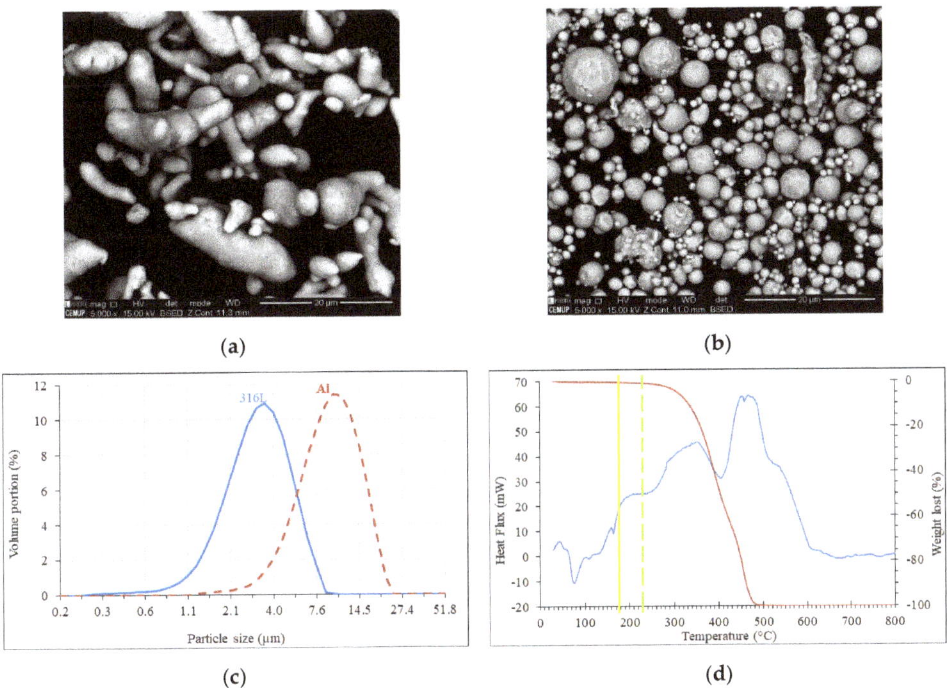

Figure 1. SEM/SE images of (**a**) Al and (**b**) 316L powders; (**c**) particle size distribution (PSD) graphics of the powders; (**d**) differential scanning calorimetry and thermogravimetric analysis graphs of the binder; the solid and dashed lines illustrate the mixing and shaping temperatures, respectively.

For this study, Al and 316L feedstocks, with and without 1 vol.% of MWCNT, were used. The binder and powder were mixed inside the Plastograph chamber. First, the binder (in a proper volume concentration) was melted inside the chamber (at 175 °C for 5 min) and then metal powders (or powder-MWCNT mixtures) were added. What regards the determination of CPVC, this started with the addition of 50 vol.% of powder to the pre-melted binder and continued through the increments of 1 vol.% powder until the chamber became full; the minimum mixing time for each sequence was considered five min (similar to previous study [14]). Since the Plastograph equipment showed the torque value in real time, it was only proceeded for the next batch when the torque value stabilized (aluminum powders mixing time frequently exceeded the previously fixed five min). After selecting the optimized composition (CPVC value subtracted by one) through this method, the feedstock preparation involved the addition of three batches of powders (or metallic powders-MWCNT mixtures) to the pre-melted binder. What regards the time required for feedstock preparation, it was decided to stop mixing when the mixture torque values reached the steady state. The effect of adding stearic acid (SA) as a diluting agent was also analyzed. The SA was mixed with the powder on a Turbula shaker for 30 min before being added to the binder. Pristine MWCNTs were highly entangled (Figure 2a) and were broken apart as much as possible with dispersing conditions that avoid damaging the nanotubes. A one step dispersion method was performed involving simultaneously sonicating (20,400 rpm) the powder and MWCNT in isopropanol for 15 min, draining and drying at

80 °C for 1 h [15]. Al-MWCNT and 316L-MWCNT mixtures (Figure 2b,c) were produced through this method before being added to the Plastograph chamber. Torque mixing was carried out at 30 rpm and 175 °C until stabilization (constant torque value) was achieved. Three batches were prepared for each aluminum feedstock, and two batches for each stainless steel. Extruding the Al feedstocks was also performed at 15 rpm and 175 °C. All aluminum and stainless-steel feedstocks were granulated and sieved before shaping. Granulation was performed by a blender equipment through three sequences of blending and sieving (in a total of one-minute blending), granulated particles were sieved in each sequence by a laboratory test sieve of 500 μm aperture size (Retsch®Haan, Germany).

(a)

(b)

(c)

Figure 2. SEM/SE images of (**a**) pristine MWCNTs, (**b**) Al-MWCNTs and (**c**) 316L-MWCNTs mixtures for feedstock preparation.

The homogeneity of feedstock was evaluated by the stabilization of the torque mixing value and by SEM observations. In SEM analysis, the homogeneity of feedstocks and green parts is evaluated by the distribution of dark regions (representing higher binder concentration).

Feedstocks were shaped into two different green parts: a microblind-flange and microchannel-half-flanges (Figure 3 and Table 2). The shaping facilities included a uniaxial press LIOYD LR 30K equipment (AMETEK (GB) Ltd., West Sussex, UK) assembled with an infrared radiation furnace, it was carried out at 230 °C, 8.5 MPa or 11.3 MPa for 15, 30 or 45 min in air. This shaping temperature ensures maximum replication without severely damaging the elastomer die. Two elastomer materials were used: a dark blue one with 46 ± 1 Shore A hardness (HB FLEX 5550 A + B), and a transparent one with 38 ± 1 Shore A hardness (HB FLEX RTV2 T4 S A + B). These materials were supplied by HB Quimica, LDA (HB Química, Matosinhos, Portugal). The elastomer hardness was measured by Teclock GS-719N equipment (Teclock Corporation, Nagano, Japan) at room conditions.

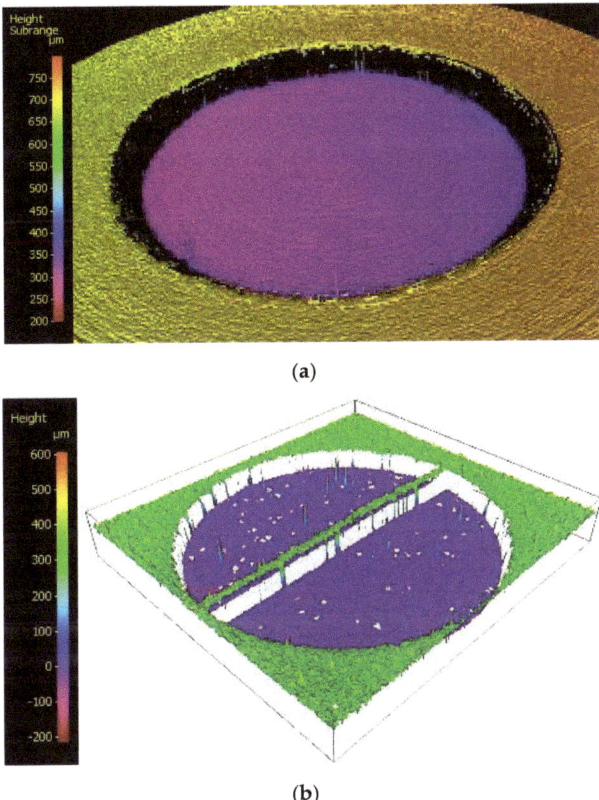

(a)

(b)

Figure 3. Configurations and roughness maps (obtained by infinite focus microscopy (IFM) analyses) of the dark blue elastomer die and the transparent elastomer die used for replicating (**a**) microblind-flange and (**b**) microchannel-half-flanges green parts.

Table 2. Dimensional characteristics (in μm) of the dies illustrated in Figure 3.

Die	Diameter	Height	Maximum Width	Minimum Width	Sa	Sz
Dark blue elastomer	4918	~<400	-	-	2	36
Transparent elastomer	4896	355 ± 9	326 ± 10	204 ± 13	9	680 [6]

[6] This value is affected by the column-like features appeared on the blue surface of the elastomer die in Figure 3b and these should be artifacts.

3. Results and Discussion

3.1. Optimization Process: Effects of Powder Type

According to Figure 4 the graphs demonstrate 59 and 61 vol.% as the CPVC values for the Al and 316L feedstocks because at these points the slope of the torque growth is greater than for the previous points, being this more pronounced for the 316L system. Although the 316L powder has a smaller PSD than the Al, it obtained a larger CPVC. This should be the consequence of having a shape factor very close to one (spherical) unlike the Al powder which has a higher factor (see Figure 1). This is consistent with a mostly accepted statement concerning the effect of powder shape on viscosity [16] (p. 73). Another study further showed that a powder size reduction (by half) did not influence the CPVC [17].

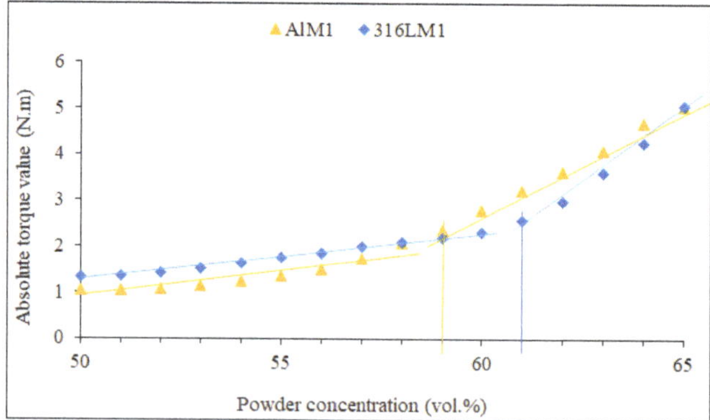

Figure 4. Determination of critical powder volume concentration (CPVC) through the optimization process (common approach): torque values function of powder increments (Al and 316L) to the M1 binder, using Z-blade mixer at 30 rpm, 175 °C and the atmospheric pressure (101 kPa). Yellow and blue intersects represent CPVC values.

Regarding the torque mixing value as the function of powder concentration, the graph illustrated in Figure 4 presents two regions: in the first region, below 59 vol.% powder concentration, the torque values of the 316L mixture are higher than those of the Al, while in the second one this order is inverted. In the first region, the difference can be attributed to the particle size, i.e., the finer powder (316L) has higher viscosity [16] (p. 73). This effect was expected to be pronounced for higher powder concentrations due to the increase in the friction caused by powder interactions [18]. However, this difference is practically constant up to a powder concentration close to 59%, after which the Al torque value grows faster with the powder concentration and exceeds the 316L value. This behavior can be attributed to the influence of particle shape at larger powder contents, this being the case, irregular shaped particles will cause viscosity increase due to presenting larger friction [16] (p. 73). Moreover, the experiment time for the determination of CPVC was longer for Al (120 min) than for 316L (80 min). This determination process was monitored visually in real time, and it was not needed to extend the 5 min mixing time for the incremental sequences of the 316L. This difference could be attributed to a shorter stabilization time required for the 316L powders. Although the 316L powders have higher surface area, meaning the 316L powders need larger mixing time, to interact with binder and stabilization, than the Al powders, the particle shape should have diminished the particle size effect [18]. Thus, it is expected that the 316L feedstock can even be homogenized faster than that of the Al.

From the CPVC points onwards, the torque increase rate of the 316L-binder system exceeded that of the Al-binder. This can be attributed to the particle interactions at larger powder concentrations, being affected by the powder's characteristics.

The difference between the CPVC value of the 316L in this study and related studies, using similar PSD, which reported values close to 66% [7,14,17,19], can be attributed to the use of different binder type or processing conditions (mixing temperature or speed). The influence of these differences is seen in a related study in which a CPVC of 61% was reported for a 316L stainless-steel powder [20].

For the Al powder the determination of the CPVC was not as easy as that of the 316L. In fact, for this material the determination of a point at which a sudden torque increase occurred was difficult. To confirm the CPVC values based on Plastograph approach (Figure 4), a complementary study was performed. Figure 5 illustrates the absolute torque increases (*Y* axes) for the continuous increments of 1 vol.% of metal powder (*X* axes). Each square highlights two increments for which the increase in torque is evident for the intermediate concentration. In relation to the Al system, the increase in torque shows some fluctuations thus, two orange squares are indicated as the zones where the possible values of CPVC are detected (59 and 63 vol.%). For stainless steel, the CPVC value is 61 vol.%. What regards the optimized compositions, that are smaller than the corresponding CPVCs, the optimum powder volume concentration of Al should be validated through the production of green parts by feedstocks with 58 and 62 vol.% of powder concentration. Regarding the 316L powder, feedstock with 60 vol.% powder can guarantee the replicability of green part, this volume is also consistent with previous study [21].

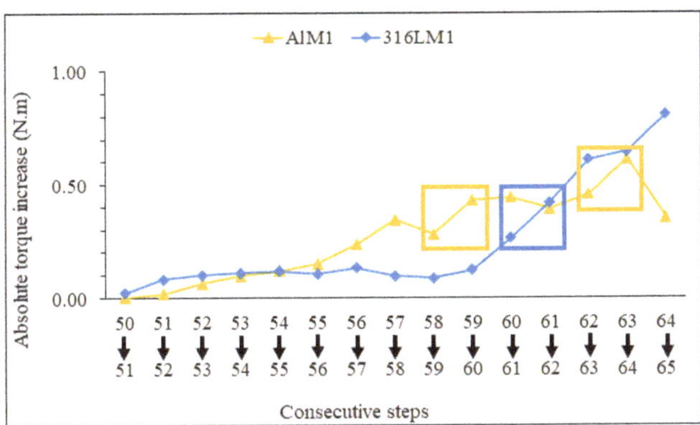

Figure 5. Determination of CPVC through the optimization process (complementary evaluation): torque increases between every two consecutive steps versus the corresponding powder addition steps.

3.2. Feedstocks: Effect of Composition on Torque Values

Table 3 presents the acronym and some characteristics of the Al and 316L feedstocks. The major result concerns the effect of increasing powder concentration or adding MWCNT onto the increase in the feedstocks' torque values, which is consistent with other studies [22–24]. As expected, the increase of powder concentration lead to higher torque value [25]; the feedstocks with 65 vol.% of powders presented the maximum torque values.

Table 3. Acronyms, compositions and mean torque values required for mixing of feedstocks. The mean values and the standard deviations were measured from the stabilized regions (after 12 min of time) of the torque mixing curves obtained by the feedstock preparation.

Acronym	Constituents	Concentration (vol.%)	Torque (Nm)	Standard Deviation
Al58M1	Al & M1	58:42	2.66	0.06
Al-MWCNT58M1	Al, CNT & M1	58(1):42	3.92	0.04
Al-MWCNT58M1SA5	Al, CNT, M1 & SA	58(1):42(5)	1.97	0.07
Al62M1	Al & M1	62:38	3.63	0.08
Al62M1SA1.6	Al, M1 & SA	62:38(1.6)	2.00	0.05
Al62M1SA5	Al, M1 & SA	62:38(5)	1.58	0.04
Al65M1	Al & M1	65:35	5.05	0.13
316L60M1	316L & M1	60:40	2.76	0.05
316L-MWCNT60M1	316L, CNT & M1	60(1):40	3.79	0.07
316L65M1	316L & M1	65:35	5.06	0.09

For the Al-binder system the effect of 1 vol.% MWCNT is more pronounced than the increase in the powder content by 4 vol.%. The addition of SA decreased the viscosity of the feedstocks. However, the increase in the SA from 1.6% to 5.0% did not result in a strong viscosity reduction, which is consistent with related studies [21,26,27]. According to the torque mixing values (Table 3), it is expected that the feedstocks with smaller values will lead to a better shaping. Table 3 also shows no significant differences between the standard deviations, except for Al65M1. The microstructural analyses of feedstocks, prepared by the torque mixer, show the presence of large porosities when

SA was added to the feedstock (Figure 6). This indicates that the porosity may assist the reduction in viscosity.

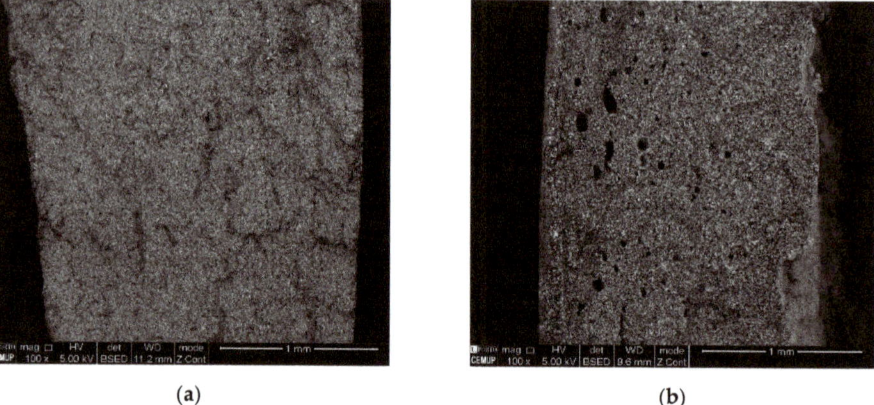

(a)   (b)

Figure 6. SEM/backscattered electron (BSE) images of (**a**) Al62M1 and (**b**) Al62M1SA1.6 feedstocks after torque mixing.

3.3. Feedstocks, Torque Stability and Homogeneity

Figure 7a illustrates the torque variations of Al58M1, Al-MWCNT58M1 and Al-WCNT58M1SA5 feedstocks. The unstable initial period belongs to Al-MWCNT58M1 and the great fluctuation could be attributable to the dispersion and interaction of the MWCNT with the binder during the earliest mixing period. The addition of SA delayed the stabilization stage, however, all Al feedstocks reached a nearly constant torque value, usually called a steady state, after 17 min which will be considered as a reference mixing time in this study.

Microstructural observations from Al58M1 feedstock prepared by Plastograph after mixing (Figure 8a) highlighted the need for further mixing, which was performed by a single screw extruder and led to better homogeneity and reduction of porosities (Figure 8b). Subsequent microstructural analysis confirmed no differences in the dispersion of the constituents of feedstocks. Although some authors reported a good wettability of CNTs in some polymeric materials feedstocks, such as polypropylene and polyethylene glycol [28], in the present study the MWCNT agglomerations still existed after torque mixing and extrusion, with or without SA (Figure 7b,c), due to the different polymeric materials constituting the binder.

The torque changes during the mixing of the 316L60M1 and 316L-MWCNT60M1 feedstocks are shown in Figure 9a. The addition of the MWCNT increased torque values and delayed torque stabilization. This last difference can be attributed to the untangling of MWCNTs during torque mixing. The untangling is not complete after 25 min, since the microstructure revealed the presence of nanotube clusters (Figure 9b). The steady state was achieved after mixing for 17 min and it was maintained (which is more pronounced for 316L60M1 than 316L-MWCNT60M1) for an additional eight more min (25 min in total) to improve homogenization. Microscopy observations confirmed that the 316L feedstocks made by the torque mixer were homogeneous and did not need any extruding treatment. The comparison of torque values for Al and 316L, with and without MWCNTs, shows a similarity in values between them (Figure 9c), which highlights the effect of using the same binder system for all feedstocks [29].

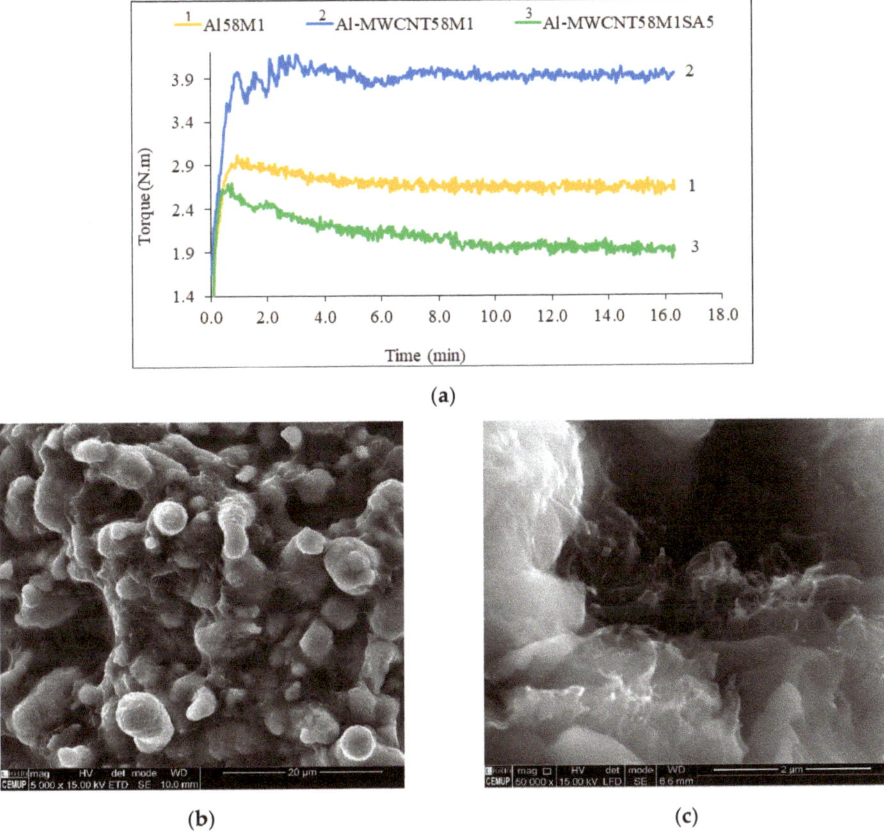

(a)

(b)   (c)

Figure 7. (a) Torque variation during mixing Al feedstocks (Al58M1, Al-MWCNT58M1 and Al-MWCNT58M1SA5) during mixing; SEM/BSE images showing (b) the distribution of the powder-binder-MWCNTs in Al-MWCNT58M1 and (c) the presence of the MWCNT aggregations in the binder of same feedstock.

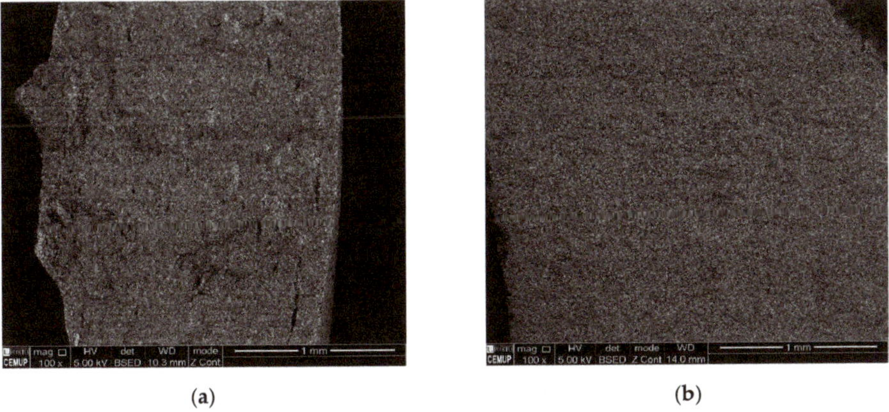

(a)   (b)

Figure 8. Microscopy observations (SEM/SE) from AL58M1 after preparation by (a) Plastograph and (b) extrusion for porosity elimination.

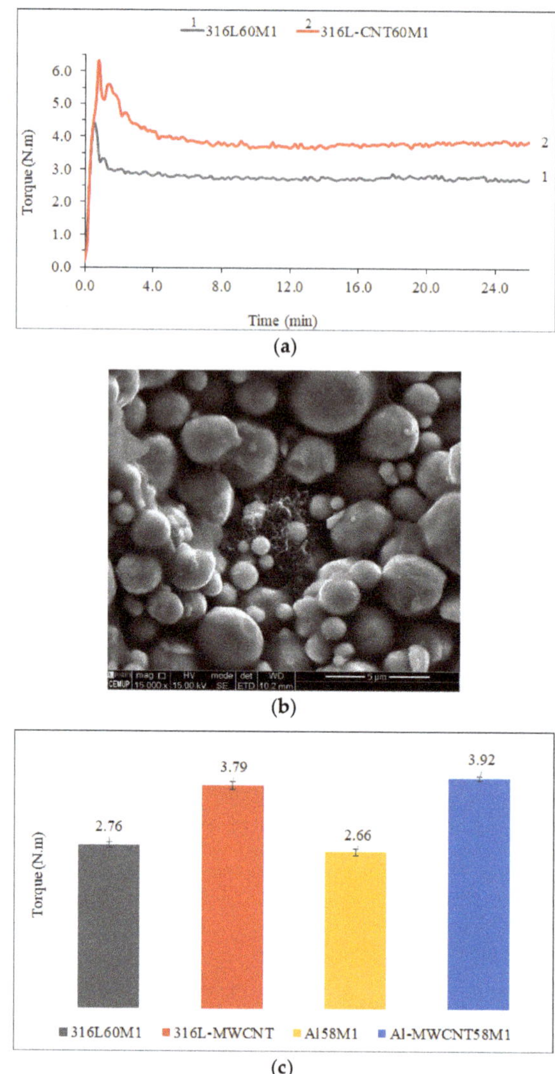

Figure 9. (**a**) Torque variation during mixing 316L and 316L-MWCNT feedstocks; (**b**) SEM/SE image of the 316L-MWCNT60M1 feedstock; (**c**) comparison of the torque values of the optimized Al and 316L systems.

3.4. Shaping Feedstocks

Microscopy observations of green microblind-flange revealed that the sharp edge and concentric rings were replicated by the Al58M1, with and without MWCNT (Figure 10a,b), though the Al-MWCNT58M1 feedstock required a larger torque mixing value, these rings were not replicated by Al62M1 feedstock and this found a strong waviness surface (Figure 10c). These waviness surfaces represent non-replicated feedstock particulates. Therefore, the metallic powder concentration of 62 vol.% is not suitable for micro-hot embossing. Although the addition of SA resulted in a decrease in viscosity (Table 3), the shaping did not increase and the waviness structure were appeared on the surfaces of Al-MWCNT58M1SA5, Al62M1SA1.6 and Al62M1SA parts (Figure 10d–f). Therefore, in this microfabrication process the powder concentration, rather than the viscosity, determines the success of the replication.

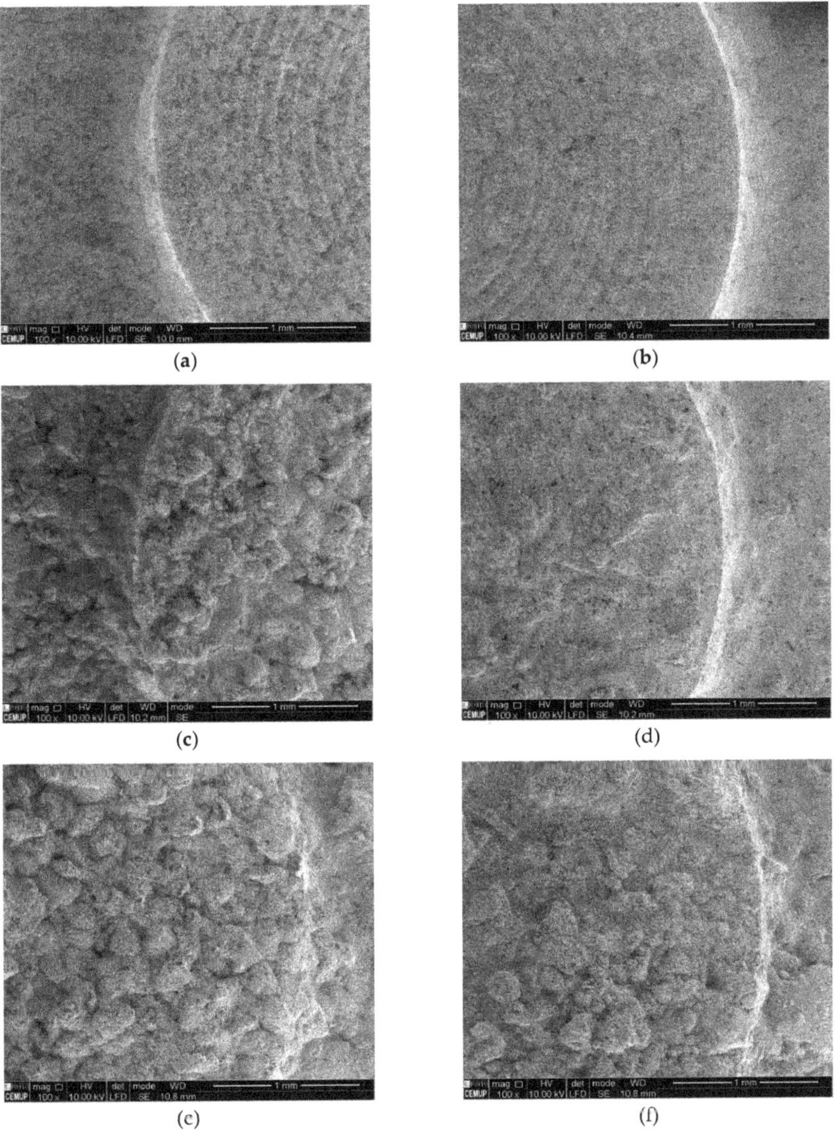

Figure 10. SEM/SE images of green microblind-flange parts made of (**a**) Al58M1, (**b**) Al-MWCNT58M1 illustrating the replication of concentric ring structure, (**c**) Al62M1, (**d**) Al-MWCNT58M1SA5, (**e**) Al62M1SA1.6, and (**f**) Al62M1SA5 without the replicability of the rings (shaped at 230 °C, 8.5 MPa for 30 min by dark blue elastomer die).

Microstructural analysis showed that there are more dark regions in Al58M1 green parts (Figure 11a) than in Al-MWCNT58M1 ones (Figure 11b,c), i.e., the homogeneity of green parts containing MWCNT increased. Meanwhile, the microcrack formation (illustrated in Figure 11a by a white arrow inset) in the green parts after demoulding diminished. This improvement could be attributable to the nanoreinforced binder material that reduced the demolding stresses, i.e., the strengthening effect of the nanotubes in the polymeric materials [30]. A longer holding time also helped in homogenizing the green parts (Figure 10c). The problems associated with the SA addition (Figure 10d–f) could be due to the separation of the powder-binder at very low viscosities [16] and this was more pronounced in the presence of MWCNT (Figure 10d) rather than for higher powder content. With regard to shaping and homogeneity of 316L green parts, it was obtained and also increased with MWCNT addition (Figure 11e,f).

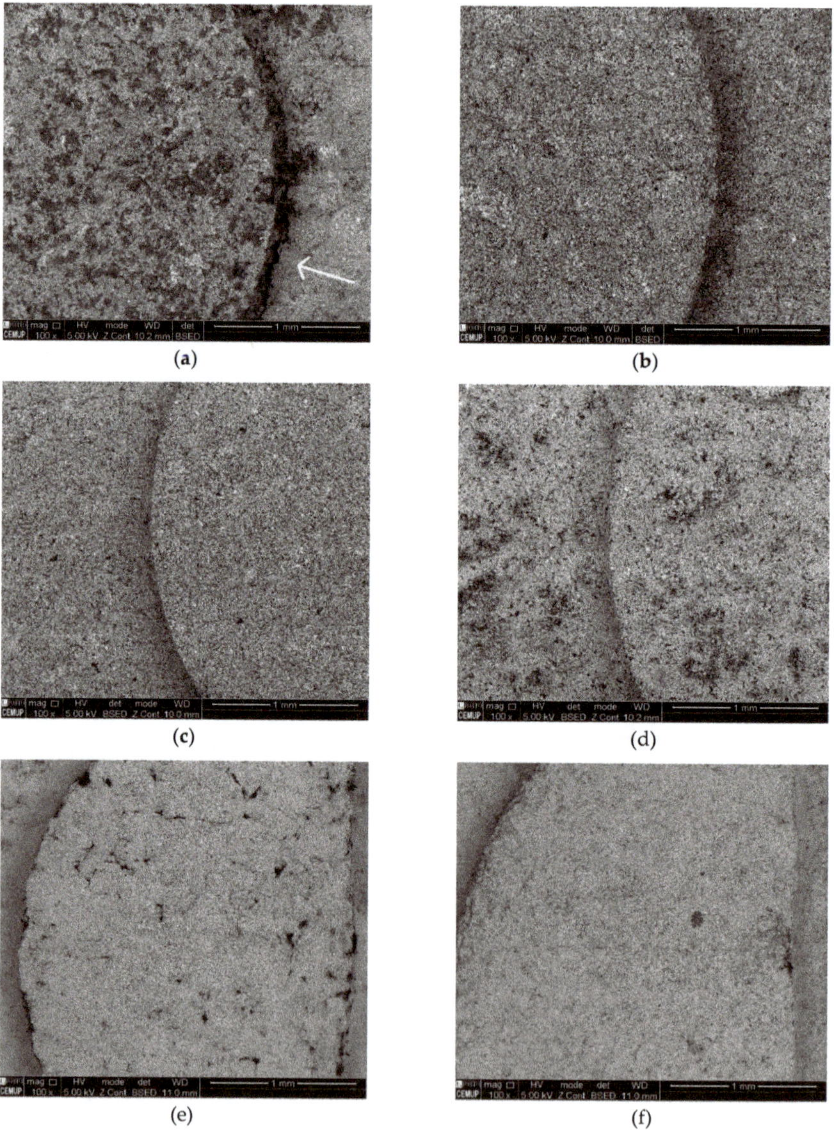

Figure 11. SEM/BSE images showing the distribution of powder-binder in green microblind-flange parts made of: (**a**) Al58M1 and (**b**) Al-MWCNT58M1, shaped at 230 °C, 11.3 MPa for 15 min; (**c**) Al-MWCNT58M1 and (**d**) Al-MWCNT58MSA5, shaped at 230 °C, 11.3 MPa for 45 min; (**e**) 316L60M1 and (**f**) 316L-MWCNT60M1, shaped at 230 °C, 11.3 MPa for 30 min by transparent elastomer.

The IFM analyses (Figures 3 and 12) revealed that the surface roughness parameters (Sa and Sz) of the green parts are larger than those of their dies (Table 4). Dimensional changes are attributed to the plastic deformation of the elastomer die during the shaping process and/or the feedstock shrinkage after cooling. Moreover, this is seen that the roughness of the green parts (Figure 12b,c) was also influenced by the number of times the elastomer die was used, (Figures 3 and 12) this change either has happened due to the die erosion or other changes occurred in the elastomer die. The reason for the latter was confirmed by an increase in its hardness (from 38 ± 1 Shore A in the pristine condition to 40 ± 1 Shore A after shaping), being affected by temperature and compaction during shaping process. This being the case that, the elastomer die has become stiffer than its pristine state leading to new dimensions and roughness (Table 4). This indicates that micro hot embossing process with elastomer die is suitable only for very small series production.

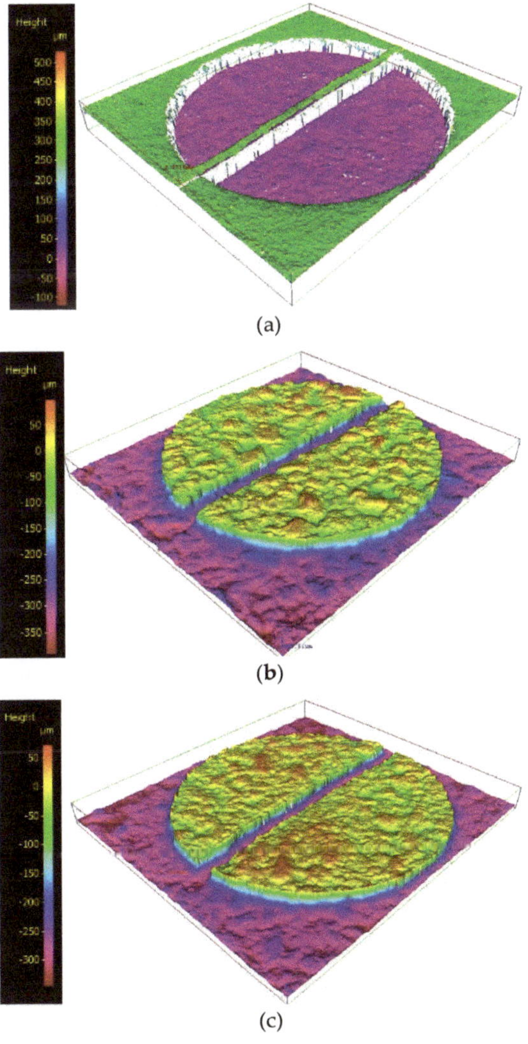

Figure 12. Roughness maps obtained by IFM analyses: (**a**) transparent elastomer die used twice for replicating green microchannel-half-flanges parts, (**b**) green part shaped by pristine die and (**c**) by the used die (Al-MWCNT58M1 parts shaped at 11.3 MPa and 230 °C for 30 min).

Table 4. Dimensional characteristics (in μm) of the used die and green parts illustrated in Figure 12.

Part	Diameter	Height	Maximum Width	Minimum Width	Sa	Sz
Used elastomer die in Figure 12a	4870	324 ± 10	347 ± 21	248 ± 12	6	106
Green part in Figure 12b	4856	266 ± 6	403 ± 14	246 ± 12	17	161
Green part in Figure 12c	4843	271 ± 54	362 ± 33	283 ± 27	13	107

Regarding this study, the authors ensures the reproducibility of these results indicating that green Al and 316L (with and without MWCNT) parts without defects were produced by optimized feedstocks through micro hot embossing, demonstrating that the presence of MWCNT led to a better homogeneity. Moreover, dimensional and roughness changes of green parts in respect to their elastomer dies are characteristics of micro hot embossing. The future work of this study will involve debinding (in a controlled atmosphere) and sintering (in low pressure atmosphere) of these green parts, and evaluating the densification, topography and strengthening.

4. Conclusions

Feedstocks based on Al and 316L powders, with and without MWCNT, were prepared. The optimized powder concentrations were based on their CPVC values. Although the Al and 316L powders have quite different powder characteristics, these powders presented close CPVC values. The addition of 1 vol.% of MWCNT increased the viscosity of the feedstocks, while the SA addition reduced it. The control of the torque variations during mixing showed that the different compositions achieved a steady state during mixing for at least 17 min. Microstructural analysis confirmed the apparent homogeneity of the feedstocks; however, MWCNT clusters in the binder were observed. Prolonging the torque mixing time for the 316L stainless steel feedstock led to a homogeneous dispersion and it did not require the extrusion process. In this study, the optimized Al and 316L feedstocks with 1 vol.% MWCNT required almost equal torque mixing values, and the feedstocks with 65 vol.% powder concentrations as well. Producing green parts confirmed that shaping was strongly affected by the powder concentration while the viscosity reduction in the Al feedstocks, by stearic acid, was ineffective. The addition of 1 vol.% MWCNTs improved shaping, increased the homogeneity of the green parts and the microcrack formation was eliminated without considerable topographical changes. The use of elastomer dies in the micro hot embossing process caused changes in roughness of the green part which increased according to the number of times the process was repeated.

Author Contributions: O.E. produced and characterized the feedstocks and green parts, and hardness measurement; O.E., M.T.V. and M.F.V. discussed the graphics, microstructures, topography images and hardness values. All the authors participated in the design of the experiments and cooperated in writing this paper.

Funding: This study was supported by the Fundação para a Ciência e a Tecnologia, Portugal, through PD-BD-52674 2014 grant, and also the Add.Additive project (POCI-01-0247-FEDER-024533).

Acknowledgments: The authors would like thank Centro de Materiais da Universidade do Porto for SEM assistance and Instituto Pedro Nunes for IFM and hardness measurement.

References

1. Peng, L.F.; Deng, Y.J.; Yi, P.Y.; Lai, X.M. Micro hot embossing of thermoplastic polymers: A review. *J. Micromech. Microeng.* **2014**, *24*. [CrossRef]

2. Heckele, M.; Bacher, W.; Muller, K.D. Hot embossing—The molding technique for plastic microstructures. *Microsyst. Technol.* **1998**, *4*, 122–124. [CrossRef]

3. Fu, G.; Tor, S.; Loh, N.; Hardt, D. Micro-hot-embossing of 316L stainless steel micro-structures. *Appl.Phys. A: Mater. Sci. Process.* **2009**, *97*, 925–931. [CrossRef]

4. Sequeiros, E.W.; Neto, V.C.; Vieira, M.T.; Vieira, M.F. Hot micro-embossing: Effect of pressure on 316L metal parts. *Powder Metall.* **2014**, *57*, 241–244. [CrossRef]

5. Sahli, M.; Millot, C.; Gelin, J.C.; Barrière, T. The manufacturing and replication of microfluidic mould inserts by the hot embossing process. *J. Mater. Process. Technol.* **2013**, *213*, 913–925. [CrossRef]

6. Sahli, M.; Millot, C.; Gelin, J.C.; Barrière, T. Development and characterization of polymers-metallic hot embossing process for manufacturing metallic micro-parts. *AIP Conf. Proc.* **2011**, *1315*, 677–682. [CrossRef]

7. Sahli, M.; Gelin, J.C.; Barriere, T. Characterisation and replication of metallic micro-fluidic devices using three different powders processed by hot embossing. *Powder Technol.* **2013**, *246*, 284–302. [CrossRef]

8. Barreiros, F.M.; Vieira, M.T. PIM of non-conventional particles. *Ceram. Int.* **2006**, *32*, 297–302. [CrossRef]

9. Hausnerova, B.; Mukund, B.N.; Sanetrnik, D. Rheological properties of gas and water atomized 17-4PH stainless steel MIM feedstocks: Effect of powder shape and size. *Powder Technol.* **2017**, *312*, 152–158. [CrossRef]

10. Bakshi, S.R.; Lahiri, D.; Agarwal, A. Carbon nanotube reinforced metal matrix composites—A review. *Int. Mater. Rev.* **2010**, *55*, 41–64. [CrossRef]

11. Ye, H.Z.; Liu, X.Y.; Hong, H.P. Fabrication of metal matrix composites by metal injection molding—A review. *J. Mater. Process. Technol.* **2008**, *200*, 12–24. [CrossRef]

12. Romero, A.; Herranz, G. Development of feedstocks based on steel matrix composites for metal injection moulding. *Powder Technol.* **2017**, *308*, 472–478. [CrossRef]

13. Supati, R.; Loh, N.; Khor, K.; Tor, S. Mixing and characterization of feedstock for powder injection molding. *Materials Letters* **2000**, *46*, 109–114. [CrossRef]

14. Barreiros, F.M.; Vieira, M.T.; Castanho, J.M. Fine tuning injection feedstock by nano coating SS powder. *Met. Powder Rep.* **2009**, *64*, 18–21. [CrossRef]

15. Emadinia, O.; Vieira, M.F.; Vieira, M.T. Effect of Reinforcement Type and Dispersion on the Hardening of Sintered Pure Aluminium. *Metals* **2018**, *8*, 786. [CrossRef]

16. Enneti, R.K.; Onbattuvelli, V.P.; Atre, S.V. Powder binder formulation and compound manufacture in metal injection molding (MIM). In *Handbook of Metal Injection Molding*; Woodhead Publishing: Cambridge, UK, 2012; pp. 64–92.

17. Sequeiros, E.W. Microfabricação de Componentes Metálicos por Microgravação. PhD Thesis, University of Porto, Porto, Portugal, July 2014.

18. Sotomayor, M.; Várez, A.; Levenfeld, B. Influence of powder particle size distribution on rheological properties of 316L powder injection moulding feedstocks. *Powder Technol.* **2010**, *200*, 30–36. [CrossRef]

19. Kong, X.; Barriere, T.; Gelin, J. Determination of critical and optimal powder loadings for 316L fine stainless steel feedstocks for micro-powder injection molding. *J. Mater. Process. Technol.* **2012**, *212*, 2173–2182. [CrossRef]

20. Park, S.J.; Kim, D.; Lin, D.; Park, S.J.; Ahn, S. Rheological Characterization of Powder Mixture Including a Space Holder and Its Application to Metal Injection Molding. *Metals* **2017**, *7*, 120. [CrossRef]

21. Ferreira, T.J.; Vieira, M.T. Optimization of MWCNT—Metal Matrix Composites feedstocks. *Cienc. Tecnol. Mater.* **2017**, *29*, e87–e91. [CrossRef]

22. Ferreira, T.J.; Vieira, M.T. Behaviour of feedstocks reinforced with nanotubes for micromanufacturing. In Proceedings of the Euro PM 2014 Congress and Exhibition, Salzburg, Austria, 21–24 September 2014; Available online: https://www.epma.com/publications/euro-pm-proceedings/product/euro-pm2014-powder-injection-moudling-modelling-and-processing (accessed on 27 November 2018).

23. Ahmad, F.; Muhsan, A.S.; Raza, M.R. Rheological behavior of carbon nanotubes/copper feedstocks for metal injection molding. *Adv. Mater. Res.* **2012**, *403–408*, 5335–5340. [CrossRef]

24. Liang, S.; Tang, Y.; Zhang, Y.; Zhong, J. The rheologyical effect of carbon nanotubes on the iron based metal powder injection molding feedstock. *Adv. Mater. Res.* **2009**, *79–82*, 469–472. [CrossRef]

25. Gonzalez-Gutierrez, J.; Duretek, I.; Kukla, C.; Poljsak, A.; Bek, M.; Emri, I.; Holzer, C. Models to Predict the Viscosity of Metal Injection Molding Feedstock Materials as Function of Their Formulation. *Metals* **2016**, *6*, 129. [CrossRef]

26. Li, Y.M.; Liu, X.Q.; Luo, F.H.; Yue, J.L. Effects of surfactant on properties of MIM feedstock. *Trans. Nonferr. Met. Soc. China* **2007**, *17*, 1–8. [CrossRef]

27. Sahli, M.; Gelin, J.C. Development of a feedstock formulation based on polypropylene for micro-powder soft embossing process of 316L stainless steel micro-channel part. *Int. J. Adv. Manuf. Technol.* **2013**, *69*, 2139–2148. [CrossRef]

28. Nuriel, S.; Liu, L.; Barber, A.H.; Wagner, H.D. Direct measurement of multiwall nanotube surface tension. *Chem. Phys. Lett.* **2005**, *404*, 263–266. [CrossRef]

29. Ahn, S.; Park, S.J.; Lee, S.; Atre, S.V.; German, R.M. Effect of powders and binders on material properties and molding parameters in iron and stainless steel powder injection molding process. *Powder Technol.* **2009**, *193*, 162–169. [CrossRef]

30. Spitalsky, Z.; Tasis, D.; Papagelis, K.; Galiotis, C. Carbon nanotube-polymer composites: Chemistry, processing, mechanical and electrical properties. *Progress Polym. Sci.* **2010**, *35*, 357–401. [CrossRef]

Article

Experimental and Numerical Investigations of a Novel Laser Impact Liquid Flexible Microforming Process

Fei Liu, Huixia Liu *, Chenkun Jiang, Youjuan Ma and Xiao Wang

School of Mechanical Engineering, Jiangsu University, Zhenjiang 212013, China; 18452480021@163.com (F.L.); 18260623466@163.com (C.J.); myj@ujs.edu.cn (Y.M.); wx@ujs.edu.cn (X.W.)
* Correspondence: lhx@ujs.edu.cn; Tel.: +86-0511-88797998; Fax: +86-0511-88780276

Received: 26 June 2018; Accepted: 27 July 2018; Published: 1 August 2018

Abstract: A novel high strain rate microforming technique, laser impact liquid flexible embossing (LILFE), which uses laser induced shock waves as an energy source, and liquid as a force transmission medium, is proposed by this paper in order to emboss three-dimensional large area micro arrays on metallic foils and to overcome some of the defects of laser direct shock microembossing technology. The influences of laser energy and workpiece thickness on the deformation characteristics of the pure copper foils with the LILFE process were investigated through experiments and numerical simulation. A finite element model was built to further understand the typical stages of deformation, and the results of the numerical simulation are consistent with those achieved from the experiments. The experimental and simulation results show that the forming accuracy and depth of the embossed parts increases with the increase in laser energy and decrease in workpiece thickness. The thickness thinning rate of the embossed parts increases with the decrease of the workpiece thickness, and the severest thickness thinning occurs at the bar corner region. The experimental results also show that the LILFE process can protect the workpiece surface from being ablated and damaged, and can ensure the surface quality of the formed parts. Besides, the numerical simulation studies reveal the plastic strain distribution of embossed microfeatures under different laser energy.

Keywords: laser impact liquid flexible embossing; microforming; 3-D large area micro arrays; liquid shock wave; high strain rate forming; numerical simulation

---

1. Introduction

With the continuously increasing applications of the parts with microscale structures in many fields, including the chemical field [1], medical devices [2], microelectronics [3], and so on, the demand for developing the micro manufacture technologies has rapidly grown in the past few decades.

The current technologies that are used to machine the three-dimensional microstructures of the MEMS (Micro-Electro-Mechanical System) components, such as electrochemical micromachining (ECMM), show a good application prospect. ECMM provides new abilities for modern micromachining technologies because of its advantages, such as the capability of machining complicated three-dimensional structures, short machining time, and better processing precision [4]. However, some properties of the parts fabricated by ECMM should be improved by controlling and optimizing the various parameters of the ECMM process, including the surface quality and material removal. The lithography (LIGA) based technology is also one of the promising technologies to fabricate the microscale structure [5]. However, most lithography (LIGA) based technologies are mainly used to machine quasi three-dimensional shapes without irregular and complex curvilinear three-dimensional (3-D) structures or high aspect ratio micro components. The increasing complex structures of the micro parts in the MEMS field, and so forth, have posed a challenge to lithography

(LIGA) based technologies. Meanwhile, these technologies are time consuming and expensive. Therefore, a new 3-D micromachining technique must be developed in order to produce micro parts with complex 3-D curved surfaces at high efficiency, high quality, and low cost.

Microembossing, developed from macro forming technology, shows promise for fabricating 3-D microscale structures on materials with low cost and less material loss. However, due to the decreasing size of the die and the formed parts, the macroembossing mechanism cannot be treated the same as the micro-formed part, and some of the process parameters and the like cannot be reduced proportionally to be used in the microembossing [6]. Meanwhile, when the conventional microembossing is applied for metals, this technology will have some limitations, such as the poor formability of metals, because of the size effect [7]; difficulty in the alignments of the micro-die and punch; life length of micro-die and punch, because of the friction and wear; and the difficulty of micro punch manufacture. As a result, high speed forming techniques such as explosive forming [8], electrohydraulic forming [9], and electromagnetic forming [10] have been developed to be used for embossing. In recent years, a novel micro high speed forming technique, laser shock microforming based on micro-die, for microembossing, has aroused widespread attention. Zhou et al. [11] proposed the laser micro forming process to emboss large area three-dimensional microstructures on aluminium foils by single laser pulse. The results showed that the replications cost is unrelated with the microfeatures intricacy, and this technology is a low-cost, high-efficiency microforming process with wide industrial application prospect. Gao and Cheng [12] verified successfully the feasibility of microscale laser dynamic forming (μLDF) in machining three-dimensional microstructures of a high aspect ratio on polycrystalline aluminium thin films with different micro-moulds, through a series of experiments. Gao et al. [13] conducted a range of numerical simulations to research the influences of the fillet ratio, aspect ratio of the mould cavity, and laser processing parameters on the deformation characteristic of copper thin foils during the μLDF process. Li et al. [14] researched forming limits and the fracture behaviour of aluminium foils during the μLDF process through experiments and numerical simulation. They revealed that the formability of aluminium foils is much higher in μLDF than that in other forming processes through forming limit diagrams because of the ultrahigh strain rate of the μLDF process. Ye and Cheng [15] fabricated patterned microstructures on the NiTi (nickel-titanium) shape memory alloy surface using mask material with various sizes through laser shock assisted direct imprinting. They thought that this process had a good application potential to machine complex microstructures on functional materials and instruments. Ehrhardt et al. [16,17] investigated the micron and submicron embossing of the metal substrates by the laser embossing process. The micro- and submicro-meter structures can be embossed into the surfaces of solid copper foils using a laser microembossing process based on laser scanning. They also investigated the effects of laser fluence and laser pulse number on the heights of the replicated structures. Liu et al. [18] successfully fabricated circularity microchannels on a metallic foil surface using laser shock embossing technology. They studied the laser shock embossing process through simulation and experiments. They demonstrated that complex 3-D microstructures can be fabricated directly by this technique with only single pulse, and that the shape of the embossed microchannel can be properly achieved in the numerical simulation with the verified finite element model. Shen et al. [19] revealed that micro rectangular array features can be formed on the metal foil surface through overlapping the laser shock embossing, which showed the feasibility of replicating a large area of 3-D array features on metal foils using overlapping laser shock embossing. There are a lot of advantages of laser microembossing (LME) compared to the conventional microembossing technologies. Firstly, it can be widely used for various materials, is not limited to superplastic materials and amorphous [12]. Secondly, the formability of the LME technology with the ultrahigh strain rate, around 10^6–10^7 s^{-1} [13], is much better than the conventional microembossing process with a low strain rate. The necking or even failure of the workpiece can be delayed because of the inertia effects during the high strain rate forming process [20]. Thirdly, LMF can reduce the experimental cost, because it only requires one side of the rigid tool to form the foils. Of course, this process can also solve the problem of punch-die alignment, because the diameter and position of the laser spot can be controlled and adjusted easily by a laser system and three-dimensional working platform. Finally, the laser energy is more precisely controllable and more localized, so the LMF can easily realize 3-D forming for parts with complex shapes at the meso-, micro-, and nano-scales.

However, the present LME technique still has several problems. Firstly, if the laser radiates directly on the workpiece surface with high energy, the surface of the workpiece would be melted or ablated, thereby destroying the surface quality of the workpiece and causing defects in the microsystems. However, if the ablative medium is applied on the workpiece surface to improve the absorption rate of the laser and to protect the workpiece surface from being damaged, the remaining ablative medium may adhere to the surface of the workpiece and is difficult to remove, which will not only need the cleaning process, extra cost, and extra working time, but will also have an adverse influence on the use of the workpiece. Secondly, the laser induced shock wave directly acting on the workpiece may easily lead to an uneven load on the samples, and the local stress concentration will cause the fracture and failure of the workpiece. In order to solve these problems, it is necessary to introduce the force transmission medium during the LME process, which not only inherits the advantages of the LME technique, but also protects the workpiece surface from being damaged, and makes the workpiece bear pressure evenly so as to improve the quality of the embossed workpiece.

Liquid impact forming technology is an advanced manufacturing technology with a high strain rate, which uses a liquid shock wave induced from different energy sources to achieve an elastic-plastic forming of parts. The methods of using liquid as the force transmission medium to achieve the high-speed impact forming are mainly divided into explosive forming, electrohydraulic forming, and liquid impact forming with a high-speed punch. These energy sources are mainly achieved from the chemical energy of the explosive, the electrical energy of the electric capacity, and the kinetic energy of a high-speed punch or hammer. Regarding the explosive forming, Samardzic et al. [21,22] investigated the application of high-speed liquid projectiles that were accelerated by the powder explosion in a launcher (a water cannon) for parts forming. They constructed an experimental setup for water projectiles generation and a series of experiments, involving the forming of grooves; stamping; extrusion; and punching of the macro-, meso-, and micro-scale, on the metal surfaces were performed. The results showed that the application of the liquid impact for the metal forming is feasible and effective, and that the shapes of the formed parts have agreement with the geometry of the dies. Meanwhile, they found that the liquid impact has no adverse influence on the surface quality of the formed parts, or that it can even can improve the topography of them. However, using powder explosion is not better for control, and is dangerous in contrast to using a high-speed punch or electrical energy as an energy source. Regarding electrohydraulic forming, Golovashchenko et al. [9] carried out the formability testing of dual phase steels during the electrohydraulic forming (EHF) process, and put forward an explanation of the formability improvement according to an analysis of the experimental results, as well as numerical modeling involving an analysis of all stages of the process. Woo et al. [23] compared the formability of electrohydroforming and electromagnetic forming through experimental research and numerical simulation. This comparison showed that electrohydraulic forming can overcome the bouncing effect and deform sheets into complex shapes, compared to electromagnetic forming. Regarding liquid impact forming with a high-speed punch, Skews et al. [24] investigated the forming of circular disks and cylindrical tubes through the liquid shock wave pressure generated in a vertical water shock tube, including free-forming, metal forming with dies, and detail imprinting. Meanwhile, the impact of the liquid shock wave that was generated by the piston during the high velocity deformation process was researched using a high-speed video camera. They concluded that this process has the prospect to be developed as the manufacture process, because of its safety and repeatability. Marai et al. [25] combined the experiments and finite element simulation to investigate the innovative hybrid impact hydroforming (IHF) technology that introduces the liquid impact forming technology into the traditional hydroforming to improve some disadvantages of the traditional hydroforming. This process is beneficial for forming complex parts with small features, and the possibility of fracture is reduced mainly because of the inertia factor.

Wang et al. [26] showed the theory of liquid shock wave and performed some experiments regarding punching on metal sheets, using IHF technology. They demonstrated that using liquid as a punch has some advantages, including improving the forming quality of complicated products, avoiding the precise of the die, and solving the difficulty of punch-die alignment. Through studying the three types of liquid impact forming technologies, using liquid as the force transmission medium has several advantages. (1) Using liquid as the force transmission medium can improve the forming accuracy of the workpiece shape and dimension. Moreover, it can make the workpiece reduce the bouncing off the die. This is because water acts as a medium to propagate the forming pressure, which can make the workpiece bear pressure relatively evenly and retain the forming pressure during the deformation process. Meanwhile, the fluidity of liquid is good, which makes the material better flow into the cavity of the die. (2) The liquid acting on the workpiece can avoid the damage of explosive powder or a punch on the surface of the workpiece, and protect or even improve the quality of the formed workpiece surface.

In this paper, a novel micro high-speed forming technique, laser impact liquid flexible embossing (LILFE), is proposed combined with the advantages of laser microembossing (LME) and liquid impact forming technology. This process is used to overcome some of the defects of laser direct shock microembossing technology, and can emboss metallic foils into 3-D large area micro arrays under the high speed condition. In this process, the laser induced shock wave is used as the energy source, and the liquid is used as the force transmission medium that is placed into the liquid chamber. The LILFE uses the liquid shock wave to form the metal foil rather than the direct impact of the laser, thus protecting the metal foil surface from being ablated and damaged. Also, LILFE employs liquid as the force transmission medium to propagate the forming pressure, which retains the forming pressure during the forming process and makes the workpiece bear load relatively evenly compared to the laser induced shock wave directly acting on the workpiece.

The aim of this paper is to show that the fabrication of microscale structures on metallic foils is feasible with the LILFE process, and can overcome some defects of the laser direct shock microembossing technology. There are the mechanism analysis, experimental research, and numerical simulation in this study. The influences of the laser pulse energy and the workpiece thickness on the forming characteristics of the pure copper foils with the LILFE process were researched through the experiments and the numerical simulation. The experiments were performed with a concave micro-die of array features on copper foils. Then, the morphology of the processed samples was observed, and the forming accuracy, depth, surface quality, and thickness variation of the formed parts were measured to evaluate the quality of the formed parts. In addition, the numerical simulations were conducted to further understand the deformation process of the workpiece. Meanwhile, the results of the numerical simulations were compared with those of the experiments. Numerical simulation studies also revealed the axial and radial plastic strain distribution of the embossed parts.

2. Experiments

2.1. Principal of Laser Impact Liquid Flexible Embossing Process

Figure 1 shows the schematic of the LILFE process. It includes a Spitlight 2000 Nd:YAG laser, optical system, experiment mobile platform, workpiece assembly, electric box, and water cooling system. It can be seen in Figure 1 that the workpiece assembly from the top to the bottom consists of the blank holder, a confining layer, ablative medium, polyurethane rubber layer, liquid chamber, liquid, sealing ring, workpiece, micro-die core, and mould substrate. Firstly, when a high-energy focused and pulsed laser beam penetrates through the transparent confining layer and irradiates on the ablative medium, the laser energy is absorbed by the ablative medium and then instantaneously vaporizes into high-temperature and high-pressure plasma. After further absorption of the laser energy, the plasma rapidly expands a strong shock wave in the limited space between the rubber layer and the confining layer under the restraint of the confining layer and the blank holder. The high-tenacity rubber layer rapidly produces a large deformation and occupies the volume of the liquid in the liquid chamber. Then, with the rubber layer impacting on the water surface at an ultrahigh strain rate, high pressure in the form of a water shock wave is generated because the water is hard to be compressed. The high water shock wave pressure propagates downward through the water in the liquid chamber and introduces impact energy onto the workpiece. As the distance increases, the liquid shock wave pressure gradually attenuates and the duration of the action on the workpiece increases. Finally, under the shock wave pressure and the restraint of the fixed micro-die, the workpiece generates a plastic deformation to reproduce the cavity shape of the micro-die at a high strain rate, as a result of the liquid shock wave pressure greatly exceeding the plastic deformation resistance.

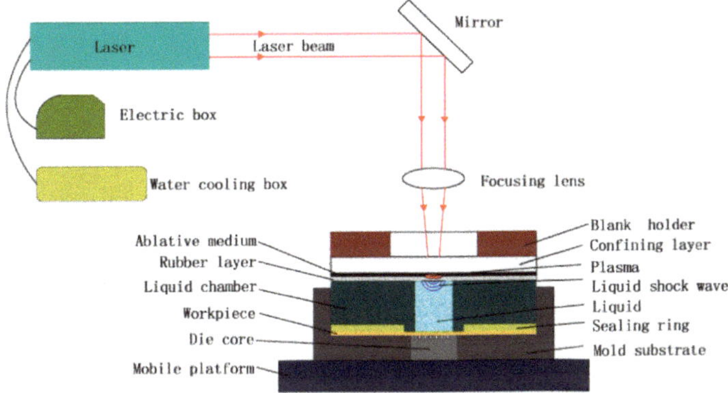

Figure 1. Schematic of the laser impact liquid flexible embossing process.

2.2. Experiment Instruments and Preparation

2.2.1. Laser Processing Conditions

A Spitlight 2000 Nd:YAG Laser that was manufactured by the German InnoLas Company (Munich, Germany) was applied in the experiments, and its maximum pulse energy is 1900 mJ/pulse. The energy distribution of the laser is the Gaussian distribution. The laser wavelength is 1064 nm, and the laser pulse width is about 8 ns. To ensure the forming accuracy of the micro-parts in the laser impact liquid flexible embossing process, the laser spot diameter was adjusted to 3 mm, which is large enough to cover the liquid chamber with the diameter of 2.6 mm. The beam diameter could be adjusted by the height of the focusing lens and micro-die. It was measured by photosensitive paper, which can record the size of the beam spot in the experiments. In the experiments, the laser pulse energy can be adjusted by the laser control system.

2.2.2. Materials and Micro-Die

In the experiments, the pure copper foils were selected as the specimen because of their relatively good properties and wide application in micro parts. Pure copper foils with the thicknesses of 30 μm and 20 μm were cut into squares of 10 mm × 10 mm, in order to prepare for the experiments. Polyurethane rubber layers with the thicknesses of 200 μm, coated with the ablative medium (black paint) were also cut into square pieces of 10 mm × 10 mm. Both of the specimen and polyurethane rubber layers should be cleaned on the surface by anhydrous alcohol before the experiments. The transparent confining layer of the organic glass (PMMA) with a side length of 45 mm and thickness of 3 mm was applied to keep the plasma from moving upward. The blank holder with about 12 N was used to press on the confining layer to restrain the expansion of the plasma. The liquid chamber with a height of 13 mm and a diameter of 2.6 mm was used to hold the liquid (water) and propagate the water shock wave pressure. Table 1 gives the detailed experimental parameters of the LILFE.

The 100# copper grating micro-die was used to investigate the deformation characteristics of LILFE. The two-dimensional (2-D) morphology of the 100# copper grating micro-die observed by the KEYENCE VHX-1000C digital microscope, is shown in Figure 2a. Figure 2 also shows the two-dimensional image, sectional view, and the main geometric dimensions of the single cavity of the micro-die.

Table 1. Detailed experimental parameters.

Parameters	Values
Laser Energy	565–1800 mJ
Copper Foil Thickness	20/30 μm
Laser Spot Diameter	3 mm
PMMA Thickness	3 mm
Rubber Layer Thickness	200 μm
Ablative Medium Thickness	10 μm
Blank Holder Force	12 N

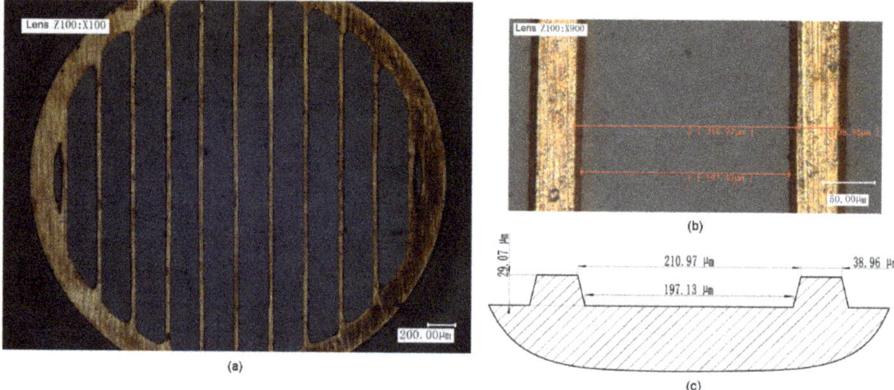

Figure 2. (**a**) Two-dimensional (2-D) morphology of the whole micro-die, (**b**) 2-D image of a single cavity of the micro-die, and (**c**) sectional view and dimension of a single cavity of the micro-die.

2.2.3. Characterization Methods

After the experiments, the KEYENCE VHX-1000C digital microscope (KEYENCE Corporation, Osaka, Japan) was used to observe the two-dimensional, three-dimensional, and cross-section morphology, as well as for measuring the accuracy, depth, and thickness distribution of the formed features. The cross-section morphology was observed and the related parameters were measured along the cross section, after the formed parts were mounted by a low viscosity epoxy, then ground by 80# to 3000# sandpapers, and polished by a polishing machine. To investigate the surface morphology and surface quality of the formed parts, the roughness of the workpiece was measured by a high resolution true colour confocal microscope (Axio CSM 700, Carl Zeiss, Oberkochen, Germany).

3. Modelling and Finite Element Simulation

In this study, a numerical simulation was performed to examine the results of the LILFE experiments in detail. The numerical modeling, meshing, sets of material parameters and laser shock wave pressure, and so on, were conducted in the Hypermesh software (17.0, Altair Engineering Inc., Troy, MI, USA), while the post-processing analysis of the forming characteristics of the pure copper foils with the LILFE process could be realized in software LS-DYNA (15.0, ANSYS Inc., Canonsburg, PA, USA). The numerical model of LILFE is composed of the rubber layer, a liquid chamber, liquid, a metal foil, and a forming micro-die. The micro-die is positioned below the workpiece, and the workpiece is placed between the micro-die and the liquid chamber, as shown in Figure 3. The laser induced shock wave exerts on the surface of the high-tenacity polyurethane rubber layer. Then, with the rubber layer impacting on the water surface at an ultrahigh strain rate, high pressure in the form of a water shock wave is generated. The high water shock wave pressure propagates downward through the water in the liquid chamber and introduces an impact energy onto the workpiece. In this study, in order to ensure accuracy and completeness, a full model needed to be employed to perform the numerical simulation.

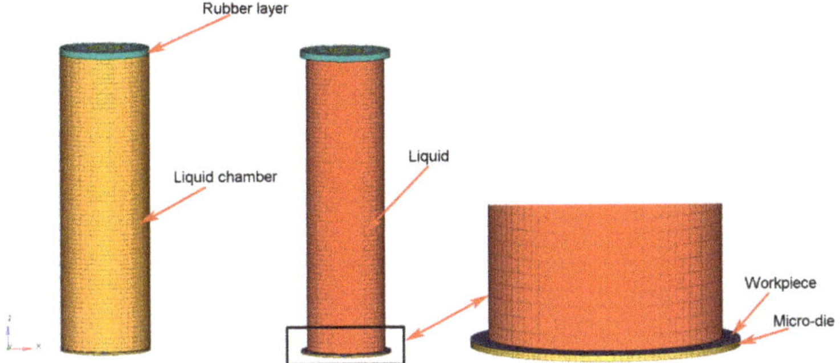

Figure 3. The numerical model of laser impact liquid flexible embossing process.

3.1. Loading

In order to simulate the LILFE successfully, the spatial and temporal distribution of the shock wave pressure induced by the laser beam should be obtained. Fabbro et al. [27] built a mathematical model to demonstrate the relation between the peak pressure of the laser shock wave in the confined mode and the laser pulse intensity, which can be defined by Equation (1), as follows:

$$P_{max}(GPa) = 0.01\sqrt{\frac{\alpha}{2\alpha+3}}\sqrt{Z(g \cdot cm^{-2} \cdot s^{-1})}\sqrt{I_0(GW \cdot cm^{-2})} \tag{1}$$

where P_{max} is the peak pressure of the shock wave induced by the laser, α is the efficiency of the interaction ($\alpha = 0.1$ in this study [28]), where αE contributes to the pressure increase, and $(1 - \alpha)E$ is devoted to the generation and ionization of the plasma, and E is the absorbed laser energy. I_0 is the incident laser power density and Z is the shock impedance, which can be calculated by Equation (2), as follows:

$$\frac{2}{Z} = \frac{1}{Z_1} + \frac{1}{Z_2} \tag{2}$$

where Z_1 is the confining layer (PMMA) impedance, Z_2 is the polyurethane rubber impedance, and $Z_1 = 0.32 \times 10^6$ g/(cm²·s), $Z_2 = 0.47 \times 10^6$ g/(cm²·s) [29]. Because the distribution of the laser beam generated by the Spitlight 2000 Nd:YAG laser in the experiments is the Gaussian distribution, the incident laser power density I_0 can be calculated by Equation (3) [30], as follows:

$$I_0 = \frac{4E}{\pi d^2 \tau} \tag{3}$$

where E is the single pulse energy of the incident laser beam, d is the laser spot diameter, and τ is the laser pulse width (8 ns). It is remarkable that the maximum laser power density is calculated as 3.18 GW/cm² by this formula, through the maximum laser energy used in this study of 1800 mJ. Thus, the maximum laser power density in this study is below the laser breakdown in the air. According to Peyre and Fabbro [31], the loading time of the laser induced shock wave pressure in the confined mode lasted two to three times longer than the laser pulse width (τ) when the glass was used as the confining medium. Wang et al. [32] revealed that the loading time is taken as three times of the laser pulse duration, because of the use of the constraint mode. Thus, in this paper, the loading time of the shock wave pressure is adopted as 24 ns, which is three times of the laser pulse width (τ) of 8 ns (which is the full width at half-maximum peak pressure). Therefore, the curve of laser induced shock wave pressure changing with the time in the simulation is plotted in Figure 4. However, as the diameter of the laser spot is 3 mm and the distribution of laser beam is Gaussian distribution in this study, it is necessary to take the spatial distribution of the shock wave pressure into consideration during the numerical simulation of the LILFE, in order to reach an agreement with the actual experiment condition. Zhang et al. [33] built the modification model that the laser induced shock wave pressure changes in the radial direction of the laser beam spot, which is based on the mathematical model built by Fabbro et al. The modification model built by Zhang et al. can be expressed by Equation (4), as follows:

$$P_{(r,t)} = P_{(t)} \exp\left(-\frac{r^2}{2r_0^2}\right) \tag{4}$$

where r is the radial distance from the centre of the laser beam, r_0 is the radius of laser beam, and $P_{(t)}$ is the shock wave pressure distribution changing with the time. The spatial distribution of the laser shock wave pressure is shown in Figure 5. Therefore, the spatial and temporal distribution of the laser induced shock wave pressure can be achieved by combining Figures 4 and 5. In the simulation, the loading region is discretized along the radial direction of the laser spot into a plurality of equidistant annular regions, shown in Figure 5, and then a time-varying and Gaussian-distributed load based on Equation (4) is applied to each region. Therefore, the laser induced shock wave pressure can change in the radial direction of the laser beam spot, and the pressure gradually decreases from the centre region to the edge region of the laser beam spot. Meanwhile, the pressure of the same divided region is uniformly loaded at each particular moment. In general, the smaller the division of the loading region, the closer it is to the actual loading condition. When the distance of the radial direction of the divided region tends to zero, the pressure distribution of the entire laser beam spot region is uniform Gaussian distribution. In the simulation of this paper, the loading area with a radius of 1.5 mm is divided into 25 annular regions to calculate the uniform load in each region, according to Equation (4), after comprehensively considering the time, efficiency, and accuracy of the solution.

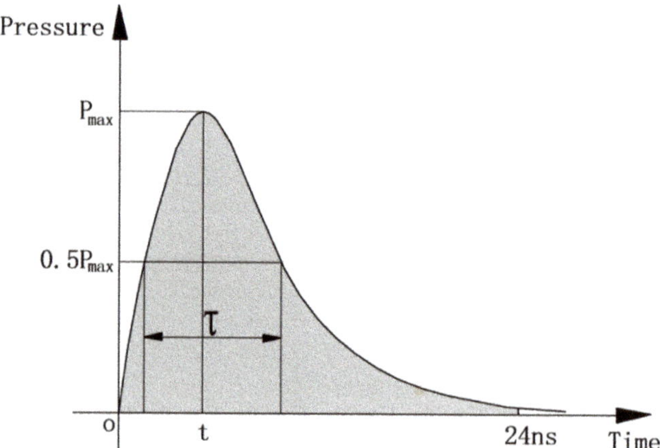

Figure 4. The curve of laser induced shock wave pressure changing with the time.

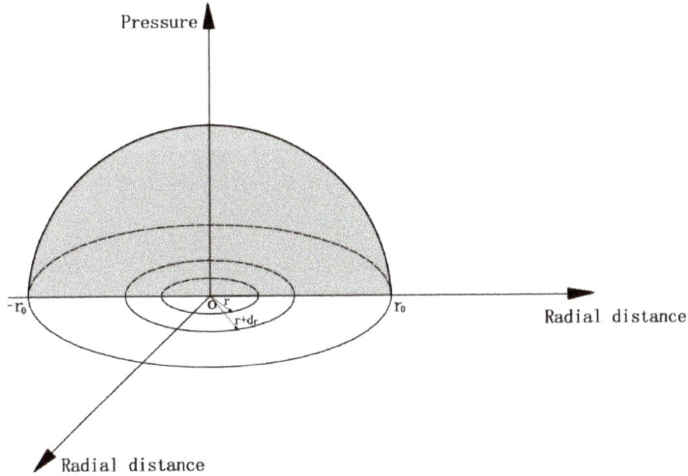

Figure 5. Spatial distribution of laser induced shock wave pressure.

3.2. Constitutive Model of the Structural Components

Structural components are composed of the liquid chamber, micro-die, rubber layer, and metal foil. The liquid chamber and micro-die are defined as rigid bodies—material model *MAT_RIGID (Material 20 in Hypermesh notation)—and they are discretized by eight-node solid hexahedron elements (SOLID 164 solid element) with the Lagrangian formulation of Hughes–Liu (elform = 1 in Hypermesh notation). The definitions of other components are as follows.

3.2.1. Metal Foil Materials Constitutive Model

Four-node shell elements (SHELL 163 shell element) are employed to define the copper foil, and the element formulation is Belytschko–Tsay (elform = 2 in Hypermesh notation), which is usually applied in a structure analysis.

The type of contact between the workpiece and the die in the numerical simulation is the keyword *CONTACT_AUTOMATIC_SURFACE_TO_SURFACE. This is a recommended contact type for the numerical models, which bears a large impact at a high strain rate. This type of contact can prevent the penetration of 2-D shell elements and external faces of 2-D continuum elements by penalty forces. Meanwhile, the parts of the slave part set can be checked whether they contact with parts of the master part set, and self contact can be checked for any part in both of the sets. The type of contact between the workpiece and the liquid chamber in the numerical simulation is also the keyword *CONTACT_AUTOMATIC_SURFACE_TO_SURFACE.

In order to achieve precise and valid results during the forming processes at a high strain rate, the constitutive model of copper foil material must be chosen properly. In the LILFE process, the copper foils are impacted by the water shock wave pressure at a high strain rate. When the metal foil is formed under the dynamic loading during the LILFE process, the inertia effect and the stress wave propagation are of great importance, because the properties of the material generate extreme change according to the grade of the strain rate and temperature [34]. The multi-axial stress state of the metal is generally defined according to the Von Mises stress, the plastic strain ε, plastic strain rate $\dot{\varepsilon}$ and temperature T are applied to express this stress in many constitutive models, as follows:

$$\sigma = f\left(\varepsilon, \dot{\varepsilon}, T\right) \tag{5}$$

Johnson and Cook [35] developed the constitutive formulation of Johnson–Cook model that is applicable for the deformation of most metals at a high strain rate. Thus, this numerical simulation chooses the Johnson–Cook model to express the flow stress of the material in terms of strain, strain rate, strain hardening, and temperature. On the basis of the Johnson–Cook model, the Von Mises flow stress can be defined by Equation (6), as follows:

$$\sigma = (A + B \cdot \varepsilon^n)\left(1 + C \cdot \ln\frac{\dot{\varepsilon}}{\dot{\varepsilon_0}}\right)\left[1 - \left(\frac{T - T_r}{T_m - T_r}\right)^m\right] \tag{6}$$

where A, B, C, n, and m are the material constants measured at or below the transition temperature; A is the initial yield stress at $\dot{\varepsilon_0}$ (Pa), B is the strain hardening coefficient (Pa), C is the strain rate-hardening coefficient, n is the strain hardening exponent, m is the thermal softening exponent. ε is the equivalent plastic strain, $\dot{\varepsilon}$ is the equivalent plastic strain rate, $\dot{\varepsilon_0}$ is the reference strain rate under quasi-static loading, T is the present temperature of the workpiece, and T_r and T_m are the room and melting temperatures, respectively. The related constitutive model parameters of copper are given in Table 2 [35,36]. In this study, the liquid is used to protect the metal foils from being influenced by the thermal field, so this deformation process can be regarded as the adiabatic process, where the deformation of the workpiece is entirely attributed to the liquid shock wave pressure induced by the laser energy function without any thermodynamics effect. The thermal effect need not be taken into account in this study. Therefore, the constitutive model can be simplified as Equation (7), as follows:

$$\sigma = (A + B \cdot \varepsilon^n)\left(1 + C \cdot \ln\frac{\dot{\varepsilon}}{\dot{\varepsilon_0}}\right) \tag{7}$$

Table 2. Constitutive model parameters of copper [35,36].

Material	A(MPa)	B(MPa)	C	n	m	T_r(K)	T_m(K)	$\dot{\varepsilon_0}(s^{-1})$
Copper	90	292	0.025	0.31	1.09	298	1356	1.0

3.2.2. Hyperelastic Material Constitutive Model for Polyurethane Rubber Material

During the LILFE process, the high-tenacity polyurethane rubber layer is used to act on the liquid surface at an ultrahigh strain rate, thus generating a high water shock wave pressure. Eight-node solid hexahedron elements (SOLID 164 solid element) are employed to define the rubber layer, and the element formulation is the Hughes–Liu (elform = 1 in Hypermesh notation).

The type of contact between the rubber layer and the liquid chamber in the numerical simulation is the keyword *CONTACT_AUTOMATIC_SURFACE_TO_SURFACE.

The polyurethane rubber can generate large deformation under the strong shock wave pressure in the LILFE and it has nonlinear stress-strain characteristics under this condition, thus the polyurethane rubber material is defined as the Hyperelastic theory (Mooney–Rivlin model) [37], as follows:

$$\sigma_{ij} = \frac{\partial W}{\partial \varepsilon_{ij}} \tag{8}$$

$$W = \sum_{k+m=1}^{n} C_{km}(I_1 - 3)^k + (I_2 - 3)^m + \frac{1}{2}k(I_3 - 1)^2 \tag{9}$$

where W is the strain energy per unit of reference volume, I_1, I_2, and I_3 are the strain invariants, and k is the bulk modulus. The polyurethane rubber is incompressible, thus setting $I_3 = 1$. C_{km} is the hyperelastic constant, which determines the material response. Two Mooney–Rivlin parameters of C_{10} and C_{01} are usually used for describing the deformation behaviour of the hyperelastic rubber material. Table 3 gives the mechanical property parameters of the polyurethane rubber material.

Table 3. Mechanical property parameters of polyurethane rubber material.

Material	Hardness A (°)	M–R Constant C_{10} (MPa)	M–R Constant C_{01} (MPa)	Poisson's (v)
Polyurethane Rubber	70	0.736	0.184	0.49997

3.3. Models for Fluid Components

In this study, the fluid components are composed of the water and void, as shown in Figure 6. When the liquid component is built, it is necessary to build a void component in the fluid components, which can allow the liquid elements to flow inside the liquid chamber and interact with the workpiece to flow into the cavity of the micro-die. In the calculation, the constitutive properties of the fluid material employed as the void must be the same as those of the material filling the voided elements. The *INTIAL_VOID card in the Hypermesh is used to define the initial voided part set ID or part numbers, and the void component it defines will not affect other components.

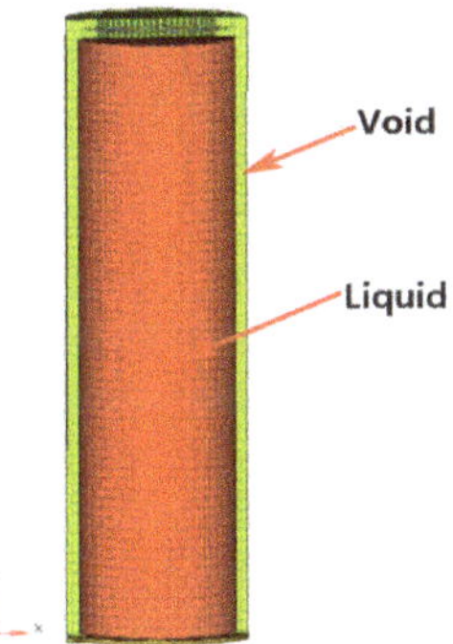

Figure 6. The image of the model of fluid components.

During the LILFE process, the fluid in the liquid chamber undergoes too large deformation, thus, the pure Lagrangian algorithm or Eulerian algorithm cannot solve this problem well. For this reason, the arbitrary Lagrangian–Eulerian (ALE) formulation, that the motion of mesh is independent of the material motion being analysed, is selected for the treatment of the fluid. The ALE method combines the advantages of the Lagrangian and Eulerian methods, which allows the mesh of the fluid domain to move in an arbitrary way, although the mesh velocity must still satisfy the boundary conditions. The ALE method requires an automatic and continuous re-zoning of the mesh and usually this is done in such a way that the mesh displacement is dependent on the displacement of the moving body [38,39]. It can improve the accuracy of the simulation under large deformation conditions, and can make the free surface of the material flow automatically without mesh distortion errors. The mesh motion can be chosen arbitrarily, providing additional flexibility and accuracy. The philosophy of the arbitrary Lagrangian–Eulerian methodology is to exploit this degree of freedom so as to improve the robustness, accuracy, and efficiency of the simulation [40]. Therefore, the simulation of this paper adopts the ALE method. The fluid filled in the liquid chamber is discretized through eight-node solid hexahedron elements (SOLID 164 solid element) with an ALE formulation of the plane stress (elform = 12 in Hypermesh notation).

The response of the structures to the impact loading is of concern in the simulation of the LILFE [41]. In the process, understanding the interaction between the fluid flow and the structural deformation is a technical challenge of vital importance. This interaction significantly affects the results and accuracy of the simulation [42]. Thus, it is pivotal to correctly define the fluid–structure interaction (FSI), including workpiece/fluid, liquid chamber/fluid, and rubber layer/fluid, for building the finite element model of the LILFE process. If the FSI is not applied correctly, the penetration of the fluid material may generate in the structure mesh, and the results will be not accurate and valid. The research of the FSI contains the various behaviors of the solids under the action of the flow field and the influence of the solid deformation or motion on the flow field. An important feature of the FSI is the interaction between the two-phase media, namely the structure deformation or motion under the action of the fluid dynamic loads, and the deformation or motion of the structure in turn affects the flow field, thereby changing the distribution of the fluid loads and size. In general, the fluid–structure interaction problem can be divided into two categories according to the interaction mechanism: one type of characteristic is that the fluid-structure interaction only occurs at the interface of flow and structure, and the interaction on the equation is the balance of the two-phase interaction surface. The other characteristic is that the flow and solid phases are partially or completely overlapped, and the interaction utility is realized by the differential equation describing the problem. The type of interaction involved in this paper is the former one. Therefore, in this study, the *CONSTRAINED_LAGRANGE_IN_SOLID keyword was applied to define the fluid–structure interaction. This command can provide a coupling mechanism for defining the FSI. For this command, the structure can be built from the Lagrangian formulation, and the multi-material fluid is modeled by the ALE formulation. This command can allow the fluid material to flow around the structure, however, the penalty coupling force stops the leakage of the fluid material across the Lagrangian surface. Once a fluid node penetrates the Lagrangian surface, a force of recall is applied to both the fluid and structural node to make the fluid node withdraw to the Lagrangian surface. The penalty coupling force behaves like a spring system and it can avoid the penetration for the fluid material.

The liquid (water) was modeled using the material constitutive model *MAT NULL (Material Type 9 in the Hypermesh notation), which is usually applied to describe the material with a fluid-like behaviour (such as air and water). The null material has no shear stiffness or yield strength and behaves in a fluid-like mode. It can be applied to describe the (deviatoric) viscous of the form, as follows:

$$\sigma'_{ij} = 2\mu \dot{\varepsilon}'_{ij} \tag{10}$$

where σ'_{ij} is the Cauchy stress tensor, $\dot{\varepsilon}'_{ij}$ is the deviatoric strain rate, and μ is the dynamic viscosity. Moreover, the null material must be applied with an equation-of-state (EOS) in order to describe the relation of the pressure and volume deformation of the material. The equation-of-state of *EOS_GRUNEISEN [43], which is based on a cubic shock velocity (V_s)—particle velocity (V_p) is commonly used to represent the water, and it defines the pressure for compressed materials, as follows:

$$P = \frac{\rho_0 C^2 \mu \left[1 + \left(1 - \frac{\gamma_0}{2}\right)\mu - \frac{\alpha}{2}\mu^2\right]}{\left[1 - (S_1 - 1)\mu - S_2\frac{\mu^2}{\mu+1} - S_3\frac{\mu^3}{(\mu+1)^2}\right]^2} + (\gamma_0 + \alpha\mu)E \tag{11}$$

and for expanded the materials, as follows:

$$P = \rho_0 C^2 \mu + (\gamma_0 + \alpha\mu)E \tag{12}$$

where S_1, S_2, and S_3 are the coefficients of the slope of the V_s–V_p curve; C is the intercept of the V_s–V_p curve, which agrees with the speed of sound on water; γ_0 is the Gruneisen gamma; α is the first order volume correction to γ_0; $\mu = \rho/\rho_0 - 1$, in which ρ and ρ_0 are the current and initial densities of the material respectively; E is the initial internal energy. The water properties and parameters for the equation-of-state of the Gruneisen are obtained from the investigations of Boyd et al. [44] and Shah [45], and they are given in detail in Table 4.

Table 4. Water properties and parameters.

Material	$\rho_0(kg/m^3)$	$\mu(Pa \cdot s)$	$C(m/s)$	S_1	S_2	S_3	γ_0	α	$E(J/kg)$
Water	1000	1.002×10^{-3}	1484	1.979	0	0	0.11	0	0

4. Results and Discussions

4.1. The Typical Morphology of the Formed Parts

The two-dimensional, three-dimensional, and cross-section images of the replicated structures of 30 μm thick pure copper foils after the LILFE process were examined by a KEYENCE VHX-1000C digital microscope. The die of the 100# copper grating arrays with ten channels was used to investigate the capability of this process fabricating the microscale structures.

Figures 7 and 8 show the two-dimensional, three-dimensional, and cross-section images of the micro-features of the formed samples with the thickness of 30 μm, after the LILFE at the experimental condition where the laser energy was 1800 mJ and the number of the pulse was one. In Figure 7, the two-dimensional and cross-section images of the workpiece indicate that the die of the 100# copper grating arrays with ten channels was embossed into the metal foils surface. It can be seen that the ten array channels can be obtained by this process. The three-dimensional morphology and cross-section morphology of the one microchannel (channel 5) at the middle position of the workpiece are clearly shown in Figure 8. The deformation depth is about 27.63 μm. The two-dimensional, three-dimensional, and cross-section morphology of the formed parts show that the amount and shape of the formed features generally have an agreement with the micro-die. Meanwhile, the dimension and shape of each feature of the same workpiece essentially keep consistent with each other. Each feature can be formed evenly and smoothly without any fracture or failure, because this process uses liquid as the force transmission medium. The fluidity of liquid is fine, which makes the material better flow into the cavity of the die. Meanwhile, the liquid acting as a medium to propagate the forming pressure makes the workpiece bear pressure evenly, so the shape of each formed feature is smooth and symmetrical.

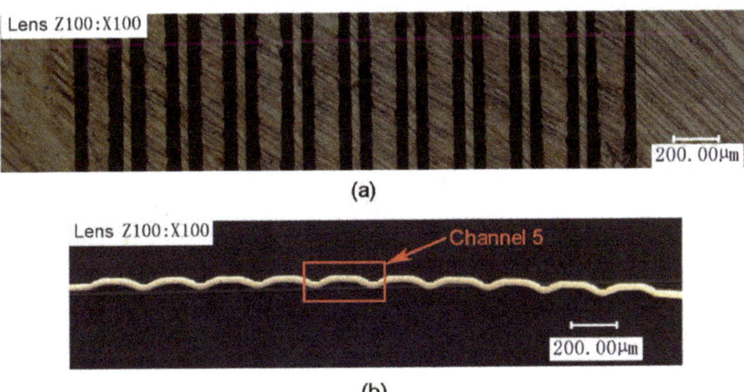

Figure 7. Image of the formed part under the laser energy of 1800 mJ: (**a**) 2-D morphology of the whole formed part and (**b**) cross-section morphology of ten array channels.

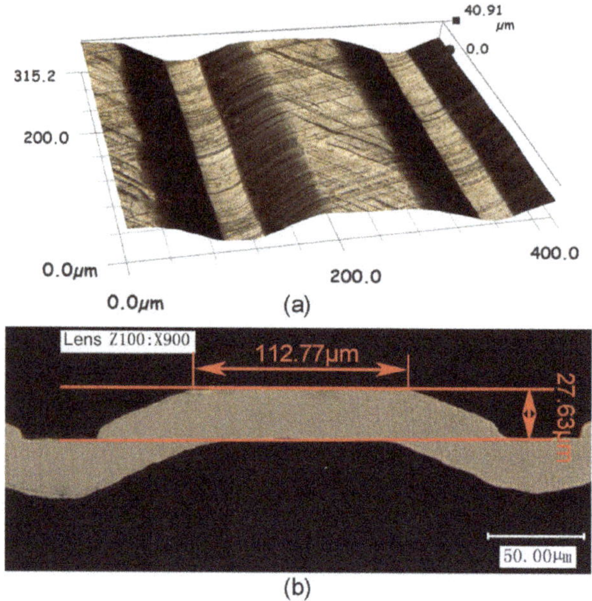

(a)

(b)

Figure 8. (**a**) 3-D morphology of one formed microchannel and (**b**) cross-section morphology of one microchannel.

4.2. Finite Element Model Verification

The deformation process of the workpiece with the thickness of 30 μm under the laser energy of 1800 mJ can be achieved in the numerical simulation, as shown in Figure 9. From 0.0036 ms to 0.0048 ms, the liquid shock wave pressure has propagated to the surface of the workpiece in the liquid chamber, and the workpiece begins to occur a bending deformation. At 0.0062 ms, the workpiece achieves further deformation due to the high water shock wave pressure continuing to transfer downward, however, it still does not reach the bottom of the die. During the time from 0.0062 ms to 0.0074 ms, the workpiece generates a greater deformation, then it collides with the bottom of the cavity and the material is forced to flow transversely toward the bottom corner of the cavity under the water shock wave pressure and the restriction of the fixed micro-die.

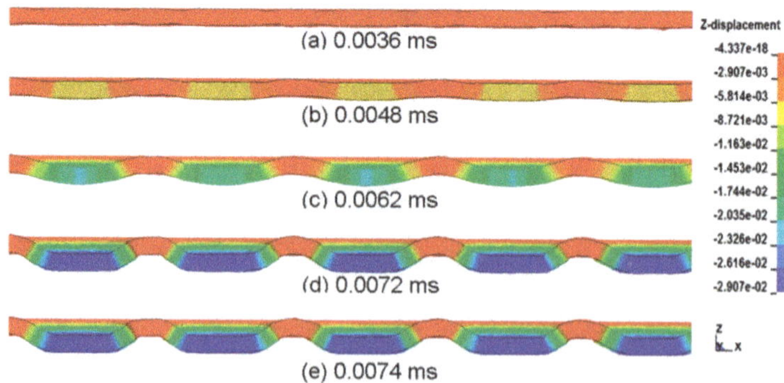

Figure 9. Deformation stages of the workpiece in the laser impact liquid flexible embossing process.

Figures 8b and 10 present the cross-section morphology and cross-section curve of one microchannel (channel 5) located in the middle region of the specimen under the laser pulse energy of 1800 mJ in the experiment. It is obvious that the width ΔL of the flat bottom region of the microchannel is about 112.77 µm and the depth reaches approximately 27.63 µm. Figures 10 and 11 show the surface morphology of the workpiece with the thickness of 30 µm, as well as the cross-section morphology and cross-section curve of one microchannel (channel 5) under the laser energy of 1800 mJ and at the time when the forming process has finished in the numerical simulation. It can be found that the deformation depth of the channel is 27.9 µm, 0.98% larger than the experimental value, and the width ΔL of the flat bottom region is 107.2 µm, 4.9% smaller than the experimental value. The error between the experiment and numerical simulation is in the acceptance range in the practical application of engineering. In general, the result of the numerical simulation is approximately consistent with that achieved from the experiments. Thus, the finite element model built in this study is useful and valid.

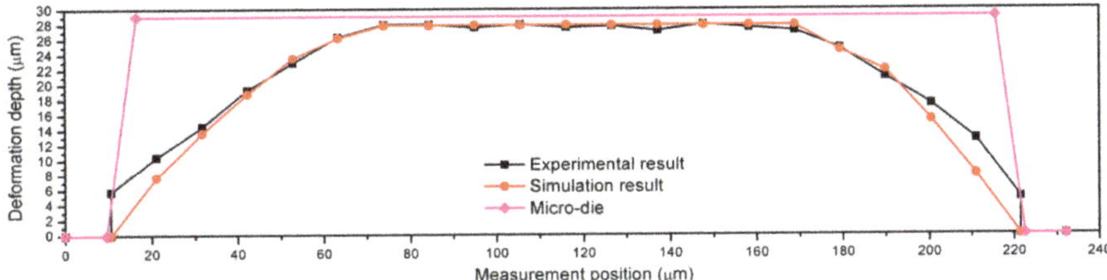

Figure 10. Cross-section curves of one microchannel under the laser energy of 1800 mJ in the experiments and numerical simulation.

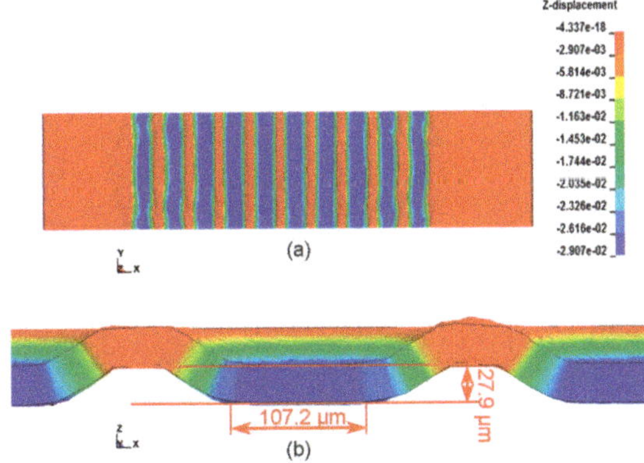

Figure 11. Image of the formed part under the laser energy of 1800 mJ in the numerical simulation: (**a**) surface morphology of the whole formed workpiece and (**b**) cross-section morphology of one microchannel.

4.3. Forming Accuracy

4.3.1. Effect of the Foil Thickness on the Forming Accuracy

The forming accuracy of the workpiece is of great importance to evaluate the forming quality, and it can measure whether the workpiece can achieve the reproduction of the rigid micro-die during the forming process. To study the effect of the foil thickness on the forming accuracy, the copper foils with the thickness of 20 µm and 30 µm were formed using the grating meshes micro-die of 100#. It can be seen from Figures 12 and 13 the 3-D morphologies and cross-section morphologies of the microchannel (channel 5) of the formed parts under different laser energy parameters of 565 mJ, 1200 mJ, and 1800 mJ are analysed. The section curve comparison of the formed parts with the thickness of 20 μm and 30 μm under each laser energy is shown in Figure 14. The width ΔL of the flat bottom region of the formed microchannels under different laser energy is depicted in Figure 15. From Figures 12–15, it can be seen that the forming accuracy of the formed parts in a thickness of 20 μm is much better than that of the formed parts in thickness of 30 μm under the same laser energy. When the workpiece in the thickness of 30 μm is formed by the laser energy of 565 mJ, it is deformed into the cavity with a smooth dome shape from a cross-sectional perspective. This is because the water wave pressure is too low to make the workpiece flow into the bottom of the micro-die cavity. When the workpiece in thickness of 30 μm is formed by the laser energy of 1200 mJ and 1800 mJ respectively, they can be deformed with the flat bottom region that is owing to the collision with the cavity bottom. However, the forming depth of the workpiece in thickness of 30 μm formed by the laser energy of 1200 mJ and 1800 mJ is less than the depth of micro-die with 29.07 μm, which can be explained by the workpiece that may generate a spring back during the forming process. The spring back behaviour will be introduced in detail in Section 4.4. In addition, the surface layer model [46] can be used to explain the phenomenon that the forming accuracy of the workpiece in the thickness of 20 μm is much better than that of the workpiece in the thickness of 30 μm under the same laser energy for the polycrystalline material. It is defined as the overall material flow stress that is jointly determined by the flow stress in the internal grains and surface layer grains of the workpiece. It can be defined as Equation (13), as follows:

$$\sigma_f = \alpha_i \sigma_{f,i} + \alpha_s \sigma_{f,s} \tag{13}$$

where the σ_f is the overall material flow stress, α_i and $\sigma_{f,i}$ are the volume fraction and flow stress of the internal grains respectively, and α_s and $\sigma_{f,s}$ are the volume fraction and flow stress of the surface layer grains, respectively. This model demonstrates that the volume fraction of the surface layer grains increases with the thickness of the workpiece decreasing. The surface layer grains are constrained less and the flow stress of the surface layer grains is smaller than that of the internal grains, so the surface layer grains are easier to generate the rotation of grains and slip deformation of the grain boundaries. Therefore, the flow stress of the thinner foils is smaller, and the thinner foils can flow into the micro-die cavity more easily under the same laser energy. Copper foils with the thickness of 30 μm would require much higher laser intensities to achieve a more accurate reproduction of the die cavity.

The numerical simulation results regarding the width ΔL of the flat bottom region of the formed channel (channel 5) with different thicknesses are illustrated in Figure 15. It can be seen from Figure 15 that the trend of the width ΔL of the flat bottom region changing with the workpiece thickness in the numerical simulation is in good accordance with the experimental results.

From Figures 12 and 13, it is interesting to observe that the strip structures were embossed on the bottom surface of the formed workpiece, which is face-to-face with the micro-die. This may be attributed to the following reason: the surface quality of the micro-die bottom is not great and some machining marks exist. The workpiece replicates the machining marks during the forming process, which causes the workpiece bottom surface that is face-to-face with the micro-die to not be smooth. This phenomenon will be studied in Section 4.5 in detail.

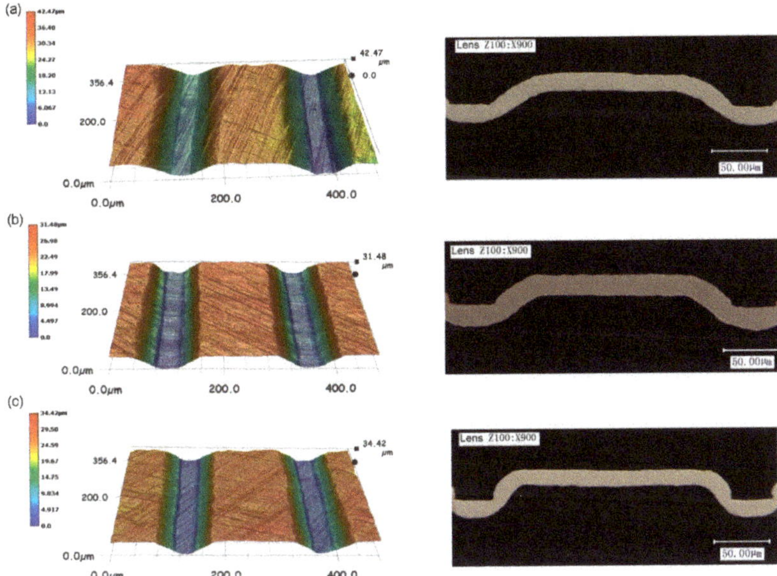

Figure 12. The 3-D morphologies and cross-section morphologies of the microchannel (channel 5) of the formed parts with the thickness of 20 μm under different laser energy parameters: (**a**) under the laser energy of 565 mJ, (**b**) under the laser energy of 1200 mJ, and (**c**) under the laser energy of 1800 mJ.

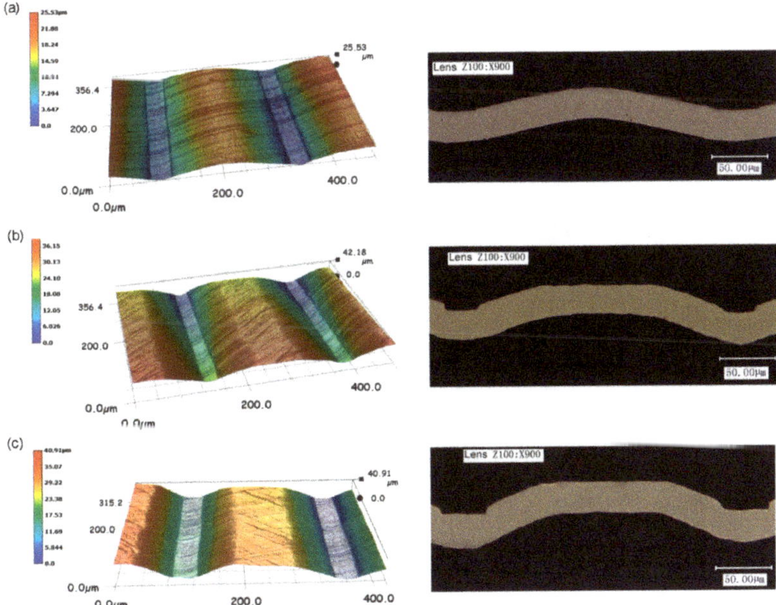

Figure 13. The 3-D morphologies and cross-section morphologies of the microchannel (channel 5) of the formed parts with the thickness of 30 μm under different laser energy parameters: (**a**) under the laser energy of 565 mJ, (**b**) under the laser energy of 1200 mJ, and (**c**) under the laser energy of 1800 mJ.

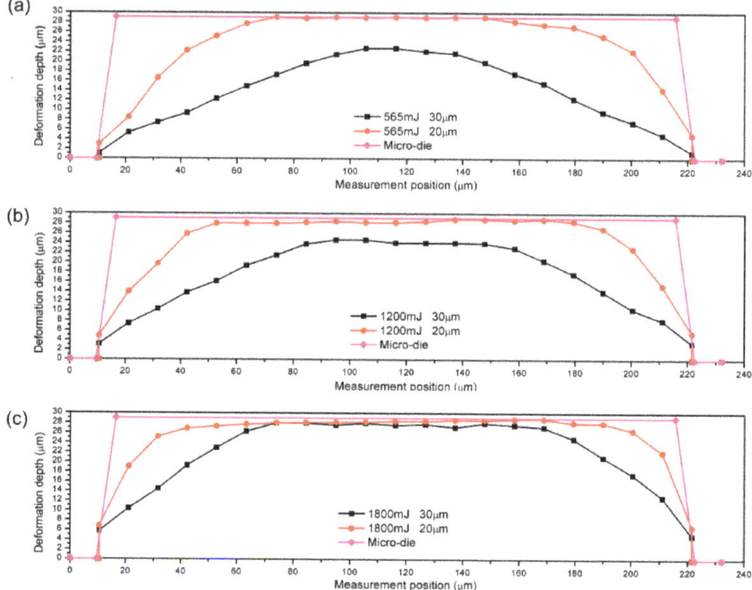

Figure 14. Sectional curves of the formed parts with the thickness of 20 μm and 30 μm under different laser energy: (**a**) under the laser energy of 565 mJ, (**b**) under the laser energy of 1200 mJ, and (**c**) under the laser energy of 1800 mJ.

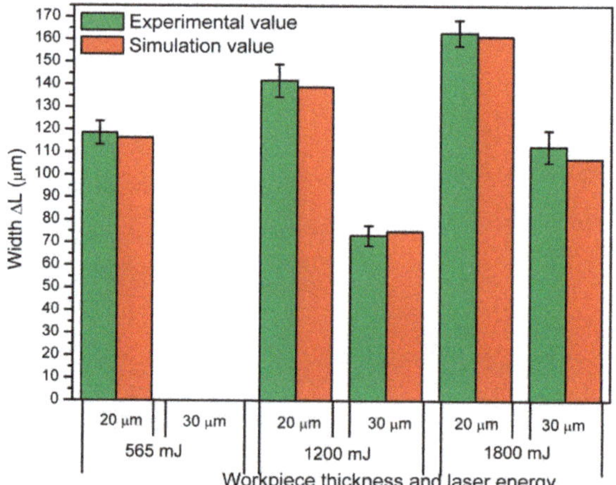

Figure 15. The width ΔL of the flat bottom region of the formed parts with different thicknesses and under different laser energy in the experiments and numerical simulation.

4.3.2. Effect of the Laser Energy on the Forming Accuracy

Pure copper foils with the thickness of 20 μm were formed using the grating meshes micro-die of 100# under different laser energy. The magnified 3-D morphology and cross-section morphology of the microchannel (channel 5) located in the middle region of the formed parts under different laser energy parameters of 565 mJ, 1200 mJ, and 1800 mJ are shown in Figure 12. It can be seen in Figure 12 that with the increase of the laser energy, the metal foil generates more deformation and the shapes of the replicated feature and the die are in better agreement. For further study clearly, the sectional curves of the replicated features under different laser energy were investigated to research the forming fitability and accuracy in Figure 16. Meanwhile, the width ΔL of the flat bottom region of the formed parts under different laser energy was measured to further estimate the forming accuracy, as shown in Figure 15. It is obvious that the width ΔL of the flat bottom region increases with the laser energy increasing from 565 mJ to 1800 mJ. By comparing the replicated structures of the samples under different laser energy with the shape of the micro-die, it shows obviously that the forming accuracy of the replicated feature under the laser energy of 1800 mJ is much better than that of the formed parts under the laser energy of 565 mJ and 1200 mJ. This phenomenon can be explained as follows, that with the increase of the laser energy, more plasma induced by a laser pulse would be generated. Then, the pressure induced by the water shock wave also gradually increases, and the higher pressure can transmit more momentum to the workpiece and increase its initial velocity. The higher momentum can make the workpiece collide the bottom of the cavity more severely, and more material can be forced to flow transversely toward the corner of the cavity under the higher water shock wave pressure and the restraint of the fixed micro-die. Thus, the higher laser energy can increase the flat bottom region of the formed parts, which makes the workpiece replicate the micro-die cavity shape better. Furthermore, it can be seen from Figure 15 that the trend of the width ΔL of the flat bottom region of the formed parts changing with the laser energy in the numerical simulation is in good agreement with the experimental results.

However, some adverse phenomena exist in the experiments when the laser energy increases to a certain extent. Firstly, the micro-die may be damaged by the strong pressure, and it may influence the forming quality and precision of the sample. Secondly, it is possible for the micro-die to be stuck on the sample, and it is hard to take out from the sample. If the workpiece and the micro-die are forced to separate, they will both suffer destruction, which affects not only the quality of the workpiece, but also the repetitive utilization of the micro-die. Therefore, the laser energy should be neither too low nor too high for the good reproduction of the samples.

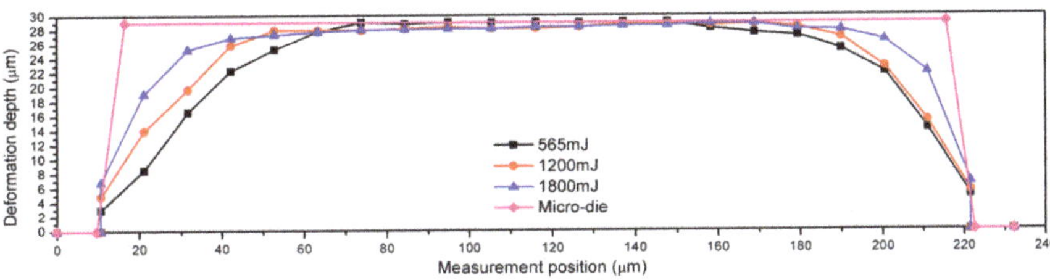

Figure 16. Sectional curves of the replicated features under different laser energy.

4.3.3. Effect of the Impact Location on the Forming Accuracy

The replicated structures of the samples with the thickness of 30 μm were embossed by the micro-die of the copper grating arrays under the laser energy of 1800 mJ, in order to study whether the forming accuracy of each feature located in the different regions of the workpiece can keep consistence. Figure 17 indicates the cross-section morphologies of the whole ten array channels. As ten array microchannels can be embossed using the liquid as the soft punch during the LILFE process, in order to understand whether the water shock pressure induced by the laser with the Gaussian distribution makes the middle region of the workpiece bear more pressure and then generate more deformation, it is necessary to understand the forming accuracy of each channel on one workpiece.

To compare the forming accuracy of each channel more intuitively, the width ΔL of the flat bottom region of each channel in the same section under the laser energy of 1800 mJ is described clearly in Figure 18. It can be seen from Figure 18 that there are a few changes in the forming accuracy. It can be found that the width ΔL of the channels located at the edge region is relatively shorter than that of the channels located at the middle region, thus, the forming accuracy of the channels located at the edge region is lower than that of the channels located at the middle region. However, the difference between the forming accuracy of the channels at the edge region and at the middle region is not very large. This phenomenon can be explained by the following, that the laser with the Gaussian distribution causes the water shock wave pressure to increase from the edge region to the center region gradually, which makes different regions of the workpiece subject to different forces, and then generate different deformation. Meanwhile, using the water as the force transmission medium can weaken the Gaussian distribution to a certain extent, and make the distribution of the pressure on the entire workpiece relatively uniform. This phenomenon was further explained by the numerical simulation. As shown in Figure 17, the cross-section image of the workpiece under the laser energy of 1800 mJ in the numerical simulation is depicted. It is clear that the width ΔL of the flat bottom region of the channels located at the edge region is relatively shorter than that of the channels located at the middle region. To compare the results of the experiment and simulation, and to illustrate this phenomenon more intuitively, the width ΔL of the flat bottom region of each channel in the experiments and the numerical simulation is shown in Figure 18. It can be seen that the trend of the results in the simulation has a good agreement with that in the experiments. The deviation is mainly attributed to the difference between the actual laser induced shock wave pressure and the ideal pressure calculated by the constructed model in the simulation.

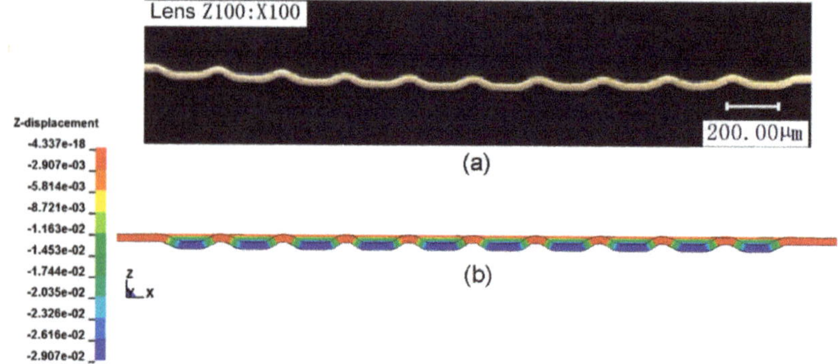

Figure 17. The cross-section images of the whole ten array channels: (**a**) experimental result and (**b**) simulation result.

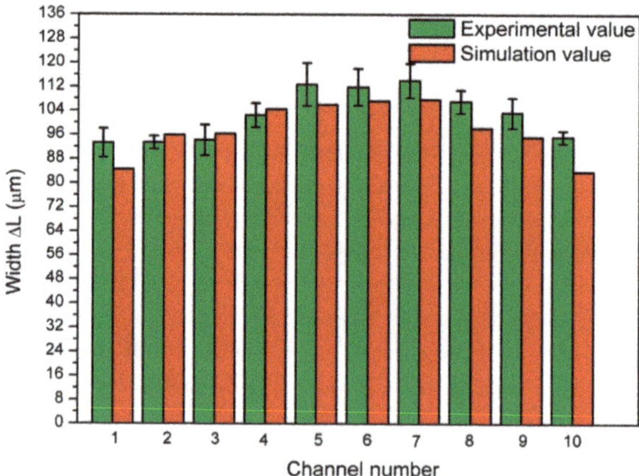

Figure 18. The width ΔL of the flat bottom region of each channel in the same section under the laser energy of 1800 mJ.

4.4. Deformation Depth

The copper foils with the thickness of 20 μm and 30 μm were deformed under different laser energy parameters of 565 mJ and 1800 mJ, respectively, so as to study the effects of laser pulse energy and the workpiece thickness on the forming depth of the replicated features, and to compare them with the results of the numerical simulation. To reveal the influence of the specimen thickness and the laser energy on the deformation depth of the formed parts, the cross section of the microchannel (channel 5) at the middle location of the specimen (shown in Figure 7) was measured through the KEYENCE VHX-1000C digital microscope. The forming depth of the microchannel (channel 5) of the formed parts was measured, as shown in Figure 19. Meanwhile, the cross-section curves of the microchannels with different thicknesses and under different laser energy in the experiments are obviously presented in Figure 20. It is clear that the forming depth of the microchannel increases, with the laser energy rising when the workpiece thickness is 30 μm. The formed parts in the thickness of 30 μm reach the depth of 27.63 μm under the laser energy of 1800 mJ. This behaviour is attributed to the increase in the water shock pressure for the increase in the laser pulse energy, which makes the workpiece generate more deformation. However, the depth of the replicated channel under the highest laser energy is still less than the depth of the micro-die with 29.07 μm. The reason is that the copper foil may generate a spring back. The spring back was also found by Liu et al. [47]. They thought that a pressure that was too high would make the workpiece generate a violent collision with the rigid die bottom, and then rebound forces were generated, which caused a spring back of the formed parts. Meanwhile, Ehrhardt et al. [48] also found the spring back after laser embossing when they investigated the submicron patterning of the metal substrates by laser embossing process. They considered that the elastoplastic properties of the metals caused this phenomenon. A spring back occurred because of the elasticity of the nickel and copper foils after the laser embossing. In addition, Zhang et al. [49] also found the spring back in the simulation during the laser shock forming (LSF) process. They revealed that the workpiece will spring back to release the stored elastic energy and it acquires kinetic energy again. After several energy exchanges, the elastically stored energy gradually exhausts and then reaches the static state finally. The spring back phenomenon was further studied in the numerical simulation. Figure 21 shows the curve of the Z-displacement for the element (S1015600) at the bottom surface of the flat bottom region of the microchannel (channel 5) with the thickness of 30 μm changing with the time when the laser pulse energy is 1800 mJ. It is obvious that the workpiece generates a spring back when it reaches to the rigid die bottom, then the forming depth occurs a little fluctuation and finally reaches a certain value.

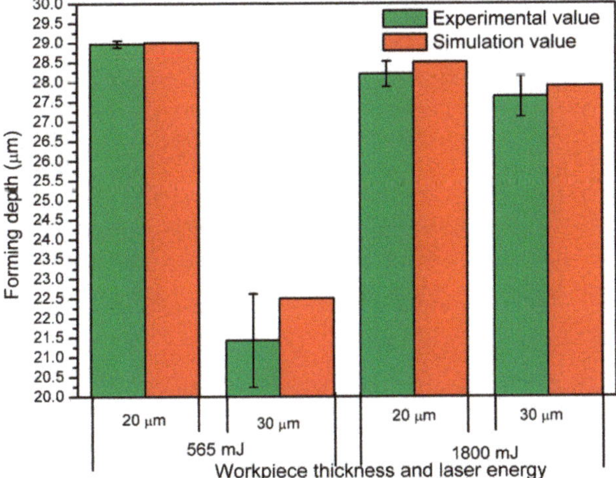

Figure 19. The forming depth of the embossed microchannel in the experiments and numerical simulation.

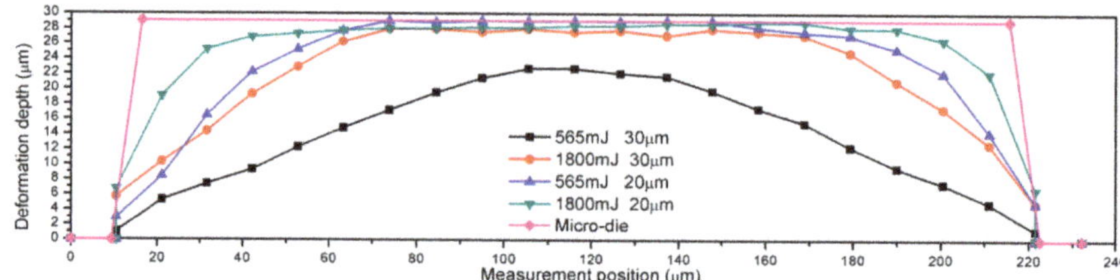

Figure 20. The experimental cross-section curves of microchannels with different thicknesses and under different laser energy.

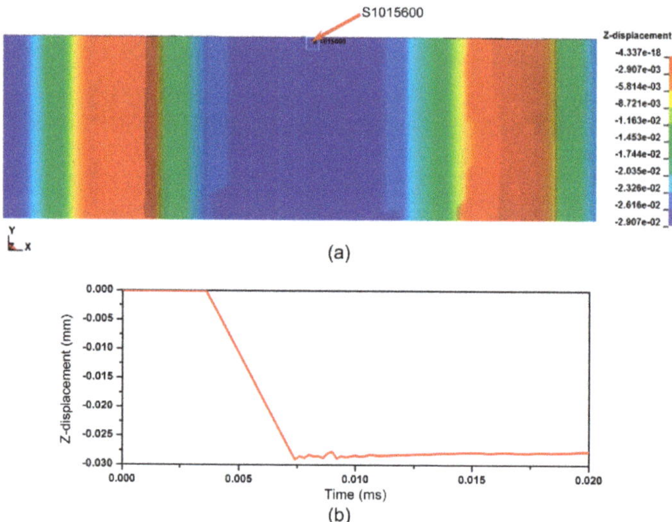

Figure 21. (**a**) The position chosen to plot the curve of Z-displacement changing with the time and (**b**) curve of Z-displacement for the workpiece with the thickness of 30 μm changing with the time under the laser pulse energy of 1800 mJ.

It can also be found from the Figures 19 and 20 that the depth of the microchannel also increases with the workpiece thickness decreasing under the same laser energy. The workpiece with the thickness of 20 μm can reach the depth of about 29 μm under the laser energy of 565 mJ, which is almost the same as the depth of the micro-die, while the forming depth of the workpiece with the thickness of 30 μm is only about 21.4 μm. This can be explicated by the surface layer model, which has been introduced in detail in Section 4.3.1.

Meanwhile, the forming depth of the microchannel (channel 5) of the formed parts in the experiments and the numerical simulation is compared, as shown in Figure 19. Through comparing the results of the experiment and the numerical simulation, it can be found that the simulation results show a good agreement with those of the experiments, yet some errors exist.

4.5. Surface Quality

The surface quality of the formed workpiece is related to the surface quality of the raw material and micro-die, as well as the process method. It also significantly affects the mechanical properties of the formed parts, such as the friction, lubricating property, abrasion resistance, corrosion resistance, and fatigue strength. Hence, it plays a significant role on the performance and durability of the MEMS or other instruments during the actual working conditions. Meanwhile, it has been put forward, in

Section 4.3.1, that the bottom surface of the embossed features may replicate the machining marks of the micro-die, which may have an influence on the bottom surface smoothness of the embossed features. Therefore, the surface quality of the samples before and after the LILFE process needs to be investigated.

In order to research the surface quality of the samples under different conditions, and to compare the surface quality of the samples before and after LILFE, pure copper foils of 30 μm in thickness used as the target material were embossed under the laser energy of 565 mJ and 1800 mJ. The high resolution true colour confocal microscope was used to observe the surface morphology of the samples before and after the LILFE and the micro-die. In order to quantitatively compare the surface quality, the arithmetical mean deviation of the profile (Ra) was measured and used in this study to indicate the surface roughness. To ensure the accuracy of the roughness measurement results, the surface roughness of the die and the samples was measured at least three times, respectively, and the average values of the measurement results were adopted in the experiment.

The surface roughness measurements were performed at the bar and bottom region of the microchannel (channel 5), located in the middle region of the formed parts, as shown in Figure 22. Figures 23 and 24 show the local magnified surface morphology at the bar and bottom region of the microchannel, which is face-to-face with the die and the liquid under the laser energy of 1800 mJ, respectively. The local magnified surface topography of the bar and bottom region of the micro-die is presented in Figure 25. The surface roughness values of the corresponding regions of the formed channel under the laser energy of 1800 mJ and the micro-die are presented in Figure 26. It is clearly seen from Figures 25 and 26 that some machining marks exist on the bottom of the micro-die, so the roughness value of the micro-die cavity bottom (about 0.265 μm) is higher than that of the micro-die cavity bar (approximately 0.065 μm). As shown in Figure 23b, the similar features (machining marks) on the surface of the micro-die cavity bottom are also found on the surface of the microchannel bottom embossed under the laser energy of 1800 mJ, which is face-to-face with the die. The machining marks, the width of which is less than 1 μm, can be successfully embossed on the workpiece surface. This can demonstrate that the LILFE process can realize the embossing of the sub-micro scale features at the same time that the larger dimension features (such as micro scale) are embossed on the workpiece. However, the surface roughness of the microchannel bottom (about 0.178 μm) becomes larger because of embossing the machining marks from the micro-die compared with the initial roughness value of the raw material (about 0.124 μm). Meanwhile, embossing the machining marks can also demonstrate that the copper foil of 30 μm can reach to the micro-die cavity bottom under the laser energy of 1800 mJ; however, the final depth is less than the depth of the micro-die cavity, because the workpiece generates a spring back, which has been explained in Section 4.4. To illustrate the increase in the surface roughness of the microchannel bottom region mainly resulting from the machining marks' embossing, the surface roughness of the microchannel bottom that is face-to-face with the die embossed under the laser energy of 565 mJ was measured to make a comparison. From Figure 27, it can be found that the microchannel bottom does not replicate the machining marks of the micro-die, because the microchannel cannot reach to the micro-die bottom under the laser energy of 565 mJ, which has been depicted in Section 4.3.1. The Ra value measured from the surface of the microchannel bottom embossed under the laser energy of 565 mJ (about 0.120 μm) is roughly same as that of the raw material, so the increase in the surface roughness of the microchannel bottom that is face-to-face with the die under the laser energy of 1800 mJ can be considered to result from the machining marks.

The Ra value at the microchannel bar region that is face-to-face with the die (about 0.105 μm), in Figure 26, is lower than that of the raw material and higher than that of the micro-die cavity bar, which illustrates that the high quality of the die surface has a smooth influence on the surface of the formed workpiece; thus, it can reduce the surface roughness of the workpiece. The defects of the residual rolled traces and some contaminations of the foil surface result in the Ra value of the microchannel bar being higher than that of the micro-die cavity bar.

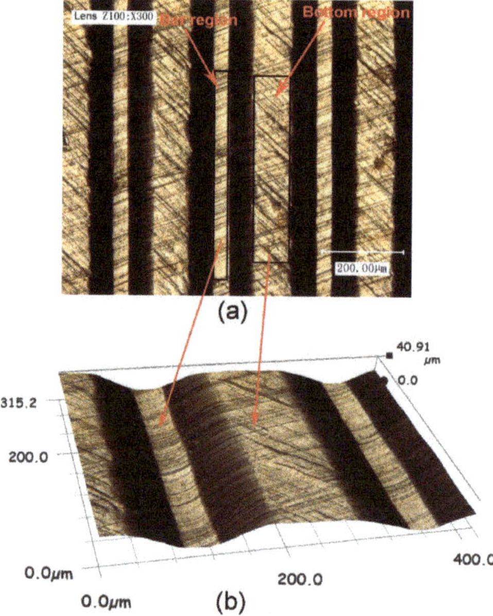

Figure 22. (**a**) 2-D surface roughness measurement positions and (**b**) local magnified morphology of roughness measurement positions in 3-D.

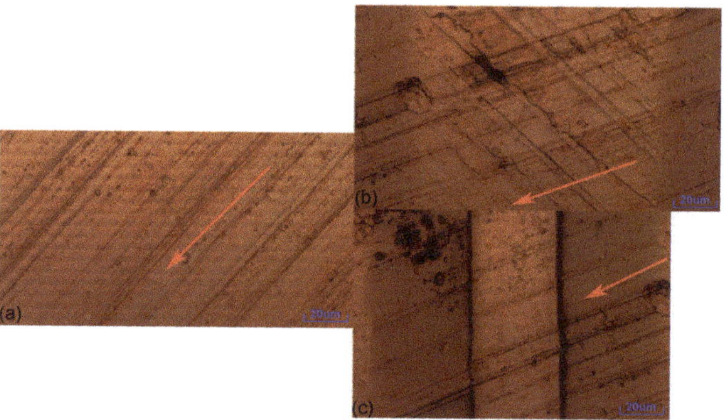

Figure 23. Local magnified surface morphology of three different regions of the microchannel that is face-to-face with the mould under the laser energy of 1800 mJ: (**a**) surface topography of the original copper foil, (**b**) surface image at bottom region, and (**c**) surface topography at the bar region. The arrows represent the rolling direction of the copper foil.

Meanwhile, the local magnified surface topography of the rear side surface of the copper foil (which is face-to-face with the liquid) at the microchannel bottom and bar region is shown in Figure 24. The Ra values of the microchannel bottom and bar region are approximately 0.123 μm and 0.126 μm, respectively, which are almost the same as the initial roughness value of the raw material. This can indicate that this process will not cause any damage to the surface of the formed parts. No laser induced thermal damage is observed on the microchannel surface at the rear side, which is face-to-face with the liquid. This can be mainly attributed to a great specific heat of water (4.2×10^3 J/(kg · °C)), which can absorb most of the heat induced by the laser. The thermal impact of the laser induced on the copper foil penetrating the water can be neglected because it is far below the melting temperature of the metal. Simultaneously, the liquid acting as the force transmission medium can completely avoid the friction and wear damage to the workpiece surface, which may be caused by some rigid dies or the punch. Thus, the LILFE process can protect the workpiece surface from being ablated and damaged, and can ensure the surface quality of the formed parts.

The standard deviation distribution of the surface roughness is also shown in Figure 26. It can be seen from the Figure 26 that the largest standard deviation (about 0.012) of the surface roughness of the formed microchannel occurs at the microchannel bottom that is face-to-face with the die. This is because the surface quality of the micro-die cavity bottom is relatively poor and the standard deviation of its surface roughness is also relatively large (about 0.01), which causes the surface roughness of the formed microchannel bottom that is face-to-face with the die to be less uniform.

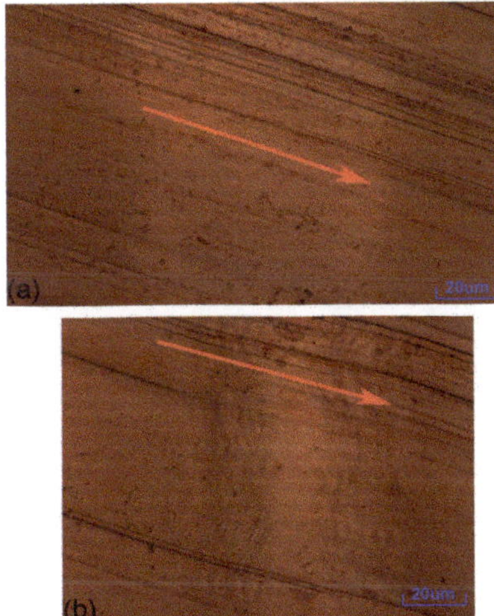

Figure 24. Local magnified surface topography of the rear side surface of the embossed microchannel (which is face to face with the liquid): (**a**) Surface image at the bottom region and (**b**) surface topography at the bar region. The arrows represent the rolling direction of the copper foil.

Figure 25. Local magnified surface topography of the micro-die: (**a**) bar region and (**b**) bottom region. The arrow represents the machining marks.

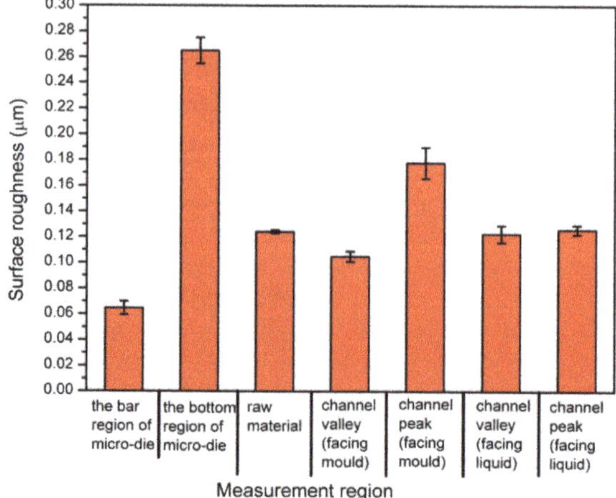

Figure 26. Surface roughness values of the corresponding regions and the micro-die.

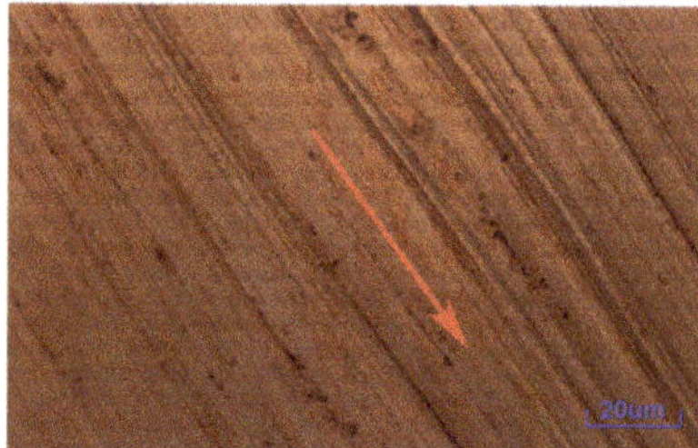

Figure 27. Surface image of the microchannel peak region, which is face-to-face with the mould embossed under the laser energy of 565 mJ. The arrow represents the rolling direction of the copper foil.

4.6. Thickness Distribution

The thickness distribution of the formed parts is one of the important characterizations during the LILFE process, because an uneven thickness distribution will cause irregular stress levels and an excessive localised thickness reduction or even fracture. In order to study the thickness variation of the samples, pure copper foils with the thickness of 20 μm and 30 μm were embossed under the laser energy of 1800 mJ. Then, the formed parts were cold-mounted with the epoxy, ground with 80# to 3000# sand papers, and polished to prepare for being examined in a cross-sectional view. The morphology and thickness distribution along the cross section of the formed parts were characterized by the KEYENCE VHX-1000C digital microscope. Twenty-two positions along the cross section of the microchannel (channel 5) located in the middle region of the formed parts were selected to characterise the thickness distribution. The measurement method of the thickness and the several positions that were measured are shown in Figure 28. It is obvious, in Figure 28, that the thickness reduction of the formed area exists and that the degree of the thickness reduction at different positions is different. To clearly and precisely reveal the thickness reduction in different points along the cross section of the formed area, the thickness thinning rate T was introduced to characterize the degree of thickness thinning. It is defined as follows:

$$T(\%) = \frac{t_0 - t_i}{t_0} \tag{14}$$

where t_0 is the thickness of the initial copper foil before the LILFE, and t_i is the thickness of the different positions of the formed region.

The thickness thinning rate distributions of the microchannel (channel 5) at the middle position of the formed workpiece under the laser energy of 1800 mJ are shown in Figure 28. As illustrated in Figure 28, the severest necking of thickness all occurs at the bar corner, regardless of the foil thickness, while the thickness thinning at the centre position of the bar is relatively slight. For example, the thickness thinning rate of 30 μm thick copper foil at the bar corner is approximately 16.3% with a thickness of about 25.11 μm under the laser energy of 1800 mJ, while the thickness thinning at the centre position of the bar is relatively slight with the thickness thinning rate of 3% and the thickness of 29.10 μm. This can be attributed to the following reasons: (1) The bar corner of the micro-die is not fabricated fillet. (2) The forming process in which the foil is firstly deformed at the bar corner, then the material of the bar corner gradually flows into the cavity to complete the whole plastic deformation. In addition, the thickness thinning of the channel bottom is more significant than that of the bottom corner and the transition region. This is because the bottom region of the microchannel impacts on the die cavity bottom with the fast speed, however, the transition region of the channel is drawn slightly. The serious impact squeezes the material to flow transversely toward the corner of the cavity bottom and the transition region under the high water shock wave pressure and the restriction of the fixed micro-die. Thus, the thickness thinning of the channel bottom is more severe.

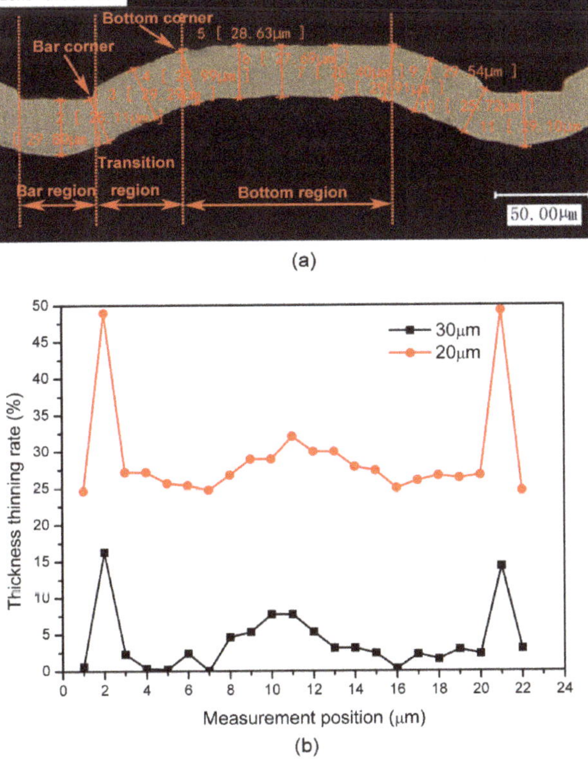

(a)

(b)

Figure 28. (**a**) The measurement method and thickness distribution of several locations and (**b**) the thickness thinning rate distribution of the microchannel with the thickness of 20 μm and 30 μm under the laser energy of 1800 mJ.

Also, the thinning increases with the decrease in the workpiece thickness. The severest thickness thinning of the 30 μm thick and 20 μm thick copper foils are approximately 16.3% and 49.3%, respectively, under the laser pulse energy of 1800 mJ. This result is consistent with the result and explanation of Section 4.3.1. The reason is that the thinner foils can flow into the micro-die cavity more easily and can generate more deformation under the same laser energy, which leads to a larger thickness thinning rate for the thinner copper foil. Furthermore, although the severe thickness thinning (49.3%) of 20 μm thick copper foils occurs at the bar corner region in the LILFE process, it still keeps without fracture and shows high formability. This behaviour was also found by Gao and Cheng [12]. They considered that the superplastic mechanical behaviour in the LDF was mainly owing to the mechanisms of inertial effects, changed constitutive model of material at high strain rates, and grain size. Thus, during LILFE, the flow stress and strain rate sensitivity of the materials can accordingly increase with the high strain rate, which is beneficial to improve the formability, and the failure of the formed samples can be delayed when the inertial forces are relatively large.

Figure 29 reveals the thickness thinning rate distribution of the microchannel under the laser energy of 1800 mJ in the numerical simulation. The maximum thinning rate in the simulation is smaller than that in the experiments. The deviation is mainly due to the meshing in the numerical simulation and the measurement error. However, the trend of the numerical simulation results is in accordance with that of the experimental results.

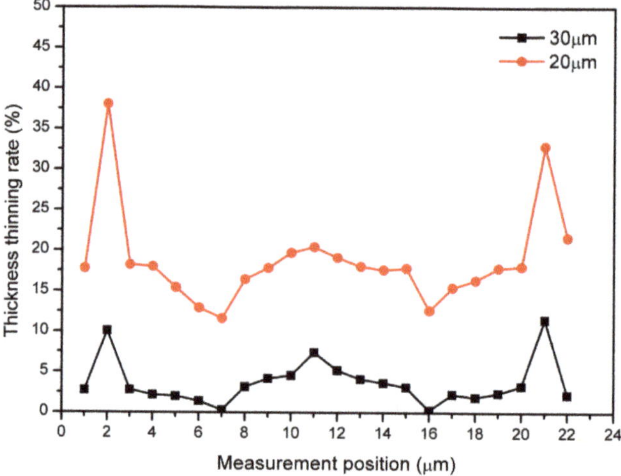

Figure 29. The thickness thinning rate distribution of the microchannel in the numerical simulation.

4.7. Strain Distribution

Figure 30 shows the contours of the axial and radial plastic strain distribution on the embossed microchannel with the thickness of 30 μm, under the laser energy of 565 mJ and 1800 mJ. It can be seen that most regions of the microchannel are subjected to tensile strain in the radial direction (X direction) and compressive strain in the axial direction (Z direction), which is quite similar to the result of Zhang et al. [50]. This is because the material is compressed by the downward water shock wave pressure, and then flows into the micro-die cavity when the liquid shock wave pressure exceeds the material yield strength. Meanwhile, the material outside the die cavity cannot flow into the cavity because it is confined by the blank-holder force, thus the material in the most deformation region is elongated in the radial direction. It can be seen in Figure 30 that the significant strain gradient occurs at the region of the bar corner, because the material at the bar corner of the micro-die cavity generates compression, stretching, and shearing owing to the bar corner of the cavity without fabricating the fillet, and the material at this region stretches into the cavity to complete the plastic deformation. Meanwhile, the compressive strain in the axial direction and tensile strain in the radial direction at the bottom region of the channel exist when the workpiece collides with the micro-die bottom under the laser energy of 1800 mJ. This can be explained by the following, that the material at the bottom region collides with the bottom of the cavity severely, and is then compressed between the high water shock wave pressure and the restraint of the fixed micro-die to flow transversely toward the corner direction of the cavity.

The contours of the axial and radial plastic strain distribution of the microchannel embossed under the laser energy of 565 mJ, without colliding with the cavity bottom, are similar to that of the laser shock micro-bulging forming [51]. It is obvious in Figure 30 that most of the regions of the channel are also subjected to tensile strain in the radial direction and compressive strain in the axial direction when the laser energy is 565 mJ, and the significant plastic strain is at the bar corner and at the bottom region. The plastic strain of the microchannel at the bottom region is larger than that at the transition region, because the material at the bottom of the deformed foil is compressed by the high water shock wave pressure, so as to elongate into the die cavity, while the transition region of the channel is drawn relatively slightly.

Figure 31 shows a group of time-varying curves of the radial strain and axial strain distribution at the bar corner region, under the laser energy of 565 mJ and 1800 mJ. It can be found that both the maximum radial plastic strain and the axial plastic strain increase with the laser energy increasing from 565 mJ to 1800 mJ. This is because the workpiece with the same thickness generates more plastic deformation under the higher laser energy, which has been introduced in the Section 4.3.2.

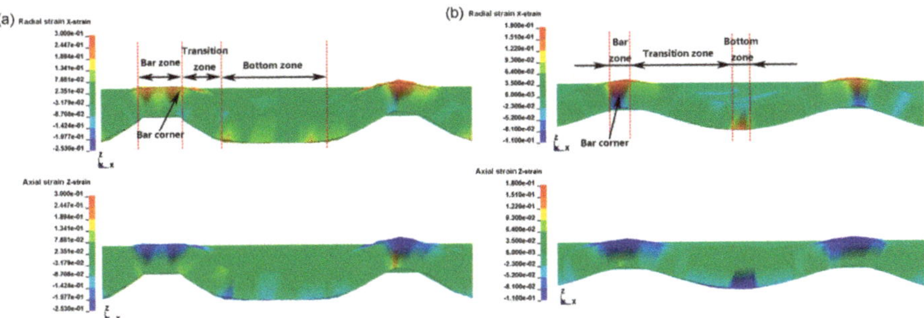

Figure 30. Plastic strain distribution of embossed microchannel with the thickness of 30 μm: (**a**) under the laser energy of 1800 mJ and (**b**) under the laser energy of 565 mJ.

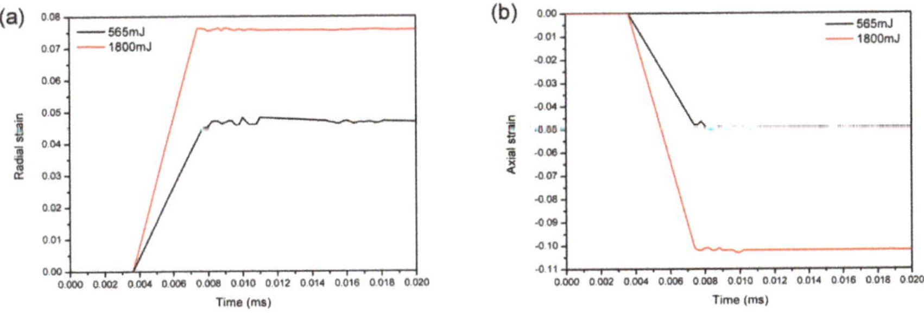

Figure 31. Time-varying curves of the strain distribution at the bar corner region under the laser energy of 565 mJ and 1800 mJ: (**a**) radial strain distribution and (**b**) axial strain distribution.

5. Conclusions

This paper investigated the influences of laser energy and workpiece thickness on the deformation characteristics of pure copper foils embossed with 100# copper grating micro-die by LILFE process, through the experiments and numerical simulation. Good consistency has been achieved between the experiments and the numerical simulation under different conditions. Meanwhile, the plastic strain distribution of the embossed parts and the typical stages of the deformation process were studied in the numerical simulation. The significant results can be obtained as follows:

1. The experiments and numerical simulation show that the forming accuracy of the formed parts becomes better with the increase in the laser energy and decrease in the workpiece thickness. However, some adverse phenomena exist in the experiments when the laser energy increases to a certain extent.
2. The forming accuracy of the channels located at the edge region is lower than that of the channels located at the middle region, and the trend of the result in the simulation has a good agreement with that in the experiments.
3. The experiments and numerical simulations show that the forming depth of the micro embossed channel increases with the decrease in the workpiece thickness and the increase in the laser energy. The formed part generates a spring back during the forming process, which will have an adverse effect on the forming accuracy of the replicated features.
4. Through measuring the surface roughness of the formed area, raw material, and micro-die, the surface quality of the formed parts is related to the surface quality of the raw material and micro-die, as well as to the process method. Meanwhile, no laser induced thermal damage is observed on the microchannel surface at the rear side, which is face-to-face with the liquid.
5. The experiments and numerical simulations show that the thickness thinning rate of the embossed parts increases with the decrease of the workpiece thickness, and the severest thickness thinning presents at the bar corner region.
6. Most regions of the channel are subjected to tensile strain in the radial direction and compressive strain in the axial direction. Both the maximum radial plastic strain and the axial plastic strain at the bar corner region increase with the increase in the laser energy.

6. Future Work

(1) A comprehensive study of the spring back phenomenon should be examined in future research. The detailed study of the spring back effect during the forming process in the numerical simulation with a longer time and a statics analysis of the steps of the spring back will be a worthy supplement for the experimental investigation, and it is useful to explain the forming behaviour of metal foils during the LILFE process.
(2) How to improve the spring back phenomenon is also important for future research. The multiple laser pulses, instead of just a single laser pulse, will be employed in the experiments and the numerical simulation to further investigate whether the spring back can be reduced, and whether the forming accuracy of the formed parts can be improved. Meanwhile, the multi-pulse approach can also be applied to overcoming the difficulty of forming the thicker workpiece.
(3) In addition, besides the process parameters, such as laser energy and workpiece thickness in this study, other parameters, including the number of laser pulse, the height and diameter of liquid chamber, the type of liquid, and workpiece with different materials, will be further researched in our future study, in order to make the research of the LILFE process more complete and systematic.

Author Contributions: F.L., H.L., and X.W. conceived and designed the experiments; F.L. and C.J. performed the experiments and numerical simulation; X.W. and Y.M. analysed the data; F.L., H.L., and X.W. wrote the paper.

Acknowledgments: This work is supported by the National Natural Science Foundation of China (Grant No. 51675243) and the 2017 Innovation Practice Fund of Jiangsu University Industrial Centre (Grant No. ZXJG201772).

Conflicts of Interest: The authors declare no conflict of interest.

References

1. Mai, J.M.; Peng, L.F.; Lai, X.M.; Lin, Z.Q. Electrical-assisted embossing process for fabrication of micro-channels on 316l stainless steel plate. *J. Mater. Process. Technol.* **2013**, *213*, 314–321. [CrossRef]

2. Wang, H.M.; Vivek, A.; Wang, Y.L.; Viswanathan, G.; Daehn, G. High strain rate embossing with copper plate. *Int. J. Mater. Form.* **2017**, *10*, 697–705. [CrossRef]

3. Vollertsen, F.; Niehoff, H.S.; Hu, Z. State of the art in micro forming. *Int. J. Mach. Tool. Manuf.* **2006**, *46*, 1172–1179. [CrossRef]

4. Schuster, R.; Kirchner, V.; Allongue, P.; Ertl, G. Electrochemical Micromachining. *Science* **2000**, *289*, 98–101. [CrossRef] [PubMed]

5. Hirata, Y. Liga process—Micromachining technique using synchrotron radiation lithography—and some industrial applications. *Nucl. Instrum. Methods Phys. Res. Sect. B Beam Interact. Mater. At.* **2003**, *208*, 21–26. [CrossRef]

6. Zhang, K.F.; Kun, L. Classification of size effects and similarity evaluating method in micro forming. *J. Mater. Process. Technol.* **2009**, *209*, 4949–4953. [CrossRef]

7. Cheng, G.J.; Pirzada, D.; Zhou, M. Microstructure and mechanical property characterizations of metal foil after microscale laser dynamic forming. *J. Appl. Phys.* **2007**, *101*, 063108. [CrossRef]

8. Zhang, R.; Zhang, T.S. Non-die explosive forming of spherical pressure vessels. *J. Mater. Process. Technol.* **1994**, *41*, 341–347. [CrossRef]

9. Golovashchenko, S.F.; Gillard, A.J.; Mamutov, A.V. Formability of dual phase steels in electrohydraulic forming. *J. Mater. Process. Technol.* **2013**, *213*, 1191–1212. [CrossRef]

10. Li, F.Q.; Zhao, J.; Mo, J.H.; Li, J.J.; Huang, L. Comparative study of the microstructure of Ti-6Al-4V titanium alloy sheets under quasi-static and high-velocity bulging. *J. Mech. Sci. Technol.* **2017**, *31*, 1349–1356. [CrossRef]

11. Zhou, M.; Huang, T.; Cai, L. The novel nanosecond laser micro-manufacturing of three-dimensional metallic structures. *Appl. Phys. A* **2008**, *90*, 293–297. [CrossRef]

12. Gao, H.; Cheng, G.J. Laser Induced High-Strain-Rate Superplastic 3D Micro-Forming of Metallic Thin Film. *J. Microelectromech. Syst.* **2010**, *19*, 273–281. [CrossRef]

13. Gao, H.; Ye, C.; Cheng, G.J. Deformation Behaviors and Critical Parameters in Microscale Laser Dynamic Forming. *J. Manuf. Sci. Eng.* **2009**, *131*, 051011. [CrossRef]

14. Li, J.; Gao, H.; Cheng, G.J. Forming limit and fracture mode of microscale laser dynamic forming. *J. Manuf. Sci. Eng.* **2010**, *132*, 061005. [CrossRef]

15. Ye, C.; Cheng, G.J. Scalable patterning on shape memory alloy by laser shock assisted direct imprinting. *Appl. Surf. Sci.* **2012**, *258*, 10042–10046. [CrossRef]

16. Ehrhardt, M.; Lorenz, P.; Frost, F.; Zimmer, K. Laser Embossing of Micro-and Submicrometer Surface Structures in Copper. *Phys. Proced.* **2012**, *39*, 735–742. [CrossRef]

17. Ehrhardt, M.; Lorenz, P.; Frost, F.; Zimmer, K. Fabrication of sub-microstructures in solid copper surfaces by inverse laser microembossing. *Appl. Phys. A* **2013**, *111*, 517–523. [CrossRef]

18. Liu, H.X.; Shen, Z.B.; Wang, X.; Wang, H.J.; Tao, M.K. Micromould based laser shock embossing of thin metal sheets for MEMS applications. *Appl. Surf. Sci.* **2010**, *256*, 4687–4691. [CrossRef]

19. Shen, Z.B.; Gu, C.X.; Liu, H.X.; Wang, X.; Hu, Y. Fabricating three-dimensional array features on metallic foil surface using overlapping laser shock embossing. *Opt. Lasers. Eng.* **2013**, *51*, 973–977. [CrossRef]

20. Balanethiram, V.S.; Daehn, G.S. Hyperplasticity: Increased forming limits at high workpiece velocity. *Scr. Metallurgica Mater.* **1994**, *30*, 515–520. [CrossRef]

21. Samardzic, V.; Geskin, E.S.; Atanov, G.A.; Semko, A.N.; Kovaliov, A. Liquid Impact Based Material Micro-Forming Technology. *J. Mater. Eng. Perform.* **2007**, *16*, 375–389. [CrossRef]

22. Samardzic, V.; Geskin, E.S.; Atanov, G.A.; Semko, A.N.; Kovaliov, A. Investigation of Liquid Impact-Based Macro-, Meso-, and Microforming Processes. *J. Mater. Eng. Perform.* **2008**, *17*, 302–315. [CrossRef]

23. Woo, M.-A.; Noh, H.G.; An, W.J.; Song, W.J.; Kang, B.S.; Kim, J. Numerical study on electrohydraulic forming process to reduce the bouncing effect in electromagnetic forming. *Int. J. Adv. Manuf. Technol.* **2017**, *89*, 1813–1825. [CrossRef]

24. Skews, B.W.; Kosing, O.E.; Hattingh, R.J. Use of a liquid shock tube as a device for the study of material deformation under impulsive loading conditions. *Proc. Inst. Mech. Eng. Part C J. Mech. Eng. Sci.* **2004**, *218*, 39–51. [CrossRef]

25. Marai, M.; Lang, L.H.; Wang, S.H.; Lin, L.J.; Yang, C.L.; Luo, X. Investigation on the effect of liquid hammer geometry to the pressure distribution of innovative hybrid impact hydroforming. *Adv. Mater. Res.* **2014**, *941–944*, 1843–1849. [CrossRef]

26. Wang, S.H.; Lang, L.H.; Lin, L.J. Investigation on the Punching Quality of Hybrid Impact Hydroforming. *Appl. Mech. Mater.* **2014**, *602–605*, 520–523. [CrossRef]

27. Fabbro, R.; Fournier, J.; Ballard, P.; Devaux, D.; Virmont, J. Physical study of laser-produced plasma in confined geometry. *J. Appl. Phys.* **1990**, *68*, 775–784. [CrossRef]

28. Zheng, C.; Sun, S.; Ji, Z.; Wang, W.; Liu, J. Numerical simulation and experimentation of micro scale laser bulge forming. *Int. J. Mach. Tools Manuf.* **2010**, *50*, 1048–1056. [CrossRef]

29. Zhou, J.Z.; Yang, J.C.; Zhou, M.; Zhang, Y.K.; Guo, D.H.; Wu, H.X. Experimental Study on the Effects of Overlay Properties on Laser-induced Shock Waves. *Chin. J. Lasers* **2002**, *29*, 1041–1044. [CrossRef]

30. Zheng, C.; Ji, Z.; Song, L.B.; Fu, J.; Zhu, Y.H.; Zhang, J.H. Variation of fracture mode in micro-scale laser shock punching. *Opt. Lasers Technol.* **2015**, *72*, 25–32. [CrossRef]

31. Peyre, P.; Fabbro, R. Laser shock processing: A review of the physics and applications. *Opt. Quantum Electron.* **1995**, *27*, 1213–1229.

32. Wang, X.; Du, D.Z.; Zhang, H.; Shen, Z.B.; Liu, H.X.; Zhou, J.Z.; Liu, H.; Hu, Y.; Gu, C.X. Investigation of microscale laser dynamic flexible forming process—Simulation and experiments. *Int. J. Mach. Tool. Manuf.* **2013**, *67*, 8–17. [CrossRef]

33. Zhang, W.W.; Yao, Y.L.; Noyan, I.C. Microscale laser shock peening of thin films, part 1: experiment, modeling and simulation. *J. Manuf. Sci. Eng.* **2004**, *126*, 10–17. [CrossRef]

34. Mangonon, P.L. *The Principles of Materials Selection for Engineering Design*; Prentice Hall: Upper Saddle River, NJ, USA, 1999; ISBN 0132425955.

35. Johnson, G.R.; Cook, W.H. A constitutive model and data for metals subjected to large strains, high strain rates and high temperatures. In Proceedings of the 7th International Symposium on Ballistics, Hague, The Netherlands, 19–21 April 1983; pp. 541–547.

36. Bae, G.; Xiong, Y.M.; Kumar, S.; Kang, K.; Lee, C. General aspects of interface bonding in kinetic sprayed coatings. *Acta. Mater.* **2008**, *56*, 4858–4868. [CrossRef]

37. Dirikolu, M.H.; Akdemir, E. Computer aided modelling of flexible forming process. *J. Mater. Process. Technol.* **2004**, *148*, 376–381. [CrossRef]

38. Takashi, N.; Hughes, T.J.R. An arbitrary Lagrangian-Eulerian finite element method for interaction of fluid and a rigid body. *Comput. Methods Appl. Mech. Eng.* **1992**, *95*, 115–138. [CrossRef]

39. Kjellgren, P.; Hyvärinen, J. An Arbitrary Lagrangian-Eulerian finite element method. *Comput. Mech.* **1998**, *21*, 81–90. [CrossRef]

40. Barlow, A.J.; Maire, P.H.; Rider, W.J.; Rieben, R.N.; Shashkov, M.J. Arbitrary Lagrangian–Eulerian methods for modeling high-speed compressible multimaterial flows. *J. Comput. Phys.* **2016**, *322*, 603–665. [CrossRef]

41. Shams, A.; Lopresto, V.; Porfiri, M. Modeling fluid-structure interactions during impact loading of water-backed panels. *Compos. Struct.* **2017**, *171*, 576–590. [CrossRef]

42. Hou, G.; Wang, J.; Layton, A. Numerical Methods for Fluid-Structure Interaction—A Review. *Commun. Comput. Phys.* **2012**, *12*, 337–377. [CrossRef]

43. Varas, D.; Zaera, R.; López-Puente, J. Numerical modelling of the hydrodynamic ram phenomenon. *Int. J. Impact. Eng.* **2009**, *36*, 363–374. [CrossRef]

44. Boyd, R.; Royles, R.; El-Deeb, K.M.M. Simulation and validation of UNDEX phenom-ena relating to axisymmetric structures. In Proceedings of the sixth international LS-DYNA users conference, Dearborn, MI, USA, 9–11 April 2000.

45. Shah, S. Water Impact Investigations for Aircraft Ditching Analysis. Master's Thesis, RMIT University, Melbourne, Australia, 2010.

46. Engel, U.; Eckstein, R. Microforming-from basic research to its realization. *J. Mater. Process. Technol.* **2002**, *125–126*, 35–44. [CrossRef]
47. Liu, H.X.; Sha, C.F.; Shen, Z.B.; Li, L.Y.; Gao, S.; Li, C.; Sun, X.Q.; Wang, X. Fabrication of Dish-Shaped Micro Parts by Laser Indirect Shocking Compound Process. *Micromachines* **2016**, *7*, 105. [CrossRef]
48. Ehrhardt, M.; Lorenz, P.; Lotnyk, A.; Romanus, H.; Thelander, E.; Zimmer, K. Pattern transfer of sub-micrometre-scaled structures into solid copper by laser embossing. *Phys. Proced.* **2014**, *56*, 944–950. [CrossRef]
49. Zhang, X.Q.; She, J.P.; Li, S.Z.; Duan, S.W.; Zhou, Y.; Yu, X.L.; Zheng, R.; Zhang, B. Simulation on deforming progress and stress evolution during laser shock forming with finite element method. *J. Mater. Process. Technol.* **2015**, *220*, 27–35. [CrossRef]
50. Zhang, X.Q.; Zhang, Y.; Zhang, Y.W.; Pei, S.B.; Huang, Z.L.; Deng, L.; Li, S.Z. Numerical and experimental investigations of laser shock forming aluminum alloy sheet with mold. *Int. J. Mater. Form.* **2018**, *11*, 101–112. [CrossRef]
51. Liu, H.X.; Sun, X.Q.; Shen, Z.B.; Li, L.Y.; Sha, C.F.; Ma, Y.J.; Gau, J.T.; Wang, X. Experimental and Numerical Simulation Investigation on Laser Flexible Shock Micro-Bulging. *Metals* **2017**, *7*, 93. [CrossRef]

Review

Review on Advances in Metal Micro-Tube Forming

Christoph Hartl

Faculty of Automotive Systems Engineering and Production Engineering, TH Köln-University of Applied Sciences, Betzdorfer Str. 2, 50679 Köln, Germany; christoph.hartl@th-koeln.de; Tel.: +49-22-8275-2550

Received: 25 March 2019; Accepted: 6 May 2019; Published: 10 May 2019

Abstract: Metallic tubular micro-components play an important role in a broad range of products, from industrial microsystem technology, such as medical engineering, electronics and optoelectronics, to sensor technology or microfluidics. The demand for such components is increasing, and forming processes can present a number of advantages for industrial manufacturing. These include, for example, a high productivity, enhanced shaping possibilities, applicability of a wide spectrum of materials and the possibility to produce parts with a high stiffness and strength. However, certain difficulties arise as a result of scaling down conventional tube forming processes to the microscale. These include not only the influence of the known size effects on material and friction behavior, but also constraints in the feasible miniaturization of forming tools. Extensive research work has been conducted over the past few years on micro-tube forming techniques, which deal with the development of novel and optimized processes, to counteract these restrictions. This paper reviews the relevant advances in micro-tube fabrication and shaping. A particular focus is enhancement in forming possibilities, accuracy and obtained component characteristics, presented in the reviewed research work. Furthermore, achievements in severe plastic deformation for micro-tube generation and in micro-tube testing methods are discussed.

Keywords: micro-tubes; micro-tube drawing; micro-hydroforming; laser assisted; severe plastic deformation; micro-tube testing

1. Introduction

The application of micro-components in technical systems and devices today represents a crucial value-adding factor for numerous industrial sectors, such as medical engineering, healthcare, mobility or communication. Micro-manufacturing techniques for the fabrication of these components therefore play a key role in facilitating corresponding innovative and affordable products. The increasing trend in miniaturization over the past few years was a driver for the development of a broad range of novel manufacturing techniques and scaled-down conventional processes. In particular, this concerns cases where traditional methods, which are typically applied for silicon-based components of micro-electro-mechanical-systems (MEMS), reach their limits in terms of the materials to be processed or the aspect ratios, etc. [1,2]

Due to their numerous advantages, forming processes provide a promising approach for manufacturing relevant micro-parts from metal materials [3]. Micro-forming can enable near-net-shape characteristics, an enhanced shaping of complex geometries, the possibility to produce micro-parts with a comparatively high stiffness and strength [4], and the applicability of a wide spectrum of materials. Moreover, in general, forming processes provide a high-volume production for micro-components, which is required today [1,4].

These issues also apply for micro-tube forming techniques, as tubular micro-components play an important role in a broad range of industrial products for economically significant markets. Products based on micro-tubes and tubular micro-components can be found, for example, in the fields of medical engineering, electrics and electronic systems, optoelectronics, chemical systems, sensors, and thermal management, as summarized in Table 1 from [5,6].

Table 1. Examples of products based on micro-tubes and tubular micro-components.

Industries	Examples of Products	Typical Materials
Medical engineering	Painless injection needles Catheters Stents Needles for cosmetic surgery Implants	Stainless steel Titanium/titanium alloys Magnesium/magnesium alloys
Electronic and electrical engineering	Contact probes Electrode tubes for EDM	Nickel/nickel alloys Copper/copper alloys
Optoelectronics	Optical fibers Laser accelerators	Nickel/nickel alloys Kovar
Chemical technology	Micro-reactors Micro-nozzles	Stainless steel Titanium/titanium alloys
Sensing technology	Differential gas pressure detector Micro-quantitative analysis device	Platinum/platinum alloys Tungsten/tungsten alloys
Heat transfer technology	Micro-heat exchanger	Titanium/titanium alloys Aluminum/aluminum alloys

As an example, for the market of medical engineering, where tubular micro-components are strongly represented, a global revenue of approximately USD 595 billion is expected in 2024, with a compound annual growth rate of about 5.4% between 2015 and 2024 [7].

Despite the importance and industrial presence of products with tubular micro-parts and the advantages forming processes can offer for their manufacture, a review of advances in micro-tube forming, in conjunction with the related technologies for tool design and consideration of size effects, is not available at present.

There exist comprehensive overviews of the latest achievements in micro-manufacturing technology in general [1,2,6,8–11], which represent the state of the art of various micro-manufacturing processes applied to metallic and non-metallic materials, including selected forming techniques. Reviews concentrated specifically on developments in micro-forming predominantly deal with bulk and sheet metal forming at the microscale [3,12], considering industrial applications in micro-sheet metal forming [13] and techniques for the micro-forming of bulk micro-parts from sheet metal material [14]. Additionally, surveys concerning research on size effect-affected deformation behavior [15,16] and friction in micro-forming [17] mainly focus on bulk and sheet metal forming.

However, there are no reviews available that consider the micro-tube forming technology and corresponding innovations as a whole, although the economic significance of micro-tubular products and the advantages of their manufacture with the forming technology is obvious, as described above. On this background, this review paper aims to provide an overview about the latest advances in micro-tube forming technologies and to complement existing reviews dealing with micro-forming.

Following the general definitions of micro-parts within the field of micro-forming, the term "micro-tubes" includes here tubular semi-finished products and components, with typical part-dimensions or part-features in the range of sub-millimeters up to a few millimeters [11]. This review surveys forming technologies for the fabrication of metallic micro-tubular products and considers the usual classification of forming processes as processes from the group of manufacturing processes in which plastic deformation is used to change the shape of metalwork pieces. However, for the sake of completeness, reference is made at this point to related methods, for which the efficient fabrication of micro-tubular components from metals was also demonstrated, but which are based on deposition techniques, such as electroforming [18,19] and sputtering [20].

The structure of this review is based on the core processes investigated and the forming results envisaged in the individual reviewed research work that is dealing with the plastic forming of micro-tubes. This leads to a subdivision into (a) drawing processes of semi-finished products,

(b) forming processes with pressurized media of three-dimensional products, (c) the laser-assisted forming of three-dimensional products, (d) severe plastic deformation processes to achieve a defined material microstructure, and (e) expansion processes to obtain material data. Accordingly, this review provides summaries of new developments in the fabrication of micro-tubes as a semi-finished product (Section 2), research results in innovative micro-tube hydroforming processes (Section 3), published achievements in laser-assisted micro-tube forming (Section 4), new strategies in severe plastic deformation for the fabrication of micro-tubes (Section 5), and investigations into testing methods of micro-tubes (Section 6). Achievements in enhanced shaping possibilities, accuracies, component properties, process stability, and modeling techniques for the individual groups of forming techniques are considered and discussed, where corresponding results were available.

2. Micro-Tube Fabrication

Recent developments in the fabrication of micro-tubes aim in particular to provide improved semi-finished products, with more economic manufacturing processes, for use in microsystem-based applications of medical engineering and microfluidics for heat exchangers. When primarily considering the fabrication of seamless metallic tubes, the investigated technologies can be divided into (a) dieless drawing techniques [5,21–35], (b) process chains, applying scaled-down and partly modified tube drawing [36–47], and (c) manufacturing techniques of grooved micro-tubes for applications in heat exchange [43,48–54]. Apart from this research into tube manufacturing processes, notable progress was also achieved in the design of the process and tools for the profile extrusion of tubular micro-profiles with a number of channels [55–59]. Tables 2–5 summarize the details of this research concerning the investigated technologies, micro-tube dimensions, with an outer diameter of d_o and tube wall thickness of t, materials, and designated applications in experimental and practical works.

Table 2. Reviewed experimental research on the dieless drawing of micro-tubes.

		Tube Dimensions			
Listed Shaping Techniques	**Material**	d_o (mm)	t (μm)	**Designated Application**	**Reference**
Machining, hot extrusion, dieless drawing (IH)	Superplastic alloy (Zn-22Al)	0.19	49.5	-	[35]
Machining, hot extrusion, multi-pass dieless drawing (IH)	Superplastic alloy (Zn-22Al)	0.343	91	MN, SP	[32]
Rotary laser multi-pass dieless drawing	β Titanium alloy (Ti-22V-4Al)	<0.5	<130	CA, ME, ST	[23]
Rotary laser dieless drawing	Stainless steel (SUS304)	<0.5	<130	-	[25,27]
Dieless drawing (IH)	Stainless steel (SUS304)	<0.5	<130	-	[24]
Laser dieless drawing	Stainless steel (SUS304)	<0.52	<100	MN, IN, MH	[29,30]
ECAP, hot extrusion, dieless drawing (IH)	Magnesium alloy (AZ31)	<2.0	<500	BI	[31]
Extrusion, multi-pass dieless mandrel drawing (IH)	Magnesium alloy (AZ31)	3.35	690	ME, SA, CH	[21]
Extrusion, rotary laser dieless drawing	Magnesium alloy (AZ31)	<5.0	<500	–	[22]
Dieless drawing (IH)	Stainless steel (SUS304)	<6.35	<1000	–	[26]
Hot extrusion, multi-pass dieless drawing (IH)	Superplastic alloy (Zn-22Al)	a	–	MH, CO, FC, ET	[28]

a, rectangular and multi-core rectangular tubes, with a channel size of 0.533 × 0.923, 0.58 × 0.58, and 0.75 × 1.3 (mm); BI, biodegradable implants; CA, catheters; CH, chemical industry; CO, commutators; ECAP, equal-channel angular pressing; ET, tubes for electro-discharge machining; FC, fuel cells; IH, inductive heating; IN, injection needles (painless, hypodermic); ME, medical technology; MH, micro-heat exchanger; MN, micro-nozzles; SA, sanitary; SP, stepped tubes; ST, stents.

Table 3. Reviewed experimental research on scaled-down micro-tube drawing and variants.

Listed Shaping Techniques	Material	d_o (mm)	t (μm)	Designated Application	Reference
Multi-pass fluid mandrel cold drawing	Stainless steel (SUS304)	0.2	42	ME, IN	[43]
Multi-pass cold drawing (FP, MD)	Shape memory alloy (Ti (51 at. %)-Ni)	1.48	210	CA, ST	[46]
Hot extrusion, multi-pass mandrel cold drawing	Magnesium alloy (Mg-Zn-Y-Nd)	2.0	150	BS	[36]
Multi-pass cold tube sinking	Titanium alloy (Ti-0.2Pd)	2.4	400	ME	[42]
Extrusion, drilling, cold-rolling, cold drawing (SI, MD)	Zinc alloy (Zn-5Mg-1Fe) (wt. %)	2.5	130	BS	[38]
Extrusion, machining, hot indirect extrusion, mandrel cold drawing	Magnesium alloy (ZM21)	2.9 2.9	200 220	BS	[40,44]
Hot extrusion, multi-pass cold-rolling, multi-pass mandrel cold drawing	Magnesium alloys (Mg-Nd-Zn-Zr, AZ31, WE43)	3.0	180	BS	[39]
Multi-pass fluid mandrel cold drawing	Magnesium alloy (AZ31)	3.6	610	ME	[45]
Single-pass mandrel cold drawing	Shape memory alloy (Ni(56 wt %)-Ti)	4.0	400	-	[37]
Hot extrusion, cold-rolling	Magnesium alloy (AZ31)	<7.71	<900	BS	[41]
Multi-pass cold drawing (SI, FP)	Shape memory alloy (Ti (49.3%)-Ni)	<8.0	-	ME	[47]

BS, biodegradable stents; CA, catheters; FP, fixed plug; IN, injection needles (painless, hypodermic); MD, mandrel drawing; ME, medical technology; SI, sinking; ST, stents.

Table 4. Reviewed experimental research on the fabrication of grooved micro-tubes.

Listed Shaping Techniques [a]	Material	d_o (mm)	t (μm)	Number of Grooves	Reference
Tube spinning, multi-pass sinking	Copper	3.0	>223	18	[50,51]
Multi-pass fluid mandrel drawing	Copper	<5.0	>364	45	[49]
Multi-pass fluid mandrel drawing	Copper	5.1	300	55	[43]
Tube spinning, multi-pass sinking	Copper	<6.0	<480	12, 18	[53]
Tube spinning	Copper	6.0	300	60	[54]
Tube spinning	Copper	6.0	-	60	[52]

[a] all processes conduced at room temperature.

Table 5. Reviewed experimental research on the extrusion of profiles with multiple micro-channels.

Material	w (mm)	h (mm)	t (μm)	Number of Channels	Reference
Aluminum alloy (AA3003)	16.0	1.8	250	10	[57]
Aluminum alloy (AA1100)	16.0	2.0	300	10	[55,56]
Aluminum alloy (AA3003)	–	–	250	17	[58]
Aluminum alloys (A1100, A3003)	–	–	>200	12	[59]

[a] cross-section dimensions, w, width; h, height; t, wall thickness.

2.1. Dieless Drawing

The dieless drawing of micro-tubes is an innovative technique to achieve comparatively larger reductions of the tube cross-section area within a single drawing step. Research results for the investigated micro-tube materials are available for zinc-aluminum-based superplastic alloys, titanium alloys, and stainless steel, as shown in Table 2. The forming principle, represented in Figure 1, together with fabricated micro-tube examples, consists in the local plasticization of the tube material by a heat source and the application of a tensile stress to the tube. Due to the decreased yield strength in the heated area and the superimposed tensile stress state, the dimensions of the tube cross-section start to be locally reduced. A controlled relative movement between the heat source, together with a cooling unit, and the tube then allows for the transfer of these reduced dimensions to the entire tubular workpiece. A distinction can be made between continuous and non-continuous dieless drawing, depending on how this relative movement is achieved [5].

Due to the fact that dieless drawing is conducted without rigid forming dies, difficulties in the fabrication and handling of scaled-down drawing dies and the influences of size effects in friction can be avoided [34]. As an advantage of size effects, it can be expected that the increase in heating and cooling rates at the microscale will lead to an increase in feasible drawing speeds for dieless micro-tube drawing [24,33].

The forming temperature, distance between heating and cooling, tensile velocity, and the material characteristics of the formed tube are essential parameters determining a feasible reduction in the cross-section area [32]. Typical limits for the achievable cross-section reduction include an unequal plastic deformation of the produced micro-tube and its possible fracture [5]. As an example, Figure 2 shows the effect of the temperature on forming limits for different materials and micro-tube dimensions, indicating that reductions in cross-sections of up to 80% can be obtained in a single pass, considering the example of micro-tubes made from the material, Zn-22Al [23].

Research on heat sources for dieless micro-tube drawing and on the influencing parameters in the forming process predominantly concerns high-frequency induction heating systems, as shown, e.g., in [5,28,32], and laser heating, without [29,30] and with rotating relative movement between the micro-tube and the laser beam [25,27]. A rotating movement provides a more uniform temperature distribution and enhanced forming limits, compared to the use of laser irradiation, without rotation [25], but non-uniformity in the optical properties of the workpiece can lead to local overheating [22]. Independently from the type of heat source used, an effective temperature in the forming zone cannot be achieved when the drawing speed is faster than the thermal conductivity [33], whereby typical feeding velocities for dieless micro-tube drawing are in the range of about 0.06 to 0.5 mm/s [23]. It was found that increasing the length of the heating area at the formed tube led to a decrease of the mean strain rate within the formed material [60], and that, in turn, led to an enhancement of the limiting reduction in the cross-section area [60]. Further studies on the relationships between the drawing velocity, heating temperature, and forming limit were presented, for example, in [26] for the drawing of mini-tubes made of stainless steel.

The grain size of the dieless drawn micro-tubes increases slightly during forming. From the example of the drawing of tubes made from the alloy, AL78-ZN, an increase of the average grain size d_g from 0.5 to 1.1 μm was determined for a dieless drawing with three passes and a reduction in the cross-section area of 66.7% per pass [32]. In this case, the surface roughness of the formed micro-tubes did not change significantly during forming [32]. For the drawing of stainless-steel micro-tubes in the temperature range of 950 to 1100 °C in argon atmosphere and atmospheric oxygen, respectively, it was found that the oxidation and surface roughening affected the drawing limit and varied widely, depending on the temperature [24].

A specific peculiarity of dieless tube drawing processes is the fact that the ratio of the outer to the inner diameter of the drawn tube remains constant during forming, irrespective of the used tube material and tube dimension, even after a multi-pass dieless drawing [34,35], and this is also achievable for non-circular multicore tubes [28]. A theoretical explanation of this experimentally proven effect was provided in [34]. This similarity of cross-section change simplifies the process design. Additionally, the drawing process can also be conducted with the use of a wire mandrel, where the tube is drawn with this inserted mandrel to specify a selected inner tube diameter. This novel strategy was examined for the dieless drawing of magnesium micro-tubes, with obtained cross-section reductions of about 58,3% [21]. Based on this technique, a concept for the manufacture of bio-absorbable micro/mini-tubes, made from magnesium alloys for vascular anastomosis and for stents was suggested, with fabricated outer tube diameters d_o in the range of 0.4 up to 2 mm and 0.1 up to 1 mm, respectively, using the drawing with rigid dies as the final forming step [21]. Figure 3 represents this concept, together with a comparison with the fabrication of such products using solely rigid drawing dies [21]. These concepts can help to eliminate difficulties in dieless drawing associated with dimensional accuracy and surface roughness [21].

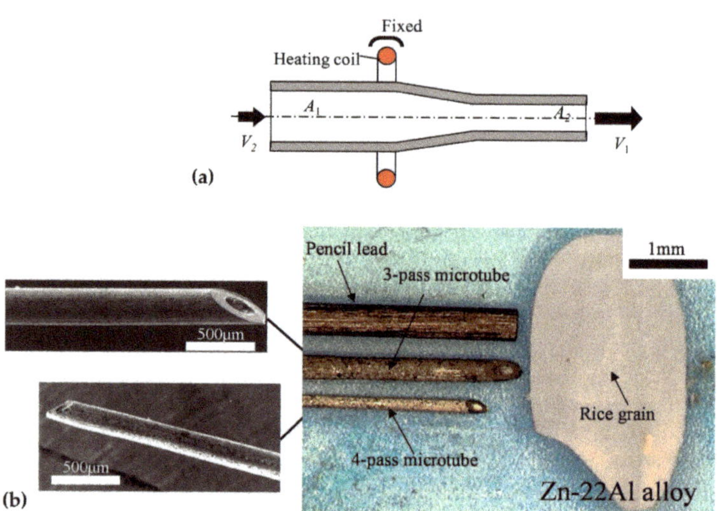

Figure 1. Dieless drawing of micro-tubes: (**a**) process principle (A_1, A_2: tube cross-section areas; V_1, V_2: drawing velocities) and (**b**) fabricated micro-tube examples. Republished with the permission of Trans Tech Publications, from reference [23], copyright 2016; permission conveyed through Copyright Clearance Center, Inc.

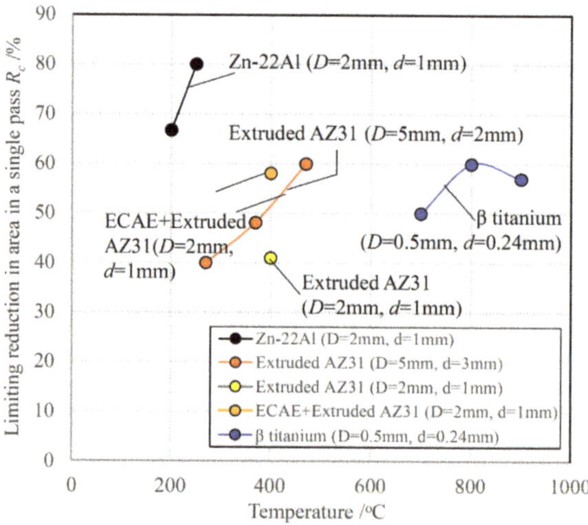

Figure 2. Effect of the heating temperature on the limiting reduction in the cross-section area (D, outer tube diameter; d, inner tube diameter; ECAE, equal channel angular extrusion). Republished with permission of Trans Tech Publications, from reference [23], copyright 2016; permission conveyed through Copyright Clearance Center, Inc.

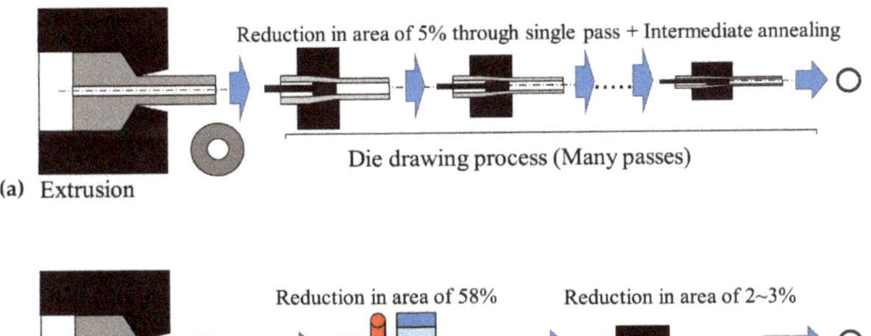

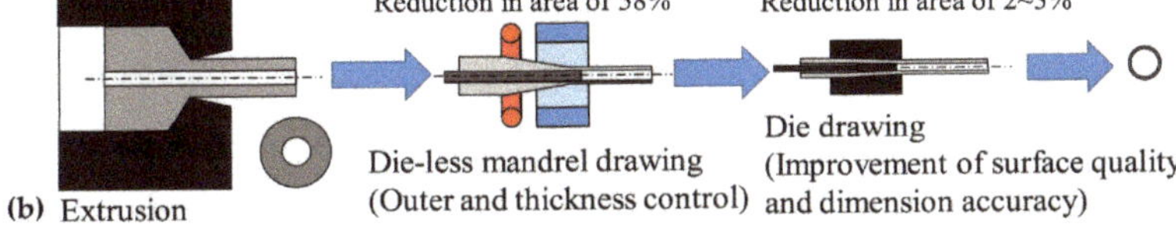

Figure 3. Conceptual illustration for the mass production of magnesium alloy micro-tubes, with an improved accuracy and quality: (**a**) conventional manufacturing process die drawing and (**b**) proposed manufacturing process using dieless mandrel drawing. Reprinted from reference [21], copyright 2018, with permission from Elsevier.

Developed models for the numerical analysis of dieless drawing processes were based on thermo-mechanical-coupled finite element simulations [5,22,33,34] and finite difference methods [60], and they have shown a good accuracy and reproducibility when material flow curves, which consider the dependencies between the flow stress and strain, and strain rate and temperature, were implemented.

2.2. Scaled-down Tube Drawing and Variants

Research into process chains using scaled-down tube drawing processes focused in large part on the processing of degradable biomedical metallic materials, such as alloys based on magnesium [36,39–41,44] or zinc [38], as shown in Table 3. These alloys have attracted extensive attention as materials for metallic stents for the treatment of cardiovascular disease, since there is no requirement for a secondary removal surgery. However, the hexagonal close-packed crystal structure of these materials significantly reduces cold formability. Furthermore, the application of the fabricated tubes for stents requires certain properties of these semi-finished products relating to, for example, strength, accuracy, and a metallic microstructure. The latter influences the biodegradability of the final component, as a sufficient degradability is related with a low density of crystallographic defects and a fine and homogenous microstructure [36].

For the manufacture of stents made from a magnesium alloy, a process chain was developed, starting with the hot extrusion of a tubular blank and subsequent multi-pass cold drawing operations, combined with heat treatment between the forming steps [36]. A dimensional error of less than 3.8% was achieved, and a mean grain size of 4.4 µm was determined for the final tubes, after annealing. It was shown that the mechanical properties and bio-corrosion improved due to the microstructure, after the achieved grain refinement [36]. The process chain described in [39] and applied to three different magnesium alloys was similar, but it had several cold-rolling steps between the hot extrusion of the blank and the multi-pass drawing operations with intermediate annealing. Dimensional errors below 2.8% were determined for the fabricated micro-tubes, and the influence of the obtained microstructure of the individual alloys on the mechanical properties and corrosion behavior was analyzed. The average grain size of the produced micro-tubes in this study were in the range of 10.9 and 15 µm [39]. Investigations into the effect of cold-rolling on the mechanical properties and evolution of the twinning of magnesium alloy tubes under varied processing parameters showed that different types of twinning dominated the mechanical behavior of the formed tube [41]. The type of twinning that appeared depended on the selected dimensional parameters for the cross-section-area changes in the rolling process [41].

A comparatively short process chain for the fabrication of micro-tubes made from magnesium was developed and demonstrated [40,44]. Extrusion processes were applied to generate a solid billet that was first machined and then formed into a tube by a hot indirect extrusion process, followed by four cold drawing steps, without intermediate annealing. The indirect extrusion process could be considered to contribute to the refining of the grain structure, which showed an average grain size of 10 μm [40] and 15 μm [44], as shown in Figure 4. A dimensional error below 5.5% was achieved [44].

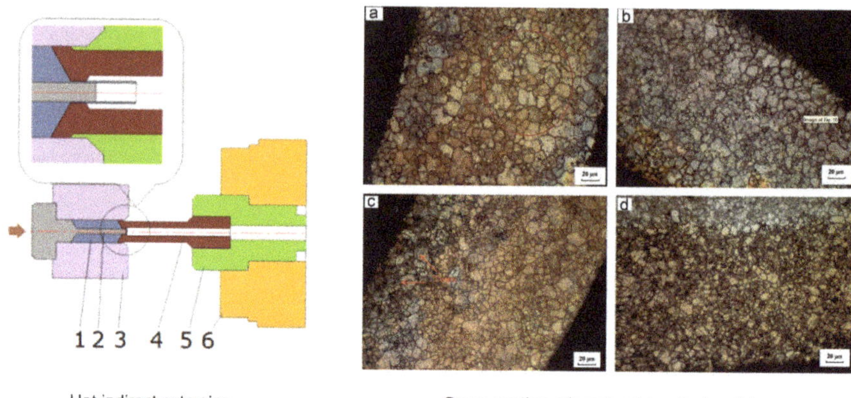

Figure 4. Schematic illustration of hot indirect extrusion (1, hollow billet; 2, mandrel; 3, container; 4, die; 5, die set; 6, front plate) and cross-section microstructure of fabricated micro-tubes, after: (**a**) hot indirect extrusion; (**b**) two-pass drawing; (**c**) three-pass drawing; and (**d**) four-pass drawing. Reprinted from reference [40], copyright 2014, with permission from Elsevier.

Investigations into the use of a novel biodegradable zinc alloy for stent fabrication were documented in [38]. The developed process chain consisted of drilling an extruded billet, several cold-rolling operations, and multi-pass drawing, without and with mandrel. Each forming operation was followed by a heat treatment process of the components, and an average grain size of 20 μm was attained, with a dimensional accuracy for the final tubes below 3.8% [38]. The mechanical properties and corrosion behavior were determined, and prototypes of the stents were tested mechanically.

Most of these studies report that mandrel drawing was used for the micro-tube drawing operations. In mandrel drawing, the tube is drawn over an inserted mandrel that passes through the die with the tube. For the fabrication of micro-tubes made from a Ni–Ti shape-memory alloy, mandrel drawing provided higher drawing limits, compared to drawing with a fixed plug or tube sinking without any plug [46]. The increase of surface roughness was found to be an additional limiting factor in producing micro-tubes through tube sinking processes [42,46]. The use of soft copper material for the mandrel could enable a maximum reduction in the cross-sectional area through mandrel drawing of about 27% for the drawing of Ni–Ti alloys [46]. The drawing of tubes from Ni–Ti alloys with fixed plugs turned out to be limited to outer diameters of the tubes above 4 mm due to difficulties in joining the plug to the supporting rod, and the forming with floating plugs led to unstable processes, caused by breaks of the drawn material [46]. Investigations based on numerical simulations using hyperelastic and the Mullins effect model, in conjunction with experiments, have shown that the dimensional accuracy of micro-tubes made from Ni–Ti alloys could be improved using brass, instead of copper, for the mandrel material, but that drawing stresses slightly increase [37]. However, the drawbacks of the mandrel drawing of micro-tubes consist in difficulties relating to the removal of the mandrel from the drawn tube and a lower surface quality, compared to that of drawing with fixed plugs [46].

A possible way of avoiding complications in removing a rigid mandrel from the drawn micro-tube consists in the use of liquids, such as water or oil, within the sealed tube during the drawing process, instead of the mandrel [45]. For the fabrication of micro-tubes made from stainless steel, with $d_o = 2.0$ mm, by performing 21 passes, without intermediate annealing, the application of this fluid-mandrel drawing technique resulted in an increase of the wall thickness that was about 42% lower compared to that of the use of sinking [43], as shown in Figure 5.

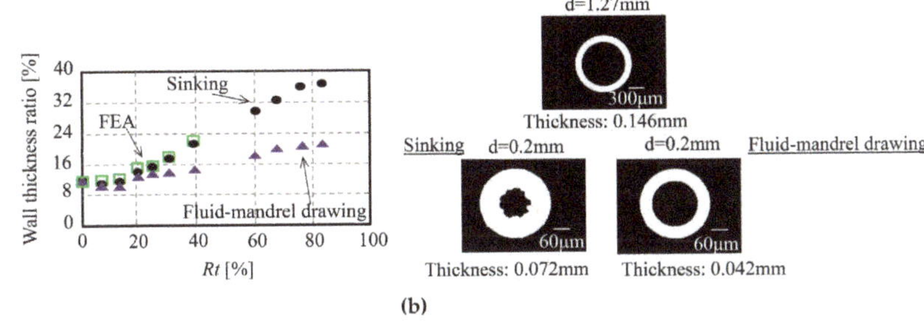

Figure 5. Comparison of fluid-mandrel drawing and sinking: (**a**) development of the wall thickness ratio versus the ratio R_t of the reduction of the outer tube diameter and (**b**) the cross-section shapes of the initial tube (above) and the drawn micro-tubes (below). Republished with the permission of Trans Tech Publications, from reference [43], copyright 2014; permission conveyed through Copyright Clearance Center, Inc.

2.3. Grooved Tubes and Non-Circular Tubular Products with Micro-Channels

Metallic tubular micro-products are gaining importance for applications in heat exchange technology [49]. Micro-tubes that are ribbed or grooved on the inside are of interest, for example, in relation to micro-heat pipes in electronic products [51], as shown in Table 4, and extruded profiles with micro-channels are suitable for improving the heat transfer efficiency in items, such as automobile air conditioners, radiators, or gas/fluid coolers [58], as shown in Table 5. For example, it was reported that a reduction of a heat exchanger channel, from diameters between 2 and 5 mm to less than 1 mm, can increase the heat transfer efficiency by more than 200% [61].

A conventional method—described in the corresponding literature as a spinning process—to manufacture grooved tubes, with outer diameters above 5 mm, consists in applying pressure using rotating steel balls on a tubular blank, which is supported inside with a ripped plug and drawn additionally in the axial direction [49,50]. However, similar to the plug drawing of smooth tubes, this method is limited to tube outer diameters above 4–6 mm due to difficulties in fabricating the plug and in its handling [53]. Specified dimensions of the grooves and of the intermediate ribs have to be obtained by the forming process to ensure the necessary properties of the final product in the heat transfer. Important parameters determining the quality and feasible dimensions of the formed inner grooves for this minimum producible diameter range are the number of spinning balls, the position of the plug, and the drawing speed [52,54].

Within certain limits, smaller diameters can be achieved by applying multi-pass tube sinking to a grooved tube, fabricated by the spinning process, described above [50,51,53]. However, this method is limited, because the ratio of the width of grooves to the width of the ribs reduces with each drawing pass, and after a certain number of passes, the grooves are closed by the deformed ribs [49,53]. Furthermore, an excessive increase in the wall thickness could be observed [49,53]. Investigations into drawing such grooved tubes with a liquid medium inside—according to fluid mandrel drawing [45]—have shown that these drawing limits of tube sinking could be extended considerably [43,49]. Figure 6 illustrates the change in wall thickness for the multi-pass drawing of a grooved micro-tube with different media, as an example. Finite element models were developed for the analysis and optimization of the multi-stage sinking of straight- [53] and spiral-ribbed micro-tubes [48], considering three-dimensional forming conditions.

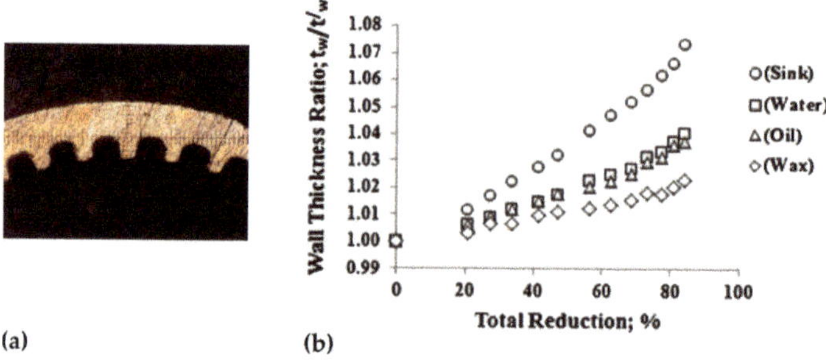

(a)   (b)

Figure 6. Fluid-mandrel drawing of grooved micro-tubes: (**a**) cross-section of the initial tube made from copper, with d_o = 5 mm and t = 364 μm, and (**b**) the development of the wall thickness ratio versus the total reduction area for different drawing media. Reprinted with permission from Springer Nature, reference [49], copyright 2014.

Achievements in the fabrication of extruded tubular profiles made from aluminum alloys with micro-channels (Figure 7) by hot extrusion processes involved investigations that led to a deeper understanding of the microstructure evolution during extrusion and of influences on weld seam strength.

Figure 7. Examples of cross-sections of a micro-channel tube and a conventional tube. Reprinted from reference [58], copyright 2014, with permission from Elsevier.

In [55], a viscoplastic self-consistent model, considering crystal plasticity, together with finite element analysis and a flow line model, was employed to investigate the microstructure evolution during the extrusion process of a micro-channel tube, with ten square channels of a width of 1400 μm. It was found that the rotation of shear planes caused a texture change during the material flow, and relationships between the grain refinement and shear effects were determined [55].

A detailed study on the influence of the extrusion process and subsequent rolling and brazing operations on the microstructure and mechanical properties of this profile type was conducted [56]. Experimental comparisons of the extrusions of micro-channel tubes made from different aluminum alloys have shown that the alloy, A3003, provided a more reduced grain growth, due to the following brazing process, and a higher strength than the alloy, A1001 [59]. The effect of the extrusion die design on the quality of the weld seam strength of micro-channel tubes, with a channel size of 500 μm × 500 μm, was the subject of research in [57] and [58], supported by experiments, numerical simulations, and physical simulations.

2.4. Discussion

The reviewed research on dieless drawing for micro-tube fabrication shows that clear advantages of this technology consist in: (a) the possible large degree of cross-section reduction in a single drawing pass, which allows the number of manufacturing steps to be crucially reduced and, with this, production costs; (b) the feasibility to process materials that are difficult to form, but that are of great interest for important fields, such as medical technology; and (c) a comparatively large knowledge base concerning fundamentals, documented in numerous publications. The need for development is seen in measures for improving the manufacturing accuracy, where possible strategies were already pointed out, for example, the combination with conventional drawing processes.

Compared to dieless drawing, scaled-down drawing processes, summarized in Section 2.2, can provide a comparatively high and feasible manufacturing accuracy. However, for the processing of materials that are difficult to form, such as materials with a hexagonal lattice structure, manufacturing is typically accompanied by a greater number of manufacturing steps and annealing operations. This usually results in rising manufacturing costs and more difficulties in guaranteeing a consistent product quality. Therefore, it is useful to intensify research on strategies that also consider a reduction of process steps, as already demonstrated in isolated cases. The technique of fluid mandrel drawing has been shown to be able to provide advantages in facilitating micro-tube manufacturing, but there are still several outstanding issues, such as its capability in relation to high-volume production and correlations between the forming result and varying tube dimensions and tube materials.

The techniques discussed in Section 2.3 are of importance for industries that provide heat exchanger systems for a broad range of consumer goods and industrial installations. The miniaturization of extruded micro-channel tubes can be considered to be at a comparatively advanced stage, and existing research results deal with a holistic view of correlations between the manufacturing chain and product properties. A significant advantage of the extrusion process can be seen in its suitability for mass production, with a low number of manufacturing steps. For improvements in the miniaturization of grooved micro-tubes, approaches were developed and investigated based on the concept of fluid mandrel drawing. As in the case of the drawing of smooth micro-tubes, the need for investigations into the capability of high-volume production is also seen here as an outstanding issue.

3. Micro-Tube Hydroforming

In tube hydroforming, a tubular blank is plastically formed into a three-dimensional shape within a forming die cavity by applying pressurized liquid or gaseous media on the inside of the workpiece [62]. Additional mechanical loads on the formed component can be utilized to contribute to the plastic flow of the formed material or supporting areas of the component during expansion [62]. Applications of hydroforming at macro-scale predominantly concern the industrial production of thin-walled, complex-shaped components for numerous sectors, such as the automotive and aerospace industry or sanitary and piping manufacture [63–67]. Scaled-down hydroforming offers an economic potential for the micro-manufacturing industries, for example, for the production of metallic components for medical technology, elements of fuel cells, micro-fluidic devices, micro-heat exchangers, or hollow micro-shafts [6,68,69].

Research into scaling-down hydroforming for the manufacture of micro-tubes started about 10 years ago and has predominantly focused on the issues of (a) process development [69–84] and (b) appropriate tool and machine design [68,77,85–91]. In both cases, size effects due to miniaturization played an important role. This concerned not only size effects in the forming behavior of the micro-tubes that influence the forming process and the result [76,79,80,83,84], but also the effects on friction [86] as well as difficulties in scaling the conventional hydroforming tooling to the microscale [86]. A particularity of hydroforming consists in the comparatively large number of necessary forming loads [68]. This requires a tool design that ensures its sufficient high durability, with a minimum of elastic deflection and a reliable sealing of the formed components under the applied fluid pressure.

Fluid pressures of the order of 50–120 MPa were applied within the context of published investigations into the hydroforming of micro-tubes [68,70,75].

3.1. Process Development

For the design and optimization of hydroforming processes and components, knowledge about forming limits is required to estimate the feasibility of the process and to determine the necessary process loads for the forming operation. Corresponding research on these limits and suitable load paths have been published for the micro-hydroforming of tubular blanks with outer diameters in the range of 500 to 2000 μm, summarized in Table 6, which contains corresponding information about the tube material, dimensions, process type, and envisaged workpiece shape.

Table 6. Reviewed experimental research into the fabrication of micro-hydroforming parts.

Hydroforming Process Type	Material	Tube Dimensions d_o (mm)	t (μm)	Designated Application	Reference
T-shape forming (AF)	Copper (1220-H)	0.5	100	–	[71,72]
T-shape forming (AF)	Stainless steel (SUS304)	0.5	100	–	[70]
Cross-shape forming (AF)	Copper (1220-H)	0.5	100	–	[70,71]
Non-axisymmetric expansion	Stainless steel (AISI 304)	0.8	40	MF	[68,89]
Rotationally symmetrical expansion	Stainless steel (AISI 304)	0.8	40	MS	[89,91]
Rotationally symmetrical expansion. (AF)	Stainless steel (SS 304)	1.0	100	–	[75]
Rotationally symmetrical expansion (AF)	Platinum (PtIr10)	1.65	76	ME	[78,79]
Rotationally symmetrical expansion (AF) Axisymmetric expansion (AF) T-shape forming (AF) Y-shape forming (AF)	Stainless steel (SS 304)	2.0	160	–	[75]
Rotationally symmetrical expansion (AF) Axisymmetric expansion (AF) T-shape forming (AF) Y-shape forming (AF)	Stainless steel (SS 304)	2.0	200	–	[85]

AF, with axial feeding; ME, medical technology; MF, micro-fluidic demonstrator; MS, micro-shaft demonstrator.

The results of the investigations into the micro-hydroforming of tubes, with an outer diameter of $d_o = 500$ μm and a wall thickness of $t = 100$ μm, into T-shaped (Figure 8a) and cross-shaped components were presented in [70–73]. The investigated load paths in these studies consisted of the controlled application of an axial movement of the tube ends, generating an axial compressive stress and an increase of the internal pressure. Different crack formations within the expanded component area and wrinkling and buckling in non-expanded zones were determined as the plastic instabilities that limited the range of applicable loads for the hydroforming of T-shaped and cross-shaped micro-parts [70–73], as shown in Figure 8b. For the expansion of T-shaped micro-components, an excessive local increase in the wall thickness, opposite to the expansion, also limited the process [72]. Within the frame of the research on a novel micro-hydroforming tool concept, described in Section 3.2, the micro-hydroforming of Y-shaped, T-shaped, axisymmetric, and rotationally symmetrical components, with $d_o = 1.0$ and 2.0 mm, was tested, based on load paths determined using finite element simulations [74,75].

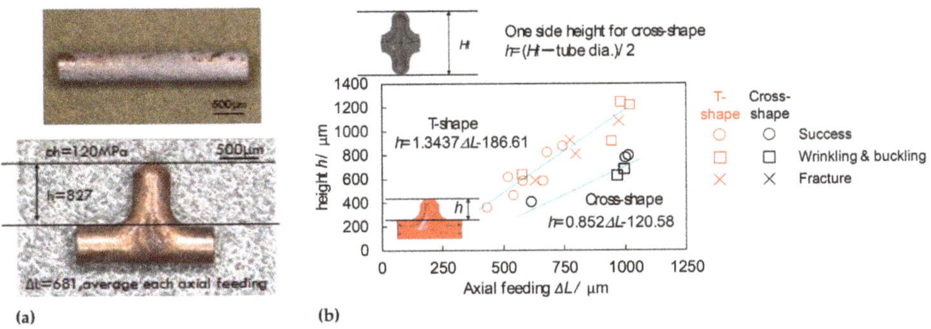

Figure 8. Micro-hydroforming of tubes made from copper: (**a**) the forming result, using fluorocarbon spray as a lubricant, and (**b**) the relation between the protrusion height and the axial feeding. Reprinted from reference [70], copyright 2017, with permission from Elsevier.

Besides the selected amount of forming loads and their control during the forming operation, the grain size of the tube material can crucially influence the feasible expansion in micro-hydroforming and the scattering of the formability. For the micro-hydroforming of stainless-steel tubes, with $d_o = 800$ μm, $t = 40$ μm and an average grain size d_g between 16 and 26 μm, it was found that the possible expansion of the micro-tubes, until the appearance of crack initiation, was lower than that for the forming of conventionally sized macro-tubes, made from the same material [83]. This could be attributed to size effects, resulting from the reduced ratio of the tube wall thickness to the average grain size t/d_g of the tube material. In [80,84], it was predicted theoretically, using a crystal plasticity finite element-based modeling technique, that the localized necking of the micro-tube wall was related to the angle between the crystal slip systems and the hoop stress direction.

As a result of the small number of grains in the tube wall direction, necking and subsequent bursting progresses faster in the expansion of micro-tubes than in macro-tubes, which can be regarded as a homogeneous continuum due to their larger number of grains. Furthermore, with a decreasing ratio t/d_g, an increase in the scattering of process parameters and forming results can be observed. This was found by [76,78] in comparative studies of micro-tubes made from stainless steel, with a ratio of $t/d_g = 3.3$, and tubes made from a platinum alloy, with $t/d_g = 1.1$. For the example of a rotationally symmetrical expansion, where merely the internal pressure and no axial compressive stress was applied, the scattering of the internal pressure at the occurrence of crack formation turned out to be 16 times higher for the tubes with the smaller ratio t/d_g [76,78]. With the application of a superimposed axial stress, a reduction of this scattering by about 77% was achieved, combined with an increase in feasible expansion by about 114% [79]. For the micro-hydroforming of T-shaped and cross-shaped micro-components, it can be assumed that, when more than approximately 20 grains are located along the wall thickness of the formed micro-tube, grain-size-based size effects do not occur [70].

The application of heat energy to the micro-hydroforming process by direct electric resistance heating was tested by [69,81,82], with the objective of reducing the required deformation energy for micro-tube hydroforming and to increase the material formability. Experiments with stainless steel tubes, with $d_o = 2.1$ mm and $t = 150$ μm, were conducted in a tool for free expansion, without axial forces, using water as a pressurizing medium and a current density of up to 26.7 A/mm^2. As essential results of these investigations, it was shown that the necessary forming pressure could be reduced [82], that the microstructure influenced the achievable forming temperature [69], but that non-uniform deformations of the formed micro-tubes were to be observed [81].

The lubrication aspects of micro-hydroforming were part of the investigations into fabricating T-shaped and cross-shaped components [70]. In particular, galling between the forming die and the micro-tube, caused by a high amount of friction, limited the process [72]. According to the presented results, the coating of the tubular blanks with a fluorocarbon spray enabled a bulging height that was more than two times higher, compared to forming without any lubricant [72]. Teflon sheets were used in the micro-hydroforming experiments conducted in [75].

3.2. Tool and Machine Design

Currently, three different concepts of micro-hydroforming systems have been proposed, differing, in particular, in the way they achieve the pressure supply from the pressurizing medium to the workpiece, together with the sealing of the tube ends and the design of the axial movement of the sealing punches. One specific issue that had to be solved with the individual systems consisted in the design of reliably working sealing punches. The punch-tube-die junction has to fulfill several requirements: the material feed, the sealing, and the high-pressure supply, which present a huge challenge in scaling down the conventional hydroforming process to the micro level [85]. Furthermore, the micro-hydroforming process is also affected by the ability to manufacture small and accurate features on the punch [86].

The device described in [87] was based on a bolted setup for the forming dies, a pressure intensifier that was driven by a press, and sealing punches that were positioned and moved via screw threads. An internal pressure for the micro-tube forming of 400 MPa could be achieved using the system. To conduct the pressurizing media to the inside of the workpiece, the sealing punches were notched axially around their outer circumferences and radially at their end faces. In this way, the pressurizing medium could be transported along the outside of the punch to the inside of the micro-tube. A radius of 25 μm was selected for the notches [87].

A floating die assembly concept was presented in [74,75] that was based on decoupling the requirements at the die-tube-punch junction, as applied in conventional tube hydroforming systems. This tooling assembly was divided into two levels: a high-pressure housing and a floating micro-tube hydroforming die assembly. With this, difficulties in developing a reliable sealing system were eliminated, the system decoupled the interdependence of the material feed and the tube sealing, and punches of various geometric configurations could be designed [75]. To clamp the pressure housing chamber, the tool system was integrated into a 1500 kN press. It was equipped with feed actuators for the punch movement, a computer-controlled pressure intensifier for the pressurization of the formed tubes up to 140 MPa and feed actuators for the punch movement. Similar to the concept described before, notched sealing punches, fabricated by wire electro-discharge machining from A2 tool steel and heat treated to 60 HRC, were developed here as well. However, merely one notch was implemented on the individual punch, with a radius of 0.4 mm [75].

To investigate micro-hydroforming under production conditions, a system was developed with the method of scaling down, in large part, macro-scale hydroforming to the microscale, but considering issues concerning elastic tool deflection caused by comparatively high loads on miniaturized tool elements [68,77,90,91]. The system was equipped with a spindle-driven pressure intensifier, which allowed an internal pressure of up to 400 MPa to be applied. The closing force of the hydroforming dies was realized by a hydraulic press, with a maximum force of 20 kN, and the sealing punches were moved by linear actuators with spindle gears. For the control of all functions and recording of experimental data, the system was equipped with a computer-based control system. An important topic in tool design was avoiding the occurrence of relative deflection between the sealing punches and the die cavity in the vertical direction due to elastic deformations, generated by the tool closing force. Such deflections can cause unreliable sealing during hydroforming, an increasing wear of the sealing punches or even damage to the tool elements [68]. Against this background, a mounting of the sealing punches was designed, which ensured that the vertical position of these punches corresponds to the position of the die cavity, irrespective of elastic tool deflections [68]. Scaled-down traditional hollow tapered punches were used in the experiments with this system for micro-tubes with $d_o > 1.0$ mm and a conical punch design for smaller tubes [77].

Several design variants for micro punches were proposed and analyzed with the aid of finite element simulations [85,86], analytical models [86], and experiments using a floating die assembly concept [85]. It was shown that partially and fully notched punches induced von Mises stresses below the yield stress of conventional tool steel and could withstand the material feed load required to form Y-, T-, and bulge shapes from SS304 steel micro-tubes, with outer diameters of 1 and 2 mm [85]. For the investigated micro-tubes and the die system, it was found that the effective punch length should not exceed 10 mm to avoid the buckling of these tool elements [85]. The feasibility of the scaling traditional macro-sized hydroforming tools to the microscale was investigated in [86], in which the effects of scaling the tooling are conceptually observed, and a mathematical model was developed to determine the stresses in tapered punches, as shown in Figure 9. One major conclusion was that current manufacturing techniques are not capable of accurately creating holes that are small enough for an acceptable stress level of the punch [86].

Measures to influence the accuracy of the produced micro-hydroforming components by reducing the elastic deflection of the forming dies under a load were investigated using finite element-based simulations [89]. The study recommended the implementation of a tool design that enables the superimposition of bending stresses, in conjunction with an adapted joint-face design and controllable closing forces, to control elastic distortions of the die cavity, caused by the internal pressurization of the workpiece [89].

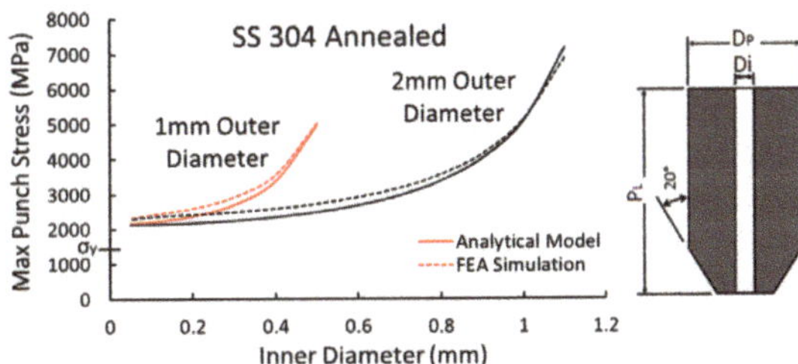

Figure 9. Maximum punch stress as a function of the inner punch diameter D_i for the micro-hydroforming of stainless-steel tubes with different tube outer diameters. Reprinted with the permission of SAGE Publications, Ltd., from reference [86], copyright 2018.

3.3. Discussion

Various achievements of conducted research on process fundamentals and die design for micro-hydroforming contributed to the well-developed status of this technology. Applicable solutions are available for the sealing of the formed micro-tubes—as one of the most critical issues of this miniaturized process—that enable the axial feeding of the tube material and are manufacturable with conventional methods. Forming limits were determined for complex geometries, but further investigations are recommended to extend these investigations to a broader range of tube and part geometries and materials in order to provide a basis for industrial component and process design. Analogous to the hydroforming of conventional-sized macro-components, micro-hydroforming also offers a considerable potential for high-volume production. Furthermore, the demonstrated feasibility of manufacturing complex-shaped components is of interest for important industries, such as medical engineering or micro-fluid technology. Hence, further research work should concentrate on the application of micro-hydroforming in mass production.

4. Laser-Assisted Forming of Micro-Tubular Components

Besides laser application in micro-tube manufacturing using dieless drawing (see Section 2.1.), there exist only a few published results on the use of laser radiation that had the objective of modifying a micro-tubular component's shape. However, strategies to assist or to conduct forming, processed by laser radiation [92], are known, since more than 20 years, and several research results on micro-sheet forming have been already presented, e.g., in [93]. Current investigations into the laser-assisted forming of micro-tubular parts were conducted (a) to generate shape elements on tubular components by applying superimposed external or internal loads to the effect of laser heating [94,95], and (b) for the bending of micro-tubes, solely due to the laser impact [96–98].

The feasibility of the laser-assisted and dieless fabrication of micro-bellows from stainless-steel tubes, with d_o = 500 μm and t = 50 μm, was successfully demonstrated [94]. Micro-bellows, commonly manufactured by electroforming, stereolithography or micro-roll forming, are used as hermetic seals, volume compensators, sensor elements, and connectors [94]. A laser, with a maximum power of 120 W and a wavelength of 808 nm, was used to heat the micro-tubes, which were set into rotation using a device that enabled a maximum rotational speed of 3000 rpm. An individual bulged shape was then generated by an upsetting deformation [94].

With the objective of providing a solution for the fabrication of grooved micro-tubes for micro-heat pipes, with diameters that are below conventionally feasible diameters, a dieless laser-assisted strategy with internal pressurization of the workpiece was investigated [95]. Inner grooves, with an average height of 209.5 μm, in a tube, with d_o = 2 mm and t = 200 μm, were fabricated using the simultaneous application of heat energy by a pulsed laser, with a power of 10 W and an air pressure of 0.01 MPa, on the inside of the tube [95].

The laser-assisted bending of micro-tubes provides a potential, for example, for the positioning of optical fibers with respect to a waveguide chip, where a submicron alignment is required [96]. In this type of bending process, the tube is heated locally on one side by the laser radiation, which results in a local thermal strain in the tube wall. This strain introduces a compressive stress that exceeds the yield strength and, through this, generates a compressive plastic strain, as shown in Figure 10a. When the tube cools down, the heated area will effectively contract, and the result is a bending of the tube toward the laser, as shown in Figure 10b. In [96], an experimental setup was presented for the bending of stainless-steel tubes, with d_o = 635 μm and t = 153 μm, with a real time measurement of the tube deformation. A laser source, with a wave length of 1080 nm, was used in these experiments, with a focused spot size of the laser beam of 400 μm. The selected laser power was between 7 and 10 W. The bending of micro-tubes made from stainless steel, with outer diameters between 457 and 711 μm, was investigated [97], and it was shown that it was possible to align an optical fiber, with an accuracy of 0.1 μm, although the laser forming process caused a significant scattering in response to the bending angle and direction [97], as shown in Figure 10c. A self-learning algorithm was proposed and tested within the frame of this research, which determined for each bending step the most suitable values of the main parameters: the axial laser spot position on the tube and the laser power [97].

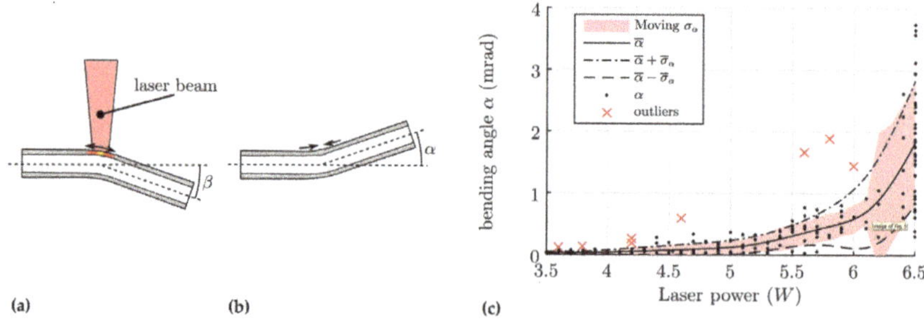

Figure 10. Laser-assisted bending of micro-tubes: (**a**) generation of local compressive plastic strains by laser radiation; (**b**) cooling and contracting of the heated area with the bending of the tube; and (**c**) measured bending angle as a function of the applied laser power for a micro-tube made from stainless steel, with d_o = 457 μm and t = 57 μm. Reprinted from reference [97], copyright 2016, with permission from Elsevier.

Studies in the laser-assisted bending of micro-tubes made from nickel, with an outer diameter of 960 μm and a wall thickness of 50 μm, were described in [98]. A particular result of this research was that the introduction of compressive pre-stresses by mechanical bending increased the final laser bending. Different laser sources were tested, and the influence of the laser power, pulse length, and pre-stress constraint on the bending deflection was investigated [98]. A high laser power and short pulse duration were found to be preferable due to the high thermal conductivity of the selected tube material [98].

It can be summarized that the laser-assisted micro-tube forming methods discussed here provide promising approaches and concern emerging industrial fields. The laser-assisted forming of micro-bellows and micro-heat pipes can open-up new opportunities for flexible manufacturing, but the research is still in an early stage. Further investigations into the influence of the forming parameters, tube dimensions, and materials on the forming result, as well as on the forming limits, appear necessary here. The published results concerning laser-assisted bending and strategies to cope with the scattering of bending results give the impression of a comparatively well-developed technique, with an obvious potential for industrial application.

5. Severe Plastic Deformation with a Focus on Micro-Tube Fabrication

Research into severe plastic deformation techniques (SPD) has attracted increasing interest over the past few years due to the potential of these methods to produce bulk nanostructured and ultra-fine-grained metals and metallic alloys [99–102]. Important objectives of corresponding investigations have been achievements in certain material properties that correlated with a reduced grain size, such as an enhanced material strength or superplasticity at lower temperatures [100,101]. The expected benefits of materials for micro-manufacturing, fabricated from SPD processes, are, aside from a reduction of grain size-related scaling effects, nearly homogeneous mechanical properties and a limited work hardening behavior [100]. Around 120 variants of SPD methods have been introduced up to now, providing metallic materials of different shapes, such as rods, bars, billets, tubes, or sheets [99]. Even though various SPD techniques were successfully applied to conventional tube dimensions [103], and these techniques can provide advantages for micro-components, only a few published results exist concerning research on micro-tube SPD processes.

On the background, an improvement of the mechanical properties and corrosion resistance of tubular materials for stent manufacture by grain refinement, a process sequence of equal-channel angular pressing (ECAP, or sometimes abbreviated as ECAE [99]), direct extrusion, machining, and micro-tube extrusion, was developed and presented by [104]. With this strategy, micro-tubes made from the magnesium alloy WE43, with $d_o = 3.3$ mm and $t = 220$ μm, were fabricated, showing an average grain size of $d_g = 3.5$ μm. In this study, the tensile strength could be increased from 240 MPa of the initial workpiece to 340 MPa of the final micro-tube. An important achievement in this context was the determined increase in ductility, with an increase from 6% to 20% for the elongation of the fracture [104]. This is worth mentioning, as in many cases, SPD processing diminishes ductility [101]. According to the authors, the significant grain refinement within the equal-channel angular pressing step, initiated by recrystallization under the selected processing temperature, was assumed to be one of the main reasons for the increase in ductility [104].

A process chain consisting of ECAP operations, micro-tube extrusion, and machining processes, developed by [105], showed the feasibility of micro-tubes, with a sub-micrometer grain-size range for the magnesium alloy ZM21. This ultrafine grain size was achieved by a two-stage ECAP, with a processing temperature of 200 °C, in the first pressing step, and 150 °C in the second step. It was found that the large shear strain, imparted by ECAP on the coarse structure in the first pressing step, exceeded the fracture limit of the crystals when the billet was processed at 150 °C in this step as well. The average grain size of about 500 nm, achieved with this ECAP processing remained in the sub-micrometer range after the micro-tube extrusion process, with a processing temperature of 150 °C. The tensile strength could be increased, keeping a fairly high tensile ductility. The extruded tubes, with d_o = 4 mm and t = 1 mm, were machined to an outer diameter of 2.4 mm, a wall thickness of 0.4 mm and a stent net generated by laser cutting [105].

The fabrication of grain-refined micro-tubes and their application to superplastic forming by dieless tube drawing was the subject of the studies presented in [31]. Grain refinement by combined ECAP processing and tube extrusion was investigated for the fabrication of micro-tubes made from AZ31 magnesium alloy. ECAP processing, with four passes at 200 °C, was applied to a cylindrical billet, with a diameter of 3.8 mm, and extruded, with a fixed mandrel at 250 °C, to a tube, with d_o = 2 mm and t = 500 μm, attaining an average grain size of 1.5 μm. Figure 11 shows results of the individual forming steps and the evolution of the material's microstructure. In particular, it was demonstrated that the application of ECAP processing improved the superplastic forming properties, compared to fabrications based merely on extrusion processes. Dieless tube drawing trials were successfully carried out, with a reduction in the cross-section area of 58.1% [31].

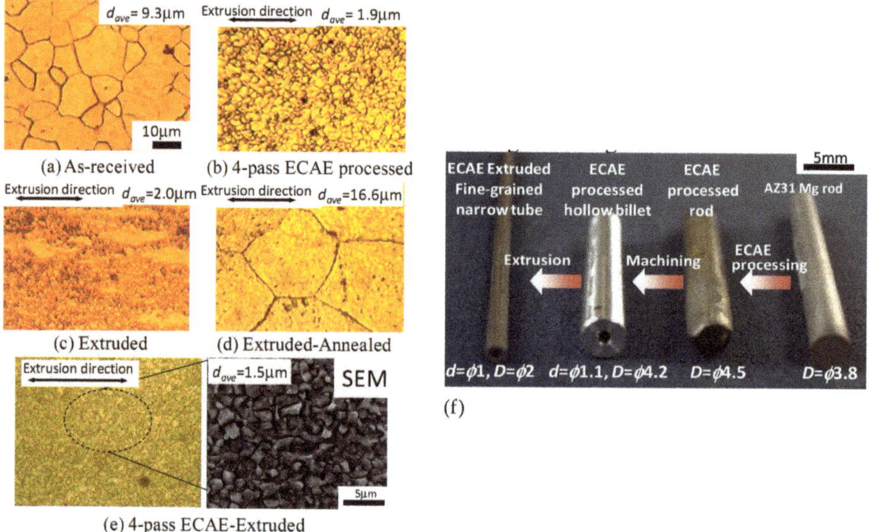

Figure 11. Micro-tubes, fabricated from ECAE-processed magnesium alloy AZ31: (**a–e**) comparison of microstructure for different treatment steps (d_{ave}: average grain size) and (**f**) specimens of the individual processing steps (D, outer tube diameter; d, inner tube diameter). Republished with the permission of Trans Tech Publications, from reference [31], copyright 2010; permission conveyed through Copyright Clearance Center, Inc.

In summary, the few available results concerning severe plastic deformation for the fabrication of micro-tubes led to the conclusion that the use of this method provides opportunities to generate components with advantageous material microstructures and properties that meet the needs of important industries, such as medical engineering. However, the state of knowledge is limited to a few types of materials based on magnesium alloys and a reduced range of fabricated tube dimensions. Therefore, more investigations dealing with further materials, which are, for example, of interest for medical implants, such as titanium or platinum, are recommended.

6. Testing Micro-Tubes Characteristics

An extensive research was carried out on investigations into the influence of size effects on the forming behavior of metal materials by scaling down standard test methods, such as micro-cylindrical compression tests or tensile tests, e.g., in [106–111]. The transfer of these standard methods for bulk and sheet materials to determine the forming behavior of micro-tubes is difficult, because their dimensions do not readily enable the preparation of corresponding specimen shapes for conducting these tests. For this reason, testing methods for micro-tubes were derived from typical methods for the evaluation of the formability and of characteristics of macro-tubes, comprising (a) flaring tests with mechanical expansion [112–116] and (b) biaxial testing under internal pressurization [83,117].

6.1. Flaring Test

In tube flaring tests, the tube end is expanded with a conical punch that is moved axially to the longitudinal axis of the specimen, as shown in Figure 12. The application of this method to determine the specific material parameters, such as the flow curve parameters or forming limits of micro-tubes, requires consideration of the fact that the selected cone angle, the friction conditions, the ratio of the wall thickness to the tube diameter, and the microstructure of the formed tube crucially influence the forming result. Investigations into the flaring of micro-tubes, with d_o = 500 μm, made from stainless steel (t = 50 μm) and copper tubes (t = 25 μm), formed with cone angles of 20°, 40°, and 60°, demonstrated that the feasible expansion increased with the increasing angle [116]. For the flaring of stainless-steel micro-tubes, with d_o = 500 μm and t = 50 μm, using a cone angle of the punch of 40°, it was established that the punch load, as well as the forming limit, increased with an increasing punch surface roughness due to mechanical adhesion [113]. Compared with dry lubrication, these values decreased when fluorocarbon resin was used as a solid lubricant [113]. Additionally, these values did not significantly change in the case of lubrication with oil in combination with a rough surface, and they decreased in cases of a finer surface roughness, which may be explained by the constellation of closed and open lubrication pockets [113].

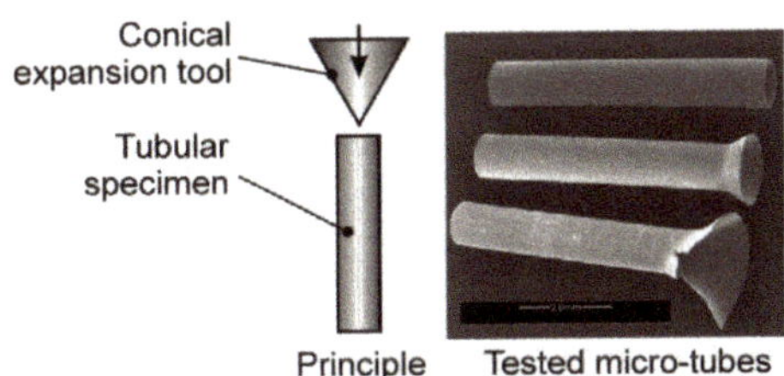

Figure 12. Principle of the flaring test for tubes and examples of micro-tubes made from stainless steel, with d_o = 800 μm and t = 40 μm, tested with a punch, with a cone angle of 60°. Reprinted from reference [77], copyright 2015, with permission from Elsevier.

Based on detailed numerical and experimental investigations into the friction mechanisms between the forming punch and the inner tube surface for the conical flaring of stainless-steel micro-tubes, the surface roughness of the conical punch should be less than R_a = 0.025 μm to reduce the effects of the tool topography and to increase the test reliability [114]. Furthermore, the interaction between the punch and inner side of the micro-tube also depends on the tube wall thickness and influences the forming behavior and forming limits [115]. Research on the influence of the material grain size on the flaring test results was conducted by [112] for stainless-steel tubes, with d_o = 700 μm and t = 150 μm, with ratios of the tube wall thickness to the average grain size t/d_g between 4.6 and 9.9, using a cone angle for the forming punch of 30.12°. According to these results, the forming limit decreased with a decreasing ratio t/d_g [112].

6.2. Expansion Test

The internal pressurization of micro-tubes allows the forming behavior of micro-tubes to be investigated under biaxial loading conditions. The pressure at the occurrence of the crack initiation and maximum feasible expansion under these loading conditions were determined with a comparatively uncomplicated device by the internal pressurization of the tubular specimen, where the tube ends were clamped, without the possibility of an axial movement [83]. However, the generated stress state in such tube expansions is always pressure-dependent and cannot be modified. In contrast, the additional superimposition of loads axially to the tube axis provides opportunities for investigations in different stress and strain states [117]. The determination of the anisotropic properties of micro-tubes was demonstrated with expansion tests, under varying axial loads for stainless-steel micro-tubes, with $d_o = 2.38$ mm, $t = 160$ μm, and an average grain size d_g between 10 and 12 μm [117].

6.3. Discussion

Among the reviewed techniques for the testing of micro-tubes, the flaring test is technically the easiest method to gather information about tube formability, but it has the drawback that numerous factors can influence the measured result. The presented research outcomes enable the consideration of the majority of these influences and the use of flaring tests for the evaluation of micro-tube formability. Expansion tests, with pressurized media and superimposed axial loads, provides more defined experimental conditions and more material data on, for instance, anisotropic behavior, but they require a higher technical effort. A challenge that needs to be addressed for both methods concerns investigations into the standardization of these tests.

7. Summary

The objective of this review was to provide a thorough overview of the latest achievements in micro-forming processes and techniques for the fabrication and shaping of micro-tubes and micro-tubular components. The focus was on novel process types, techniques with enhanced shaping possibilities, and strategies to improve the efficiency or manufactured component properties with regard to restrictions due to size effects. Figure 13 gives a graphic summary of the groups of reviewed techniques, with a range of investigated significant workpiece dimensions and materials, major fields of industrial application from the author's point of view, and the number of reviewed publications.

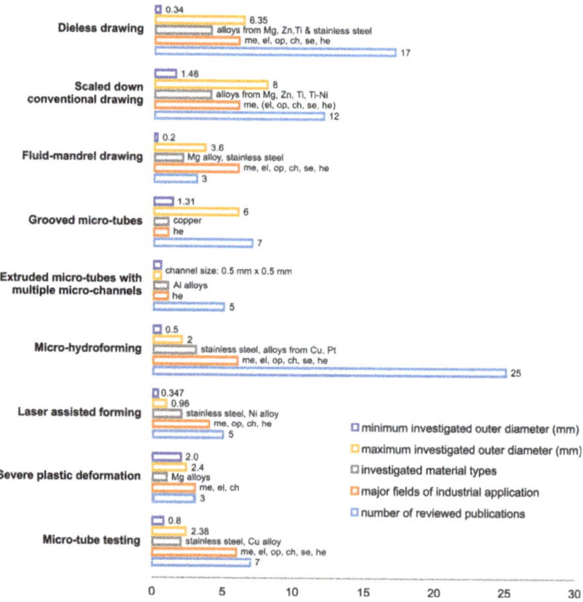

Figure 13. Graphic summary of the review (me, medical engineering; el, electronic and electrical engineering; op, optoelectronics; ch, chemical technology; se, sensing technology; he, heat transfer technology).

Research in micro-tube fabrication is concentrated to a great extent on semi-finished products for medical applications and for heat exchange technology. Dieless drawing processes have shown a considerable capability to reduce manufacturing steps and, through this, improve productivity in micro-tube manufacture. Investigations into the microstructure evolution supported the development of process chains, with scaled-down forming technologies, for the fabrication of micro-tubes with an improved manufacturing precision and product characteristics for the production of vascular stents made from materials with a reduced formability. Modifications of tube drawing processes, using liquid media as a drawing mandrel, and achievements in profile extrusion, based on studies in microstructure development of the formed material, allow semi-finished products to be manufactured efficiently, with reduced flow areas for progresses in miniaturization and thermal heat exchange.

Concerning technologies for shaping metallic tubular parts, important achievements were published for micro-hydroforming and laser-assisted micro-tube forming. For micro-hydroforming, the major issues of investigations focused on the size effects on the material behavior and appropriate design of miniaturized and highly stressed forming tool elements. New concepts in the design of these tool elements facilitated enhanced shaping possibilities for micro-tubes and the verification of forming limits in micro-hydroforming. Flexibility in the use of laser radiation was the basis for the development of a novel process to form tubular micro-bellows and micro-heat pipe-shape elements as well as to establish the applicability of laser energy for the positioning of micro-tubes by bending processes, in conjunction with computational algorithms.

Severe plastic deformation processes for grain refinement offer the potential to reduce the grain-size-related scaling effects of micro-components. A few results were published concerning research on the use of this technique for the fabrication of micro-tubes made from magnesium alloys, showing the feasibility of fine and ultrafine grain structures, with enhanced material properties and parameters for superplastic forming.

The testing methods for micro-tubes, working either with tube expansions by conical punches or by internal pressurization, were investigated, with the objective of evaluating the forming behavior. The cone angle, surface, and friction conditions, as well as tube dimensions, are the main factors influencing the measured values in tube flaring. Micro-tube expansion tests, with superimposed axial loads, have been found to be suitable for determining anisotropic characteristics.

The achievements in micro-tube forming, reviewed in this paper, can clearly contribute to enhanced manufacturing possibilities in the field of tubular micro-parts. However, the individual technologies differ in their level of development and acquired knowledge concerning the performance associated with varied tube geometries, tolerances, materials, integration into manufacturing lines, and productivity. Addressing these issues in further research work and efforts in standardization would assist in the industrial implementation of such processes and testing methods for mass production. It can be expected that a growing need for advanced micro-tubular products will come from emerging industries, such as medical technology and heat transfer technology.

Conflicts of Interest: The author declares no conflict of interest.

References

1. Qin, Y.; Brockett, A.; Ma, Y.; Razali, A.; Zhao, J.; Harrison, C.; Pan, W.; Dai, X.; Loziak, D. Micro-manufacturing: Research, technology outcomes and development issues. *Int. J. Adv. Manuf. Technol.* **2010**, *47*, 821–837. [CrossRef]
2. Fassi, I.; Shipley, D. *Micro-Manufacturing and Their Applications*, 1st ed.; Springer: Cham, Switzerland, 2017. [CrossRef]
3. Fu, M.W.; Chan, W.L. A review on the state-of-the-art microforming technologies. *Int. J. Adv. Manuf. Technol.* **2013**, *67*, 2411–2437. [CrossRef]
4. Engel, U.; Eckstein, R. Microforming-from basic research to its realization. *J. Mater. Process. Technol.* **2002**, *125*, 35–44. [CrossRef]

5. Furushima, T.; Manabe, K. Superplastic micro-tubes fabricated by dieless drawing processes. In *Superplastic Forming of Advanced Metallic Materials*, 1st ed.; Giuliano, G., Ed.; Woodhead Publishing: Cambridge, UK, 2011; pp. 327–360. [CrossRef]

6. Qin, Y. *Micromanufacturing Engineering and Technology*, 2nd ed.; Elsevier: Oxford, UK, 2015. [CrossRef]

7. *EvaluateMedTech—World Preview 2018, Outlook to 2024*; Evaluate: London, UK, 2018; Available online: https://de.statista.com/statistik/daten/studie/313462/umfrage/umsatzentwicklung-der-weltweiten-medizintechnikindustrie/ (accessed on 1 May 2019).

8. Koc, M.; Özel, T. *Micro-Manufacturing: Design and Manufacturing of Micro-Products*, 1st ed.; John Wiley & Sons: Hoboken, NJ, USA, 2011. [CrossRef]

9. Jain, V.K.; Sidpara, A.; Balasubramaniam, R.; Lodha, G.S.; Dhamgaye, V.P.; Shukla, R. Micromanufacturing: A review—Part I. *J. Eng. Manuf.* **2014**, *228*, 973–994. [CrossRef]

10. Jain, V.K.; Dixit, U.S.; Paul, C.P.; Kumar, A. Micromanufacturing: A review—Part II. *J. Eng. Manuf.* **2014**, *228*, 995–1014. [CrossRef]

11. Razali, A.K.; Qin, Y. A review on micro-manufacturing, micro-forming and their key issues. *Procedia Eng.* **2013**, *53*, 665–672. [CrossRef]

12. Vollertsen, F.; Schulze Niehoff, H.; Hu, Z. State of the art in micro forming. *Int. J. Mach. Tools Manuf.* **2006**, *46*, 1172–1179. [CrossRef]

13. Lai, X.M.; Fu, M.W.; Peng, L.F. *Sheet Metal Meso- and Microforming and Their Industrial Applications*, 1st ed.; CRC Press: Boca Raton, FL, USA, 2018. [CrossRef]

14. Fu, M.W.; Zheng, J.Y.; Meng, B. A Review of Progressive and Compound Forming of Bulk Microparts by Using Sheet Metals. *MATEC Web Conf.* **2018**, *190*, 1–10. [CrossRef]

15. Vollertsen, F.; Biermann, D.; Hansen, H.N.; Jawahir, I.S.; Kuzman, K. Size effects in manufacturing of metallic components. *CIRP Ann. Manuf. Technol.* **2009**, *58*, 566–587. [CrossRef]

16. Fu, M.W.; Wang, J.L.; Korsunsky, A.M. A review of geometrical and microstructural size effects in micro-scale deformation processing of metallic alloy components. *Int. J. Mach. Tools Manuf.* **2016**, *109*, 94–125. [CrossRef]

17. Wang, C.; Guo, B.; Shan, D. Friction related size-effect in microforming—A review. *Manuf. Rev.* **2014**, *1*, 1–18. [CrossRef]

18. Chatzipirpiridis, G.; Ergeneman, O.; Pokki, J.; Ullrich, F.; Fusco, S.; Ortega, J.A.; Sivaraman, K.M.; Nelson, B.J.; Pané, S. Electroforming of Implantable Tubular Magnetic Microrobots for Wireless Ophthalmologic Applications. *Adv. Healthc. Mater.* **2015**, *4*, 209–214. [CrossRef] [PubMed]

19. Chatzipirpiridis, G.; Avilla, E.; Ergeneman, O.; Ullrich, F.; Nelson, B.J.; Pané, S. Electroforming of Magnetic Microtubes for Microrobotic Applications. *IEEE Trans. Magn.* **2014**, *50*, 1–3. [CrossRef]

20. Tsuchiya, K.; Morishima, A.; Takamata, A.; Uetsuji, Y.; Nakamachi, E. Development of valve-less tube type micropump with PZT actuator. In Proceedings of the SPIE Conference Smart Materials Nano- and Micro Smart Systems, Melbourne, Australia, 9–12 December 2008. [CrossRef]

21. Furushima, T.; Manabe, K. Large reduction die-less mandrel drawing of magnesium alloy micro-tubes. *CIRP Ann. Manuf. Technol.* **2018**, *67*, 309–312. [CrossRef]

22. Milenina, A.; Kustra, P.; Byrska-Wójcik, D.; Furushima, T. Physical and numerical modelling of laser dieless drawing process of tubes from magnesium alloy. *Procedia Eng.* **2017**, *207*, 2352–2357. [CrossRef]

23. Furushima, T.; Manabe, K. Heat assisted dieless drawing process of superplastic metal microtubes – from Zn22Al to β titanium alloys. *Mater. Sci. Forum* **2016**, *838–839*, 459–467. [CrossRef]

24. Furushima, T.; Imagawa, Y.; Manabe, K.; Sakai, T. Effects of oxidation and surface roughening on drawing limit indieless drawing process of SUS304 stainless steel microtubes. *J. Mater. Process. Technol.* **2015**, *223*, 186–192. [CrossRef]

25. Furushima, T.; Imagawa, Y.; Furusawa, S.; Manabe, K. Development of Rotary Laser Dieless Drawing Apparatus for Metal Microtubes. *Key Eng. Mater.* **2015**, *626*, 372–376. [CrossRef]

26. Hwang, Y.-M.; Li, Z.-S.; Lin, T.-Y. Formability Discussion in Dieless Drawing of Stainless Steel Tubes. *Key Eng. Mater.* **2015**, *626*, 10–15. [CrossRef]

27. Furushima, T.; Imagawa, Y.; Furusawa, S.; Manabe, K. Deformation profile in rotary laser dieless drawing process for metal microtubes. *Procedia Eng.* **2014**, *81*, 700–705. [CrossRef]

28. Furushima, T.; Shirasaki, A.; Manabe, K. Fabrication of noncircular multicore microtubes by superplastic dieless drawing process. *J. Mater. Process. Technol.* **2014**, *214*, 29–35. [CrossRef]

29. Furushima, T.; Noda, Y.; Manabe, K. Laser Dieless Drawing Process for Metal Micro-tubes. *Key Eng. Mater.* **2010**, *443*, 699–704. [CrossRef]

30. Furushima, T.; Noda, Y.; Manabe, K. Laser Dieless Drawing Process for Metal Microtubes. In Proceedings of the International Conference on Materials Processing Technology 2010, Bangkok, Thailand, 5–6 January 2010; pp. 64–67.

31. Furushima, T.; Shimizu, T.; Manabe, K. Grain Refinement by Combined ECAE/Extrusion and Dieless Drawing Processes for AZ31 Magnesium Alloy Tubes. *Mater. Sci. Forum* **2010**, *654–656*, 735–738. [CrossRef]

32. Furushima, T.; Manabe, K.; Sakai, T. Fabrication of Superplastic Microtubes Using Dieless Drawing Process. *Mater. Trans.* **2008**, *49*, 1365–1371. [CrossRef]

33. Furushima, T.; Manabe, K. FE analysis of size effect on deformation and heat transfer behavior in microtube dieless drawing. *J. Mater. Process. Technol.* **2008**, *201*, 123–127. [CrossRef]

34. Furushima, T.; Manabe, K. Experimental and numerical study on deformation behavior in dieless drawing process of superplastic microtubes. *J. Mater. Process. Technol.* **2007**, *191*, 59–63. [CrossRef]

35. Furushima, T.; Manabe, K. Experimental study on multi-pass dieless drawing process of superplastic Zn-22%Al alloy microtubes. *J. Mater. Process. Technol.* **2007**, *187–188*, 236–240. [CrossRef]

36. Wang, J.; Zhou, Y.; Yang, Z.; Zhu, S.; Wang, L.; Guan, S. Processing and properties of magnesium alloy micro-tubes for biodegradable vascular stents. *Mater. Sci. Eng. C* **2018**, *90*, 504–513. [CrossRef]

37. Kim, Y.N.; Hwang, S.K.; Joo, H.S.; Im, Y.T. Effect of heat treatment and mandrel material on precision tube drawing of Ni-Ti shape memory alloy. *Procedia Eng.* **2017**, *207*, 1517–1522. [CrossRef]

38. Wang, C.; Yu, Z.; Cui, Y.; Zhang, Y.; Yu, S.; Qu, G.; Gong, H. Processing of a Novel Zn Alloy Micro-Tube for Biodegradable Vascular Stent Application. *J. Mater. Sci. Technol.* **2016**, *32*, 925–929. [CrossRef]

39. Liu, F.; Chen, C.; Niu, J.; Pei, J.; Zhang, H.; Huang, H.; Yuan, G. The processing of Mg alloy micro-tubes for biodegradable vascular stents. *Mater. Sci. Eng. C* **2015**, *48*, 400–407. [CrossRef]

40. Wang, L.; Fang, G.; Qian, L.; Leeflang, S.; Duszczyk, J.; Zhou, J. Forming of magnesium alloy microtubes in the fabrication of biodegradable stents. *Prog. Nat. Sci. Mater. Int.* **2014**, *24*, 500–506. [CrossRef]

41. Zhang, Y.; Kent, D.; Wang, G.; StJohn, D.; Dargusch, M.S. The cold-rolling behaviour of AZ31 tubes for fabrication of biodegradable stents. *J. Mech. Behav. Biomed. Mater.* **2014**, *39*, 292–303. [CrossRef]

42. Hong, S.-K.; Kang, J.J.; Kim, J.-D.; Kim, H.-K.; Lee, S.-Y.; Park, C.H. Finite Element Analysis for the Tube Sinking Process of the Micro Ti-0.2Pd Tube. *Appl. Mech. Mater.* **2014**, *465–466*, 693–698. [CrossRef]

43. Yoshida, K.; Yokomizo, D.T.; Komatsu, T. Production of special tubes with a variety cross-sectional shapes by bunch drawing and fluid-mandrel drawing. *Key Eng. Mater.* **2014**, *622–623*, 731–738. [CrossRef]

44. Fang, G.; Ai, W.; Leeflang, S.; Duszczyk, J.; Zhou, J. Multipass cold drawing of magnesium alloy minitubes for biodegradable vascular stents. *Mater. Sci. Eng. C* **2013**, *33*, 3481–3488. [CrossRef]

45. Yoshida, K.; Koiwa, K. Cold Drawing of Magnesium Alloy Tubes for Medical. *J. Solid Mech. Mater. Eng.* **2011**, *5*, 1071–1078. [CrossRef]

46. Yoshida, K.; Furuya, H. Mandrel drawing and plug drawing of shape-memory-alloy fine tubes used in catheters and stents. *J. Mater. Process. Technol.* **2004**, *153–154*, 145–150. [CrossRef]

47. Yoshida, K.; Watanabe, M.; Ishikawa, H. Drawing of Ni-Ti shape-memory-alloy fine tubes used in medical tests. *J. Mater. Proces. Technol.* **2001**, *118*, 251–255. [CrossRef]

48. Tangsri, T.; Norasethasopon, S. 3D FEM validation of ultra-small inner spiral ribbed copper tube using the tube sinking method. *Int. J. Adv. Manuf. Technol.* **2015**, *81*, 1949–1959. [CrossRef]

49. Tangsri, T.; Norasethasopon, S.; Yoshida, K. Fabrication of small size inner spiral ribbed copper tube by fluid mandrel drawing. *Int. J. Adv. Manuf. Technol.* **2014**, *70*, 1923–1930. [CrossRef]

50. Ou, D.S.; Tang, Y.; Wan, Z.P.; Lu, L.S.; Lian, B. Effect of drawing parameters on elongation of micro copper tube with straight grooves. *Key Eng. Mater.* **2012**, *499*, 62–67. [CrossRef]

51. Tang, Y.; OU, D.S.; Wan, Z.P.; Lu, L.; Lian, B. Influence of drawing process parameters on forming of micro copper tube with straight grooves. *Trans. Nonferrous Met. Soc. China* **2011**, *21*, 2264–2269. [CrossRef]

52. Wan, Z.P.; Li, Y.C. Forming Process and Mechanism of Microcapillary Grooves of Miniature Cylindrical Heat Pipe. *Adv. Mater. Res.* **2011**, *328–330*, 22–26. [CrossRef]

53. Tang, Y.; Long-sheng Lu, L.; Yuan, D.; Wang, Q.; Zhaoa, X. Experimental and FEM study on sinking of miniature inner grooved copper tube. *J. Mater. Process. Technol.* **2009**, *209*, 5333–5340. [CrossRef]

54. Li, Y.; Xiao, H.; Bin, L.; Tang, Y.; Zeng, Z. Forming method of axial micro grooves inside copper heat pipe. *Trans. Nonferrous Met. Soc. China* **2008**, *18*, 1229–1233. [CrossRef]

55. Tang, D.; Fang, W.; Fan, X.; Zou, T.; Li, Z.; Wang, H.; Li, D.; Peng, Y.; Wu, P. Evolution of the Material Microstructures and Mechanical Properties of AA1100 Aluminum Alloy within a Complex Porthole Die during Extrusion. *Materials* **2019**, *12*, 16. [CrossRef]

56. Tang, D.; Fan, X.; Fang, W.; Li, D.; Peng, Y.; Wang, H. Microstructure and mechanical properties development of micro channel tubes in extrusion, rolling and brazing. *Mater. Charact.* **2018**, *142*, 449–457. [CrossRef]

57. Tang, D.; Fang, W.; Fan, X.; Li, D.; Peng, Y. Effect of die design in microchannel tube extrusion. *Procedia Eng.* **2014**, *81*, 628–633. [CrossRef]

58. Tang, D.; Zhang, Q.; Li, D.; Peng, Y. A physical simulation of longitudinal seam welding in micro channel tube extrusion. *J. Mater. Process. Technol.* **2014**, *214*, 2777–2783. [CrossRef]

59. Cho, H.; Hwang, D.-Y.; Lee, B.-S.; Jo, H.-H. Fabrication of micro condenser tube through direct extrusion. *J. Mater. Process. Technol.* **2007**, *187–188*, 645–648. [CrossRef]

60. Furushima, T.; Ikeda, T.; Manabe, K. Deformation and Heat Transfer Analysis for High Speed Dieless Drawing of AZ31 Magnesium Alloy Tubes. *Adv. Mater. Res.* **2012**, *418–420*, 1036–1039. [CrossRef]

61. Palm, B. Heat transfer in microchannels. *Nanoscale Microscale Thermophys. Eng.* **2001**, *5*, 155–175. [CrossRef]

62. Hartl, C. Deformation mechanism and fundamentals of hydroforming. In *Hydroforming for Advanced Manufacturing*, 1st ed.; Koc, M., Ed.; Woodhead Publishing: Cambridge, UK, 2008; pp. 52–76. [CrossRef]

63. Koc, M.; Cora, O.N. Introduction and state of the art of hydroforming. In *Hydroforming for Advanced Manufacturing*, 1st ed.; Koc, M., Ed.; Woodhead Publishing: Cambridge, UK, 2008; pp. 1–29. [CrossRef]

64. Lee, M.-G.; Korkolis, Y.P.; Kim, J.H. Recent developments in hydroforming technology. *J. Eng. Manuf.* **2015**, *229*, 572–596. [CrossRef]

65. Yuan, S.J.; Liu, G. Tube Hydroforming. In *Comprehensive Materials Processing*, 1st ed.; Hashmi, M.S.J., Ed.; Elsevier: Amsterdam, The Netherland, 2014; Volume 3, pp. 55–80.

66. Lihui, L.; Kangning, L.; Cai, G.; Yang, X.; Guo, C.; Bu, G. A critical review on special forming processes and associated research for lightweight components based on sheet and tube materials. *Manuf. Rev.* **2014**, *1*, 1–20. [CrossRef]

67. Hartl, C. Research and advances in fundamentals and industrial application of hydroforming. *J. Mater. Process. Technol.* **2005**, *167*, 383–392. [CrossRef]

68. Hartl, C.; Anyasodor, G. Experimental and Numerical Investigations into Micro-Hydroforming Processes and Machine Design. In Proceedings of the 13th International Conference on Metal Forming, Toyohashi, Japan, 19–22 September 2010; pp. 1193–1196.

69. Wagner, S.W.; Ng, K.; Emblom, W.J.; Camelio, J.A. Influence of continuous direct current on the microtube hydroforming process. *J. Manuf. Sci. Eng.* **2017**, *139*, 034502. [CrossRef]

70. Manabe, K.; Sato, H.; Itai, K.; Vilotic, M.; Tada, K. Factors influencing the forming characteristics in micro tube hydroforming by ultra high-forming pressure. *Procedia Eng.* **2017**, *207*, 2334–2339. [CrossRef]

71. Manabe, K.; Itai, K.; Sato, H.; Vilotic, M.; Takahashi, S. FE analysis of micro tube hydroforming process for cross- and T-shaped components. In Proceedings of the 7th International Conference on Tube Hydroforming (TUBEHYDRO 2017), Bangkok, Thailand, 14–17 November 2017; pp. 37–42.

72. Manabe, K.; Itai, K.; Tada, K. Fabrication of micro T-shaped tubular components by hydroforming process. In Proceedings of the 20th International ESAFORM Conference on Material Forming, Dublin, Ireland, 26–28 April 2017; p. 050017.

73. Manabe, K.; Vilotic, M.; Itai, K.; Tada, K. Failures in micro cross-shaped tube hydroforming process. In Proceedings of the 14th International Conference on Fracture (ICF 14), Rhodes, Greece, 18–23 June 2017; pp. 1–2.

74. Ngaile, G.; Lowrie, J. Micro-Tube Hydroforming System Based on Floating Die Assembly. In *60 Excellent Inventions in Metal Forming*, 1st ed.; Tekkaya, A.E., Homberg, W., Brosius, A., Eds.; Springer: Berlin/Heidelberg, Germany, 2015; pp. 427–432.

75. Ngaile, G.; Lowrie, J. New Micro Tube Hydroforming System Based on Floating Die Assembly Concept. *J. Micro Nano-Manuf.* **2014**, *2*, 041004. [CrossRef]

76. Hartl, C.; Chlynin, A.; Radetzky, M. Investigations into Forming Limits and Process Control of Micro-Hydroforming. In Proceedings of the 7th International Conference on Tube Hydroforming (TUBEHYDRO 2015), Xi'an, China, 9–11 September 2015; pp. 3–16.

77. Hartl, C. Micro-hydroforming. In *Micromanufacturing Engineering and Technology*, 2nd ed.; Qin, Y., Ed.; Elsevier: Oxford, UK, 2015; pp. 323–345. [CrossRef]

78. Hartl, C.; Chlynin, A.; Radetzky, M. The influence of axial compressive stresses on the formability and scattering of process parameters in micro-hydroforming processes of tubes. In Proceedings of the 4th International Conference on New Forming Technology (ICNFT 2015), Glasgow, UK, 6–9 August 2015; p. 06001. [CrossRef]
79. Hartl, C.; Schiefer, H.; Chlynin, A. Evaluation of experimental and numerical investigations into micro-hydroforming of platinum tubes for industrial application. *Manuf. Rev.* **2014**, *1*, 1–7. [CrossRef]
80. Zhuang, W.; Wang, S.; Lin, J.; Balint, D.; Hartl, C. Experimental and numerical investigation of localized thinning in hydroforming of micro-tubes. *Eur. J. Mech. A/Solids* **2012**, *31*, 67–76. [CrossRef]
81. Wagner, S.W.; Ng, K.; Emblom, W.J.; Camelio, J.A. Effect of continuous direct current on the yield stress of stainless steel 304 micro tubes during hydroforming operations. In Proceedings of the ASME 2012 International Manufacturing Science and Engineering Conference (MSEC2012), Notre Dame, IN, USA, 4–8 June 2012; pp. 1–8.
82. Wagner, S.W.; Ng, K.; Emblom, W.J.; Camelio, J.A. Influence of continuous direct current on the micro tube hydroforming process. In Proceedings of the ASME 2011 International Manufacturing Science and Engineering Conference (MSEC2011), Corvallis, OR, USA, 13–17 June 2011; pp. 1–6.
83. Hartl, C.; Anyasodor, G.; Lungershausen, J. Formability of Micro-Tubes in Hydroforming. In Proceedings of the 14th International ESAFORM Conference on Metal Forming, Belfast, Ireland, 27–29 April 2011; pp. 529–534. [CrossRef]
84. Zhuang, W.; Wang, S.; Cao, J.; Lin, J.; Hartl, C. Modelling of localised thinning features in hydroforming of micro-tubes using crystal plasticity FE method. *Int. J. Adv. Manuf. Technol.* **2010**, *47*, 859–865. [CrossRef]
85. Ngaile, G.; Lowrie, J. Punch design for floating based micro-tube hydroforming die assembly. *J. Mater. Process. Technol.* **2018**, *253*, 168–177. [CrossRef]
86. Lowrie, J.; Ngaile, G. Scalability of conventional tube hydroforming processes from macro to micro/meso. *J. Eng. Manuf.* **2018**, *232*, 2164–2177. [CrossRef]
87. Sato, H.; Kobayashi, D.; Manabe, K. Numerical and experimental investigations of cross shaped micro tube forming. In Proceedings of the 7th International Conference on Tube Hydroforming (TUBEHYDRO 2015), Xi'an, China, 9–11 September 2015; pp. 64–69.
88. Wagner, S.W.; Ng, K.; Emblom, W.J.; Camelio, J.A. Novel high pressure sealing system for tube hydroforming. In Proceedings of the ASME 2013 International Manufacturing Science and Engineering Conference (MSEC2013), Madison, WI, USA, 10–14 June 2013; pp. 1–10.
89. Hartl, C.; Anyasodor, G.; Ptaschlik, T.; Lungershausen, J.; Lippert, S. Investigation into reduction of die-cavity deflection in micro-hydroforming processes using FEA. *Int. J. Adv. Manuf. Technol.* **2010**, *47*, 853–858. [CrossRef]
90. Hartl, C.; Anyasodor, G. Design and development of a new machine system for the mass manufacture of complex shaped tubular components by micro-hydroforming. In Proceedings of the 9th International Conference on Exhibition on Laser Metrology, Machine Tool, CMM & Robotic Performance (Lamdamap 2009), London, UK, 29 June–2 July 2009; pp. 206–214.
91. Hartl, C.; Lungershausen, J.; Anyasodor, G.; Ptaschlik, T. Development of tools and machines for micro-hydroforming with improved accuracy. *Zeszyty Naukowe Instytutu Pojazdów* **2008**, *5*, 157–164.
92. Vollertsen, F. *Laserstrahlumformen—Lasergestützte Formgebung: Verfahren Mechanismen, Modellierung*; Meisenbach Verlag: Bamberg, Germany, 1996.
93. Wielage, H.; Vollertsen, F. Increase of Acting Pressure by Adjusted Tool Geometry in Laser Shock Forming. In Proceedings of the International Conference on Advances in Materials and Processing Technologies (AMPT 2010), Paris, France, 24–27 October 2010; pp. 763–768.
94. Furushima, T. Development of Rotary Laser Dieless Forming Apparatus for Micro Metal Bellows. In Proceedings of the 7th International Conference on Tube Hydroforming (TUBEHYDRO 2017), Bangkok, Thailand, 14–17 November 2017; pp. 48–51.
95. Cheng, C.-M.; Tu, M.-Z.; Teng, T.-P.; Yu, S.-P.; Yu, C.-Y.; Chang, Y.-T. Development of inner-grooved tubes by dieless micro-forming with laser-assisted air pressure. *MATEC Web Conf.* **2017**, *95*, 1–4. [CrossRef]
96. Folkersma, G.; Brouwer, D.; Römer, G.-W. Microtube Laser Forming for Precision Component Alignment. *J. Manuf. Sci. Eng.* **2016**, *138*, 081012. [CrossRef]
97. Folkersma, K.G.P.; Brouwer, D.M.; Römer, G.R.B.E.; Herder, J.L. Robust precision alignment algorithm for micro tube laser forming. *Precis. Eng.* **2016**, *46*, 301–308. [CrossRef]

98. Che Jamil, M.S.; Imam Fauzi, E.R.; Juinn, C.S.; Sheikh, M.A. Laser bending of pre-stressed thin-walled nickel micro-tubes. *Opt. Laser Technol.* **2015**, *73*, 105–117. [CrossRef]
99. Bagherpour, E.; Pardis, N.; Reihanian, M.; Ebrahimi, R. An overview on severe plastic deformation: Research status, techniques classification, microstructure evolution, and applications. *Int. J. Adv. Manuf. Technol.* **2019**, *100*, 1647–1694. [CrossRef]
100. Bruder, E. Formability of Ultrafine Grained Metals Produced by Severe Plastic Deformation – An Overview. *Adv. Eng. Mater.* **2019**, *21*, 1800316. [CrossRef]
101. Cao, Y.; Ni, S.; Liao, X.; Song, M.; Zhu, Y. Structural evolutions of metallic materials processed by severe plastic deformation. *Mater. Sci. Eng. R* **2018**, *133*, 1–59. [CrossRef]
102. Segal, V. Review: Modes and Processes of Severe Plastic Deformation (SPD). *Materials* **2018**, *11*, 1175. [CrossRef]
103. Faraji, G.; Kim, H.S. Review of principles and methods of severe plastic deformation for producing ultrafine-grained tubes. *Mater. Sci. Technol.* **2016**, 1–20. [CrossRef]
104. Amani, S.; Faraji, G.; Mehrabadi, H.K.; Baghani, M. Manufacturing and mechanical characterization of Mg-4Y-2Nd-0.4Zr-0.25La magnesium microtubes by combined severe plastic deformation process for biodegradable vascular stents. *J. Eng. Manuf.* **2018**, *233*, 1–10. [CrossRef]
105. Ge, Q.; Dellasega, D.; Demir, A.G.; Vedani, M. The processing of ultrafine-grained Mg tubes for biodegradable stents. *Acta Biomat.* **2013**, *9*, 8604–8610. [CrossRef]
106. Chan, W.L.; Fu, M.W. Studies of the interactive effect of specimen and grain sizes on the plastic deformation behavior in microforming. *Int. J. Adv. Manuf. Technol.* **2012**, *62*, 989–1000. [CrossRef]
107. Chan, W.L.; Fu, M.W.; Lu, J. The size effect on micro deformation behaviour in micro-scale plastic deformation. *Mater. Des.* **2011**, *32*, 198–206. [CrossRef]
108. Barbier, C.; Thibaud, S.; Richard, F.; Picart, P. Size effects on material behavior in microforming. *Int. J. Mater. Form.* **2009**, *2*, 625–628. [CrossRef]
109. Chen, F.-K.; Tsai, J.-W. A study of size effect in micro-forming with micro-hardness tests. *J. Mater. Process Technol.* **2006**, *177*, 146–149. [CrossRef]
110. Raulea, L.V.; Goijaerts, A.M.; Govaert, L.E.; Baaijens, F.P.T. Size effects in the processing of thin metal sheets. *J. Mater. Process. Technol.* **2001**, *115*, 44–48. [CrossRef]
111. Geiger, M.; Messner, A.; Engel, U. Production of micro parts—Size effects in bulk metal forming, similarity theory. *Prod. Eng.* **1997**, *4*, 55–58.
112. Jiang, C.-P.; Chen, C.-C.; Wu, Y.-S.; Tsai, C.-H. Effect of annealing on grain size and flaring limit of seamless SUS 304 stainless microtubes. *J. Chin. Inst. Eng.* **2012**, *35*, 37–43. [CrossRef]
113. Mirzai, M.A.; Manabe, K. Tribological behavior and surface characteristics of metal microtube in flaring test. *J. Solid Mech. Mater. Eng.* **2009**, *3*, 375–386. [CrossRef]
114. Mirzai, M.A.; Manabe, K. Effect of tool surface topography on interfacial deformation behaviour of metal microtubes in flaring test. *Int. J. Surf. Sci. Eng.* **2009**, *3*, 391–406. [CrossRef]
115. Mirzai, M.A.; Manabe, K. FE Analysis of Size Effect on Deformation Behavior of Metal Microtube Considering Surface Roughness in Flaring Test. *Mater. Sci. Forum* **2009**, *623*, 79–87. [CrossRef]
116. Mirzai, M.A.; Manabe, K.; Mabuchi, T. Deformation characteristics of microtubes in flaring test. *J. Mater. Process. Technol.* **2008**, *201*, 214–219. [CrossRef]
117. Korkolis, Y.P.; Ripley, P.W.; Knysh, P. Failure of an austenitic stainless steel under linear and non-linear loading paths. In Proceedings of the 24th International Congress of Theoretical and Applied Mechanics (ICTAM 2016), Montreal, QC, Canada, 22–26 August 2016; pp. 2102–2103.

MDPI

St. Alban-Anlage 66

4052 Basel

Switzerland

Tel. +41 61 683 77 34

Fax +41 61 302 89 18

www.mdpi.com

Metals Editorial Office

E-mail: metals@mdpi.com

www.mdpi.com/journal/metals